POINT DEFECTS
IN MATERIALS

POINT DEFECTS IN MATERIALS

F. Agullo-Lopez
Universidad Autonoma de Madrid, Spain

C. R. A. Catlow
University of Keele, U.K.

P. D. Townsend
University of Sussex, U.K.

1988

ACADEMIC PRESS
Harcourt Brace Jovanovich, Publishers
London   San Diego   New York   Berkeley
Boston   Sydney   Tokyo   Toronto

ACADEMIC PRESS LIMITED
24/28 Oval Road
LONDON NW1 7 DX

United States Edition published by
ACADEMIC PRESS INC.
San Diego, CA 92101

British Library Cataloguing in Publication Data

Agullo-Lopez, F.
  Point defects in materials.
  1. Materials. Point defects
  I. Title II. Catlow, C.R.A.
  III. Townsend, P.D.
  620.1'12

ISBN 0-12-044510-7

Typeset by Setrite Typesetters Ltd., Hong Kong
Printed in Great Britain by
St Edmundsbury Press Ltd., Bury St Edmunds, Suffolk

Contents

Chapter 3
Energetic Methods of Defect Production

Chapter 6
Point Defects—Oxides

Chapter 7
Point Defects in Semiconductors and Metals

Chapter 8
Experimental techniques I—General

Chapter 9
Experimental techniques II—Optical

Chapter 10
Experimental techniques III—Electron and Nuclear

Chapter 11
Computer Modelling Techniques

Chapter 12
Quantum Mechanical Methods

Chapter 13
Statistical Mechanical Models

Chapter 14
Applications

Preface

Imperfections exist in all solids. Some of them occur despite our efforts and others because of them. In much of the earlier literature of studies of imperfection and impurity effects the distorted lattice sites were termed defects. This is unfortunate as it implies that perfect crystals, or amorphous materials, are an ideal that should be sought. On the contrary the presence of impurities and other "defects" generate a wider range of properties than would be available with perfect systems. It is salutary to realize that semiconductors, solid state lasers or stainless steel are just three examples of materials with desirable properties which are provided by appropriate imperfections. Defect studies are therefore an essential element of modern science and technology. Indeed the overall pervasiveness is apparent from the fact that the three authors of this book are variously working in departments of applied materials science, chemistry and physics.

The book provides a coverage of the theoretical tools required for defect study plus a strong emphasis on the wide range of experimental techniques needed for analysis. Particular attention has been given to the powerful resonance and hyperfine methods which often give more detailed data than the classical electrical or optical methods. Although in principle all types of materials have imperfections the emphasis here is on semiconducting and insulating materials, in part because these are better suited to detailed microscopic characterization.

The level of the book is suitable for final year undergraduates as well as postgraduates and other research workers. Discussion of imperfections in undergraduate courses should be encouraged as the simplistic ideas of perfect crystals can deflect interest away from the more challenging problems of the real world. One should admit that although there is an extensive literature on defects one can only cite a very limited range of examples where we fully understand the structure and properties of the defects. Classic examples exist for alkali halides and silicon but even for GaAs or simple oxides many uncertainties appear. For the more complex systems such as high temperature superconductors defect structures are totally speculative. Despite this one can make progress for technological applications with a combination of empiricism and intuition. A major aim of this book is to offer the foundation on which this intuition can develop. Our experience in the field has convinced us that the subject is interesting, poses challenging problems and for the forseeable future imperfections will dominate scientific progress in the uses of materials for research and technology.

Acknowledgements

Our thanks extend to many people for encouragement, advice, criticism, data and financial assistance. Among our colleagues special mention must include Dr U. Bangert, Dr F. Cusso, Professor J. M. Calleja, Dr F. Jaque, Professor M. Rahmani, Professor J. M. Spaeth, Dr J. San Juan and Dr A. M. Stoneham. The British Council have provided valuable assistance with travel costs. We are indebted to the draughtmanship of Mrs Carmen Rueda Escolar for many of the figures and typing was patiently taken through various versions by Miss A. Clark, Miss D. Porter and Miss B. Reilly.

We are grateful to the following publishing houses for permission to reproduce figures from books or journals:

Akademie Verlag, Fig. 8.11; American Institute of Physics, Figs 1.5, 4.14, 5.3, 8.6, 8.12, 8.15 and 8.16; American Physical Society, Figs 4.11, 5.9 and 6.2; Birkhauser Verlag, Fig. 5.11; Cambridge University Press, Fig. 8.4; Gordon and Breach Science Publishers, Figs 4.8, 9.2 and 9.6; Institute of Physics, Figs 1.6, 4.1, 4.3, 4.5, 4.7, 7.11, 8.7, 8.17, 9.12 and 10.13; Journal de Physique (Paris), Figs 4.6 and 5.2; North Holland Physics Publishing, Figs 4.15, 6.8, 10.11, 10.18, 10.19, 10.20, 10.24 and 10.25; Pergamon Press, Figs 1.11, 1.12, 4.4, 5.5 and 9.11; Society of Photo-Optical Instrumentation Engineers, Fig. 4.16; Springer-Verlag, Figs 2.1, 6.7, 7.4, 7.5, 9.8 and 9.10; Taylor and Francis, Figs 1.7, 3.6, 3.7, 4.13, 5.15 and 14.5; The Royal Society, Fig. 7.14.

1
Defects in Solids

1.1 STRUCTURAL DEFECTS

A structural defect is any fault in the long- or short-range order of a
material. This concept applies to crystalline solids having a three-dimensional periodic structure as well as to amorphous solids where only local or
short-range order exists. A vast variety of imperfections can exist in a solid
ranging from the presence of impurity atoms at normal lattice sites to
regions of completely different crystalline material. For simplicity it is
customary to discuss such imperfections as either point defects or extended
defects. The category of point defects includes imperfections which are
localized over a few atomic sites. Typical intrinsic examples are as follows:
vacancies (missing lattice atoms); interstitials (atoms occupying non-lattice
sites); antisite defects (where elements of a compound occur on the wrong
location). Extrinsic defects arising from the presence of foreign atoms can
take similar roles and may be coupled with intrinsic features.

The existence of various isotopes may be considered as one form of
disorder and in some cases, such as measurements of thermal conductivity
at low temperatures, isotopic abundance can markedly affect the data.
Different isotopes are particularly valuable in identifying defect structures
when electron−nuclear coupling effects are sensed as in electron paramagnetic resonance (see Chapter 10).

Extended defects influence large volumes of the material. For example a
dislocation line may run throughout a crystal at the boundary of an error in
the arrangement of lattice planes; domain boundaries exist where blocks of

crystal are misaligned; inclusions of different phases (both intrinsic and extrinsic) are often noted; large concentrations of defects may aggregate and form precipitates or extra lattice planes. One further example of an extended defect is the surface.

This book will primarily concentrate on the properties of point defects and the manner in which they may be used to control properties of real materials but there are of course interactions between all types of defect. The term "defect" is unfortunately somewhat emotive and implies an undesirable feature in the material. It must be remembered, however, that the skilful use of crystalline imperfections is the basis of the entire semiconductor industry. The fields of metallurgy, corrosion science, catalysis, phosphors, optical fibres and superconductors are just a few examples of areas where defect effects are paramount and can often be put to advantage to achieve effects which do not occur in "perfect" material. An understanding of defect properties thus allows many desirable materials and devices which would be unobtainable if constrained to be designed and developed in perfect systems.

The defects so far considered refer to the crystallographic structure of the solid. A generalization is possible to include disorder in the magnetic or dielectric structure of the material. In this wider sense, Bloch or domain walls, magnetic bubbles, etc., are also structural defects. In a similar definition, superlattice structures may be included in this category.

In this chapter the basic types of point defect are listed and a few examples of their properties are given. Extended defects (in particular, dislocations) are discussed. In the next three chapters the formation of defects is considered, firstly from thermodynamic considerations, secondly as a result of energetic particle excitation and thirdly from the more subtle mechanisms which have been identified in many insulators.

Chapters 5−7 are concerned with examples of defect models for ionic, covalent and metallic systems. Despite the different types of lattice bonding, there are many similarities not only for crystalline materials but also for amorphous systems such as glasses. In Chapters 8−10 the more commonly used experimental techniques are discussed together with comments on their limitations. Some newer techniques are included even though they may not yet be generally widespread. These chapters are followed by three devoted to theoretical modelling. Finally, the book closes with a chapter giving some applications of defects. Because defect physics is the basis of so much materials science, the examples picked here were chosen to demonstrate less familiar applications, and the areas such as semiconductor physics and metallurgy are mostly left to other more specialized texts.

1.2 POINT DEFECTS

1.2.1 Vacancies and interstitials

During crystal growth or as a result of irradiation damage, vacant lattice sites develop in all crystals. Such vacancies are generally stable at room temperature and may be mobile at slightly higher temperatures. Once the local bonding is disrupted, the immediate neighbourhood of the vacancy will relax so that in effect even a single vacancy will perturb several lattice sites. However, the major disturbance is generally confined to the nearest-neighbour shell of atoms. If the vacancy is produced by displacing a lattice atom, then the released atom can be trapped within the lattice or move out to the surface to form new crystal layers. The alternatives are known as Frenkel and Schottky defects. For the Frenkel defect the displaced atom is retained in the interior of the crystal at an interstitial site. In many cases the vacancy−interstitial Frenkel pair is only of a transient nature as interstitials tend to be mobile at very low temperatures (e.g. 20−50 K), and the pair may annihilate with the interstitial moving back to the vacancy site. If the pair is well separated, then the annealing stage may lead to alternative reactions with other interstitials, impurity atoms or extended defects. Hence the vacancies will survive and more complex interstitial structures can develop. The three simplest interstitial sites are at the interstices of normal lattice sites, as a pair of atoms sharing a single lattice site or as an additional atom inserted into a row of atoms. For cubic lattices these alternatives are termed a body-centred, dumb-bell and crowdion interstitial.

As defects cluster into larger units, they become less mobile and a classic example of this effect occurs in the photographic process. Light incident on a grain of silver halide releases silver atoms which individually are mobile but, if four such atoms cluster together, they are stable at room temperature and it is this cluster, termed the latent image, which is a key step in the photographic process. Indeed the entire photographic process can be viewed as a series of defect-controlled reactions.

1.2.2 Charge states of defects

Insulators and semiconductors require electrical neutrality. Hence the formation of even a simple vacancy must be accompanied by a redistribution of electrical charge. The defect sites allow electron transitions within the energy gap of the material with the result that the transparent crystals become coloured. The colour centres of alkali halides offer a well-documented set of examples for simple defects. Tables 1.1 and 1.2 list a number of optical absorption bands which are seen in alkali halides and for which

Table 1.1.
Electron centres in alkali halides

Centre	Description
F^+	A halogen ion vacancy. A lattice excitation near the F^+ centre produces the α band
F	A halogen ion vacancy that has trapped an electron. A lattice excitation near an F centre produces the β band. Absorption produced by transitions of the F centre to higher states give the K, L_1, L_2 and L_3 absorption bands
F_A, F_B	F centres bound to one or two neighbouring alkali-metal impurities
F_Z	A variant of the F centre with a divalent neighbouring metal ion
$F^-(F')$	A halogen ion vacancy that has trapped two electrons
$F_2(M)$	A pair of adjacent F centres
$F_3(R)$	A triangular array of three F centres. Although possible arrangements are a linear chain of three vacancies, an L-shaped set in the (100) plane or a triangular set in the (111) plane, the model currently believed is the latter (111) arrangement
$F_4(N)$	Four F centres linked together. Two possible configurations are a parallelogram of vacancies in the (111) plane which may produce the N_1 absorption and a tetrahedron of vacancies which may give the N_2 absorption
$F_2^-(M')$, F_3^- (R'), $F_4^-(N')$	By analogy with the F centre the F_2, F_3 or F_4 centres may trap an additional electron

Table 1.2.
Hole centres in alkali halides

Centre	Description
V_K	Self-trapped hole on a pair of halogen ions, i.e. effectively X_2^-
H	Hole trapped along four halogen ions in three adjacent lattice sites aligned along $\langle 110 \rangle$. This is a neutral crowdion interstitial, effectively X_4^{3-}
V_4	A halogen di-interstitial, probably aligned along $\langle 111 \rangle$
H_A, H_B	Variants of H centres adjacent to alkali impurity ions
I	Interstitial halogen ion, X^-

defect models are well established. Figures 1.1 and 1.2 give sketches of the atomic structures and Figure 1.3 indicates the position of their optical absorption bands in an idealized spectrum. In this set of examples all the vacancy defects are in the halogen sublattice and so act as electron traps. The H and V_k defects (Figure 1.4) give examples of crowdion and dumb-bell interstitials for the halogen ions. Both these interstitial defects are hole centres.

In the region of a defect the trapped charges may not be as tightly confined as in a perfect lattice; so even the F-centre vacancy may be slightly positive with respect to the lattice. This in turn allows it to weakly bond a second electron (the F′ defect) which is stable at low temperatures. An even greater diversity of charge states is found with less ionic bonding; for example in the covalently bound silicon the vacancy centre is able to exist in +2, +1, 0, −1, −2 charge states.

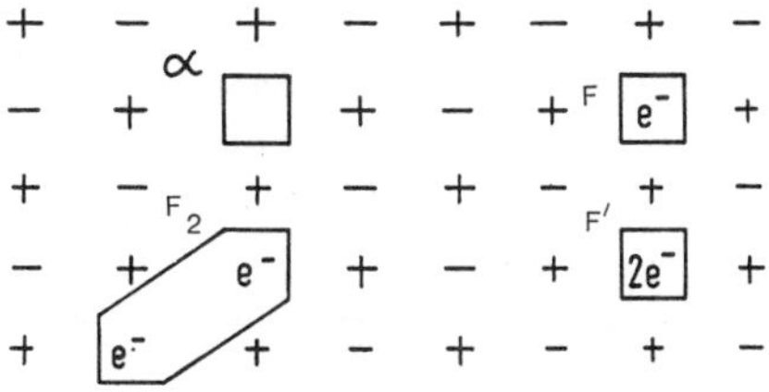

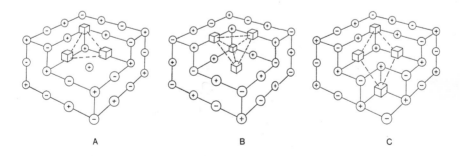

Figure 1.1.
Models of halogen vacancy defects in alkali halides.

Figure 1.2.
Sectional views of the F_3 and F_4 halogen ion vacancy defects in alkali halides. (A) F_3 has three F centres arranged on a (111) plane. F_4 may exist either as (B) a tetrahedron or (C) a parallelogram of vacancies.

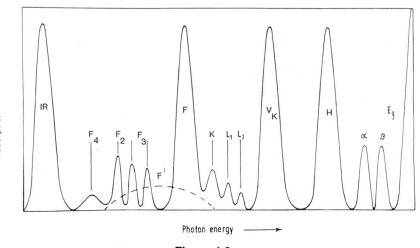

Figure 1.3.
The relative positions of the major optical absorption bands in alkali halide crystals. In perfect crystals there are no absorption features between the absorption edge E_{gap} and the infrared (IR) restrahl absorption.

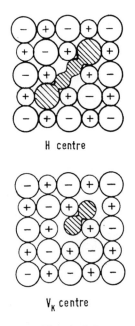

H centre

V_K centre

Figure 1.4.
Schematic structures for an H and a V_K centre in alkali halides. In each case a hole is trapped on the halogen ions.

1.2.3 Impurity ions—size and charge effects

Impurity ions can occupy normal lattice sites with minimal disturbance to the lattice if they have the same charge state and size as the original lattice ions. A familiar example of such an extrinsic defect is the Cr^{3+} ion which can replace the Al^{3+} ion in Al_2O_3 to form the red ruby. The chromium is slightly oversized and so induces strain in the crystal. At high dopant levels this would lead to fracture of the material and the system minimizes the strain energy by accommodating the impurity ions as adjacent pairs (i.e. the net strain energy of a pair of impurities is less than the sum of two separate impurities).

If the impurity has a different valence state from the host, then a number of alternative strategies are used to accommodate it into the lattice. The introduction of divalent impurities into alkali halides follows a standard pattern in which the impurity metal ion replaces the host monovalent metal ion but in order to provide local charge equilibrium a second metal ion, adjacent to the impurity, is also removed. For example magnesium in LiF is accepted as the defect $Mg^{2+}-Li_{vac}^{+}$. This is a particularly well-studied example and here it is noted that the defect is mobile by about 100°C and moves to form a ring of three units in the (111) plane of the crystal as $(Mg^{2+}-Li_{vac}^{+})_3$. To accommodate higher charge states, we might again use several intrinsic defects associated to the foreign ion, and indeed this may occur in many cases. In LiF doped with Ti^{4+} the size considerations allow the impurity to occupy a Li^+ site but in this example charge equilibrium is regained by a second set of oxygen impurities at the neighbouring fluorine sites; hence overall the defect consists of $Ti^{4+}-3O^{2-}$. Both these defects are important in the thermoluminescence radiation dosimeter described in Chapter 14.

In the alkali halide examples the impurity ions remain in their original charge state. Obviously an alternative possibility is that they will donate an electron or capture an electron to place them in the "normal" state of a lattice ion. The role of impurities in semiconductors is familiar from undergraduate texts and the substitution of P or B ions into silicon demonstrates this. The ionization of the impurity then promotes electrons to the conduction band to form the n- or p-type regions. For semiconductor devices the operating level of the doping may be as low as $10^{14}-10^{15}$ impurity ions cm^{-3}. In order for there to be reproducibility in semiconductor manufacture all other electrically active dopants must be suppressed. The simplest way to achieve this is to produce crystals of exceptional purity, prior to doping. Consequently it is now possible to obtain silicon with impurity levels measured in the range of parts per billion (ppb). From the viewpoint of defect studies this is ideal as it greatly simplifies the range of complex defects which can exist.

Gallium arsenide (GaAs) semiconductors can be doped in an analogous fashion to silicon but in this case, to produce n- or p-type material, dopants from groups VI and II are used. The ions of S and Zn replace the As or Ga ions to give the n- and p-type regions. Dopants of group IV can also be introduced into the lattice but in this case the type of site occupancy and the state of charge is less predictable. At low dopant levels Ge ions occupy Ga sites whereas at high dopant levels they move onto As sites to give p-type materials. For such changes to occur, it is probable that there must be changes both in the strain field and in the Fermi level. A further complexity in materials with covalent bond characteristics is that site occupancy of the lattice can show some imperfection. In GaAs this appears as errors in which Ga occupies As sites, and vice versa. Such errors may be stabilized by the presence of impurities, and infrared spectra have revealed the association of such antisite defects with silicon, boron and carbon impurities. The B and C ions are particularly troublesome as they originate from the furnace materials for melt-grown crystals. Antisite defects are more varied and of increasing importance in multielement compounds such as the quaternary semiconductors used in heterojunction diode lasers.

1.2.4   Defects in high concentrations

As has been noted, defects may pair or cluster into small aggregates if this reduces the free energy of the system and such processes occur both during crystal growth and as a result of diffusion. A wide range of possibilities exist including precipitation of planes of defects, formation of colloidal metal aggregates, decoration of dislocation lines, formation of new crystal phases, etc. These regions may act as both sinks or evaporation sources for point defects and even in the same crystal may play different roles depending on the temperature or the rate of excitation. Several examples will be discussed in later chapters (e.g. section 9.2.5). Extended defects are more difficult to characterize than the smaller point defects as they occur in greater variety and do not have well-defined optical or magnetic properties which allow us to distinguish between similar structures.

1.3   STRUCTURE OF POINT DEFECTS

The preceding section gave a selection of simple examples of point defects and from this it can be seen that the detailed microscopic knowledge of any point defect involves three complementary structural aspects.

(a)   Geometrical structure.   This refers to the atomic configuration at and around the defect. In general, neighbouring atoms will be shifted with regard to those in the perfect structure, an effect usually described as lattice relaxation. Far from the defect, elastic models can be used to account for that relaxation.

(b) Dynamic structure. This refers to the change in the vibrational modes induced by the presence of the defect for crystalline materials. This may give rise to the occurrence of new frequencies out of regions occupied by the acoustic and optical modes of the perfect lattice, i.e. in the gap or above the optical branches. In the former case, we are dealing with an interband mode, as illustrated by I^- replacing Cl^- in KCl. The latter case corresponds to the incorporation of an impurity of smaller mass than that of the host atoms and is termed a local mode. A typical example is the substitution of H^- for Cl^- in KCl. In other cases, defects lead to a strong vibration amplitude at the defect.

In a general way the effect of defects on the vibrational structure can be discussed by the change of the frequency distribution function $\rho(v)$ for the perfect lattice to a new distribution $\rho'(v)$.

(c) Electronic structure. This refers to the electron states and energy levels associated with the defect. As for the vibrational case, defects modify the level density $\rho(E)$ of the perfect lattice. For insulator or semiconductor materials, the change in $\rho(E)$ may give rise to the occurrence of levels in the forbidden gap between the conduction and valence bands. These levels drastically influence the electrical and optical properties of the material.

1.4 POINT DEFECTS AND STOICHIOMETRY

Point defects can be considered as modifications in the stoichiometry of the material. As an example, an atomic concentration x of cation vacancies in an AB crystal leads to a new compound whose stoichiometry is $A_{1-x}B$. These composition changes may become quite relevant at rather large concentrations of defects. In fact, a number of materials can accommodate rather wide ranges of non-stoichiometric compositions, therefore allowing for some form of crystalline engineering. Interesting examples of this behaviour are provided by non-stoichiometric compounds and intercalation materials. In intercalation materials additional ions are introduced between weakly bonded lattice planes.

1.4.1 Non-stoichiometry

A major feature of real crystals is that their composition may not match the ideal value expected from their chemical formula. This can arise in several ways. For example GaAs tends to decompose at the melt temperature as arsenic has a high vapour pressure. Therefore, in order to maintain stoichiometry, an overpressure of arsenic at about 1 atm is used. Too low or too high an arsenic pressure causes the arsenic or gallium vacancies to be frozen into the final solid. Clearly for chemical studies the sample can be viewed as GaAs but for sensitive defect problems as in semiconductor applications an imbalance of atoms even at the parts per million level is a serious problem.

A material of interest for integrated optics is $LiNbO_3$. This has numerous problems during growth in that it tends to lose Li and O ions. However, in principle a stoichiometric melt can be produced and it can be stabilized with an overpressure of Li_2O. Even for this material, growth problems remain as, during cooling, the sample frequently cracks or, when examined more closely, shows considerable variations in properties along the length of the samples. Reference to the phase diagram in Figure 1.5 shows that a 50 mol.% Li_2O-50 mol.% Nb_2O_5 mixture would lie close to a mixed crystal phase of $LiNbO_3-Li_3NbO_4$ which leads to instability. The normal practice is to avoid this problem by starting with the congruent melt composition of 48.6 mol.% $Li_2O-51.4$ mol.% Nb_2O_5. Alternatively, crystal growth can be aided by the addition of dopants such as MgO. In neither case is the material $LiNbO_3$, and the properties are strongly influenced by the built-in defects.

Oxides are well known for their ability to form in a wide range of compositions and are typified by examples of transition-metal oxides. NiO and CoO present modest $(10^{-2}-10^{-3})$ fractional deviations from stoichiometry which are accommodated by the introduction of cation vacancies. Much larger deviations are found in the similar rock-salt structure of FeO. For $T > 570°C$, the stable composition is $Fe_{1-x}O$ with $0.05 < x < 0.16$. The important question here is to understand which are the point defects (if any) responsible for the marked lack of stoichiometry. In principle, either Fe^{2+} vacancies or O^{2-} interstitials would be reasonable candidates. In fact, the situation is much more complicated. There is experimental and theoretical evidence that the basic unit of the iron-deficient structure is the so-called 4:1 cluster depicted in Figure 1.6. It is constituted by four Fe^{2+} vacancies in a tetrahedral arrangement including an Fe^{3+} interstitial at its centre. Charge compensation is accomplished by additional neighbouring Fe^{3+} interstitials. The 4:1 basic clusters can associate to form more complex 6:2 and 13:4 aggregates.

In other non-stoichiometric oxides, such as MoO_{3-x}, TiO_{2-x}, etc., a continuous range of compounds can exist. In some of these the structural vacancies are eliminated by the formation of crystallographic shear planes. In other examples the vacancies are incorporated in an ordered fashion onto lattice sites. In such a case a material may be described using a purely

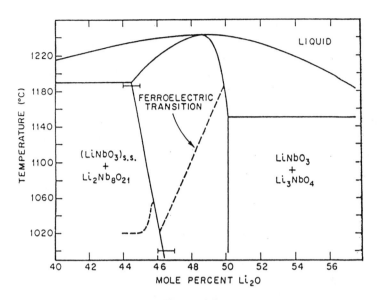

Figure 1.5.
The phase diagram of $LiNbO_3$. In practice this ferroelectric crystal is grown from the congruent melt of 48.6 mol.% Li_2O rather than from the 50 mol.% Li_2O stoichiometric melt.

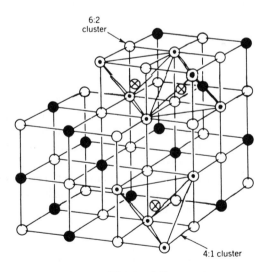

Figure 1.6.
An example of a 4:1 cluster of defects in an oxide. The 4 and 1 refer to the four M^{2+} vacancy sites, (⊙) and the central M^{3+} interstitial (⊗).

chemical nomenclature scale as $Ti_{32}O_{58}$ although this is not a helpful representation. A preferable description is $Ti_{32}O_{58}(O_{vac})_6$ which both resembles TiO_2 and emphasizes the ordered structure of the vacancies. These superlattices of ordered vacancies may similarly evolve during irradiation damage in which displacements of normal lattice ions occur. The examples of nonstoichiometry are not confined to oxides and are seen in intermetallic compounds such as AlNi (aluminides) and TiC and VC (carbides). Here again the carbon vacancies may arrange into a superlattice.

1.4.2   Intercalation materials

The most typical example of intercalation materials is graphite. It presents a structure made up of weakly bound hexagonal layers of carbon atoms. The C−C bond length inside a layer is 1.42 Å, whereas the separation between adjacent layers is 3.35 Å. This open structure permits the intercalation of a large variety of atoms and molecules between the layers. Two types of element can be intercalated.

(1)   Electron donors, such as alkali, alkaline-earth, transition and rare-earth metals.

(2)   Electron acceptors such as $FeCl_3$, Br_2 or HNO_3.

The incorporation of these elements between the carbon layers follows a definite sequence or staging characterized by a number n. A staging n means that two consecutive intercalated layers are separated by n graphite layers, constituting a periodic superstructure along the hexagonal axis. The staging $n = 1$ is schematically represented in Figure 1.7(a) for intercalation of $FeCl_3$ in graphite.

Controlled intercalation is being contemplated as a form of solid-state engineering, which allows for the preparation of new materials having a variety of structures, physical properties and applications such as energy storage batteries. The structure of the intercalation compounds depends on both the staging n and the two-dimensional structure of the intercalated layers. These two factors determine the overall stoichiometry of the material. Figure 1.7(b) illustrates two examples, one for intercalated lithium, adopting the so-called hexal structure leading to the LiC_6 stoichiometry for $n = 1$, and the other for potassium in an octal arrangement corresponding to a KC_8 composition, (also $n = 1$).

The variety of structures implies a corresponding diversity in physical behaviour. In particular, some intercalated compounds have been prepared (C_xAsF_5) with a reported room-temperature conductivity exceeding that of copper.

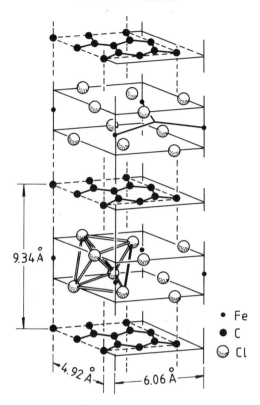

Figure 1.7a
Examples of intercalation in graphite: (a) the intercalation compound $FeCl_3$.

1.5 DEFECT ORDERING

We have already mentioned that point defects such as vacancies may become ordered and form superstructures as in VC or the ordered vacancy superstructure in oxygen-deficient MoO_3. Similarly the intercalated elements in an intercalation compound (e.g. graphite) tend to be arranged in an ordered lattice structure. Ordering is a rather general feature in defect physics, generally occurring above certain critical concentrations and under appropriate thermodynamic conditions.

Another interesting example is provided by the ordering of voids in metals or ionic crystals to form void lattices. The effect has been extensively investigated in a number of irradiated metals such as nickel, molybdenum, niobium, tantalum and tungsten. The voids are often arranged with the same lattice structures as that of the host metal. The ratio between the void

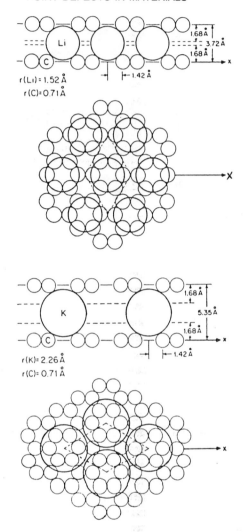

Figure 1.7b
Examples of intercalation in graphite: (b) effects of alkali ions (Li$^+$ and K$^+$) on the intercalated layer structures.

lattice parameter and void diameter has a value which is characteristic of the metal.

The phenomenon of defect ordering has been generally explained in terms of anisotropic elastic interaction among defects, although other possible models have also been advanced.

In metallurgical systems the ordering of atomic species is a well-documented phenomenon. For many binary alloys a continuous range of doping can be carried out from, say, copper to gold. A plot of the resistivity (Figure 1.8) shows that dopants raise the resistance of either material when at a low concentration. However, rather than the resistivity curve following a smooth line between the extrema, there are two pronounced dips at mixtures corresponding to Cu_3Au and $CuAu$. These special alloy mixtures correspond to an ordered stage of the two elements.

1.6 METHODS OF PRODUCING POINT DEFECTS

Typically, four methods are available to introduce point defects in materials.
   (a)   Growth and preparation.
   (b)   Thermal and thermochemical treatments.
   (c)   Plastic deformation.
   (d)   Ion implantation and irradiation.

(a)   Growth and preparation.   This mostly applies to the introduction of chemical impurities. In fact, the degree of purity of a solid critically depends on the growth or preparation method. As an example, $BaTiO_3$ when grown by the flux (KF) method presents a strong F^- and K^+ doping. In contrast, crystals grown by the Czochralski method avoid this contamination and show better purity. Techniques not requiring a crucible, such as the flame fusion growth for oxides, are considered to produce purer crystals, although often at the cost of a lower structural perfection.

(b)   Thermal and thermochemical treatments.   The simplest thermal treatments involve the heating of a sample to a high temperature followed by a fast (quenching) or slow (annealing) cooling procedure.

By quenching, we try to "freeze" the defects thermally induced at the high temperature (see Chapter 2). After quenching, the sample remains in a metastable state quite far from the thermodynamic equilibrium condition of the lower temperature. Efficient quenching procedures require thin samples having good thermal conductivity. For metals, quenching rates of about $25\,000$ K s^{-1} can be obtained by using water as a cooling fluid. Much higher rates (about 10^6 K s^{-1}) are obtained by splashing a melted metal jet onto a cool surface, and this is used for the production of glassy metals. The method is termed splat cooling. Similar glassy metallic materials have been generated after ion beam amorphization or high-power laser melting. For insulator materials, quenching treatments are less efficient, although they are also frequently used to enhance the intrinsic defect concentrations. The problem is that the low thermal conductivity of insulators produces large thermal gradients and the samples crack from thermal strain.

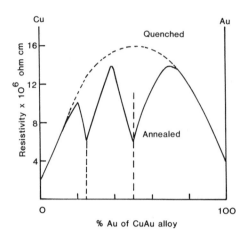

Figure 1.8.
The variations in electrical resistivity over the range of Cu−Au alloys. Note the lower resistivity for pure systems and the ordered phases Cu_3Au and CuAu. The random mixtures of quenched samples are more resistive.

Annealing treatments involve sufficiently slow cooling rates that the material approaches the equilibrium situation at any temperature, including the final temperature. In this case, thermodynamic crystal perfection should be that corresponding to the lowest temperature reached, therefore implying a reduced defect concentration in comparison with that of the pre-annealed sample.

In the above ideal treatments, a passive (non-reactive) atmosphere is assumed. This is often not the case in practice. Most treatments involve reactive atmospheres, e.g. oxidizing, reducing or additive. Heat treatments of oxides in vacuum (or even rare gas atmosphere) usually lead to oxygen losses and therefore to the formation of oxygen vacancy centres. The same result can be accomplished by using more efficient reducing atmospheres (e.g. hydrogen), although processes are often more complicated and not well understood. Sometimes, so-called additive atmospheres are utilized, such as in the additive colouring method for alkali halides and some other simple ionic materials. The alkali-halide crystal is heated in an alkali-metal atmosphere which induces the growth of the halogen-deficient compound (i.e. anion vacancies).

Other treatments also involve the application of an electric field to favour migration of some defect species and/or to inject carriers (electrons or holes) from the electrodes, during the heating. Typical examples are the electrolytic colouring of alkali halides or electrodiffusion in metals that can be used as a purification method.

Ion exchange between the host material and different chemical species can be obtained by some thermochemical treatment involving the interaction of the material with an active solution containing the ion to be substituted. This approach is suitable for modifying a surface layer and for example has been used in the production of optical waveguides. In these experiments the aim is to increase the refractive index in the region designated as a waveguide. Ion exchange of Ag^+, K^+ or Cs^+ have all been successful in soda–lime glasses by replacing sodium. The detailed analyses show that rather different depth profiles ensue with these ion species. Ion exchange has numerous limitations; for example, several ions may be involved in a single treatment. An initial attempt to form waveguides in $LiNbO_3$ by Ag^+ ion exchange with Li^+ was apparently successful but later workers found that the effect was not related to the Ag^+ but to hydrogen which entered the crystal during the treatment. In modern methods, $LiNbO_3$ is doped with protons by immersion in lithium metabenzoate.

Finally, deposition of a layer of a compound onto the surface of a material followed by heating is often used as a procedure to introduce some chemical species of the compound into the material through a diffusion process.

Thermal treatments, such as quenching, have valuable consequences as in the hardening of surfaces of steel or the precipitation of alloy phases, and the literature for metallurgy is extensive. Similar effects occur in insulators. In the example of lead-doped NaCl, there is an initial marked hardening on annealing of quenched samples. This is associated to the formation of some metastable precipitated phase which is incoherent with the NaCl matrix. Prolonged annealing causes a softening which is related to the formation of large $PbCl_2$ precipitates.

(c) Plastic deformation. Deformation treatments require understanding of both point and extended defects such as dislocations (see section 1.7) as plastic straining involves the motion and multiplication of dislocations. In metals or simple ionic crystals, multiplication rates of $10^7 - 10^8$ cm^{-2} per 1% strain are typical. In addition, plastic flow can interact with the point defect structure in the following ways.

(1) Dissolving aggregates or precipitates,
(2) Inducing clustering and precipitation.
(3) Generating intrinsic defects by dislocation intersection and jog dragging.
(4) Acting as sinks for point defects by trapping at core sites or loop coalescence.

Evidence for all these processes is scattered throughout the literature.

(d) Ion implantation and irradiation. Radiation damage will be extensively discussed in this book as a view of generating point defects. In particular the use of ion implantation to add impurity ions has become widely accepted and is a fundamental processing step in semiconductor production; this is discussed in Chapter 3. It should be noted that surface-processing treatments and the production of surface "defects" can have beneficial consequences in many materials.

1.7 EXTENDED DEFECTS

In this section, some properties of extended defects are briefly listed and, in turn, the linear features of dislocation lines, planar faults and volume imperfections are considered.

1.7.1 Linear faults—dislocation lines

Line, or one-dimensional, defects have a macroscopic length along a given direction. The most important line defect is the dislocation which essentially determines the plastic behaviour of many crystalline materials. Although they are not going to be dealt with in this book, we briefly comment on them, since they strongly interact with point defects and influence a large number of point defect processes, such as diffusion, clustering and precipitation. For example diffusion along a dislocation line may be 100 times greater than in the bulk of a crystal.

(a)   Burgers vectors.   The dislocation line is characterized by the Burgers vector b, representing the resulting lattice misfit along any closed path embracing the dislocation line. For the so-called perfect dislocations, b is a lattice vector. In mathematical language $b = \int_c \delta\mu$, where $\delta\mu$ is the displacement vector of a point moving along a path from M to Q (Figure 1.9) and b being, of course, independent of the chosen path. Depending on the relative orientation of b and the dislocation line direction l, there are *edge* dislocations ($b \perp l$) or *screw* dislocations ($b \| l$). These two types of dislocation are quite different from the point of view of the stresses or strains induced in the lattice. This can be more easily visualized if some of the ideal procedures required to create them are considered.

(b)   Edge dislocations.   An edge dislocation involves the introduction in the material of some extra lattice planes terminated at the dislocation line (Figure 1.9). The dislocation line is surrounded by compression regions of high hydrostatic stress (upper half of the material) and expansion regions of low hydrostatic stress (lower half). Thus it is easy to understand how the compression region may act as a source for vacancies and a sink for interstitials, whereas the expansion region behaves in the opposite way. The tendency of impurities to move onto the dislocation lines is termed decoration.

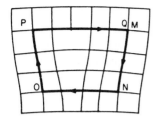 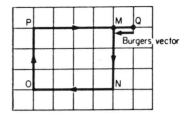

Figure 1.9.
An example of an end-on view of an edge dislocation. If the additional plane is removed, it can be seen that the Burgers vector for the edge dislocation is perpendicular to the extra plane.

(c) Screw dislocations. A screw dislocation can be created by making a cut in the material and then shearing one side of the cut over the other side (Figure 1.10). In this case, only shear and not hydrostatic stresses appear.

For both screw and edge dislocations all components of the stress (or strain) tensor decrease with distance r from the dislocation line as $1/r$. This is a long-range dependence when compared with the $1/r^3$ law associated to the stress (or strain) field of point defects.

Dislocations can take the form of closed loops or of lines ending at the crystal faces of the solid. Therefore, they mostly present a mixed character between screw or edge types. However, in many situations, long dislocation segments with a definite character (either edge or screw) can be considered.

Under applied stress, dislocations can easily move along the plane containing the Burgers vector and the dislocation line or glide plane. In fact, the plastic deformation of many crystalline materials involves the extensive "glide" of a large number of dislocations. In contrast, climb or dislocation motion out of the glide plane requires transport of mass and therefore it is a slow and thermally activated process. As is often stated, climb (at variance with glide) is a non-conservative type of motion.

Dislocation lines present atomic-size steps that may have a different character from that of the main dislocation. Steps contained in the glide plane are termed kinks, whereas those out of that plane receive the name of jogs. Kinks as well as jogs are created as a result of dislocation intersection (Figure 1.11) or can be thermally generated. Kinks can glide along with the supporting dislocations. In contrast, in order to drag jogs along during dislocation glide, vacancies have to be created or taken away. Therefore, point defect trails form behind a moving dislocation as a byproduct of plastic deformation.

For a serious study of dislocation physics the reader is referred to some of the classical introductory books, e.g. Ref. 20, in the field listed at the end of this chapter.

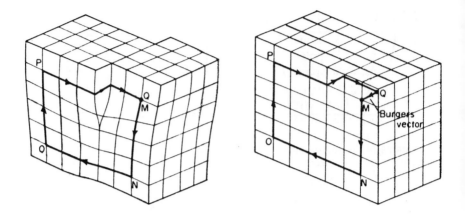

Figure 1.10.
An example of a screw dislocation and the associated Burgers vector which lies parallel to the screw axis.

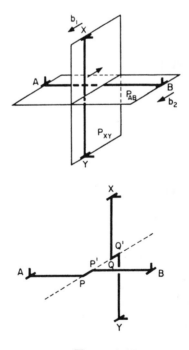

Figure 1.11.
The production of jogs in dislocation lines by the intersection of moving edge dislocations with parallel Burgers vectors.

1.7.2 Planar faults

Planar, or two-dimensional, defects include the external faces of the solid, grain boundaries in polycrystalline materials and subgrain boundaries in single crystals. Other possibilities are defects in the stacking order of lattice planes (stacking faults), or boundaries between domains with different structures, etc. These types of defect are very important in polycrystalline materials, such as ceramics, and thin films where the ratio of surface area to volume is high enough (greater than or equal to $10^{-4}-10^{-3}$) to become dominant over point defects and/or dislocations.

Two-dimensional defects, even more than dislocations, can act as sources or sinks for point defects or serve as rapid diffusion paths. However, the external surfaces of the solid may contain a variety of point and line defects as schematically shown in Figure 1.12. These defects are the active centres responsible for the high physicochemical reactivity of the surface. Unfortunately the study of surface defects is in a much-less-developed stage in comparison with that for defects in the bulk.

In many materials, relaxation of the surface atoms induces formation of a reconstruction (i.e. a new crystal arrangement). Such reconstructions are sensitive to crystal purity. They tend to be only a few monolayers in depth.

1.7.3 Three-dimensional defects

Volume, or three-dimensional, defects include inclusion and precipitated phases, colloids, voids and gas bubbles. These defects may be formed as a final stage of clustering of intrinsic or extrinsic point defects during thermal or irradiation treatments. They play a particularly relevant role in the mechanical behaviour of the material.

1.8 CRYSTAL GROWTH AND PURITY

One area which is only briefly discussed is that of crystal growth. This activity is under-represented throughout solid-state science and far too

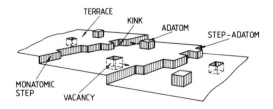

Figure 1.12.
Examples of surface defects.

many academic publications have appeared which are of dubious quality because they are based on data from poorly characterized material. It cannot be overemphasized that the preparation of crystals, or amorphous solids, of high purity and structural integrity is immensely difficult and is scientifically undervalued. Commercial materials may be less well defined if a pragmatic view is taken that they serve the purpose required of them. However, the only examples of truly high purity have arisen from commercial pressure. To prepare semiconductor material of purity measured in parts per billion, with a negligible dislocation content, in single-crystal form and with a mass of many kilograms has required immense effort. With sufficient monetary incentive this target has been routinely reached.

A similar challenge may be cited in the case of optical fibres. When fibre technology was first considered for the purposes of long-range telecommunication in 1966, the fibres used possessed lengths of tens of metres, and absorption or scatter losses were measured in decibels per metre. Once the scale of the market was appreciated, improvements in the preparation of the silica glass without impurities such as iron or water were highly competitive and now losses are measured in decibels per kilometre and single fibres have been drawn and used over more than 200 km. Purity levels of the major contaminants are now at the parts per billion level.

Without commercial pressure, few laboratories invest the necessary effort to produce high-purity material. One exception was the attempt by an Oak Ridge group to purify KCl. The results showed a continuous improvement with time; for example a plot of the logarithm of the bromine impurity content against the year of preparation gave roughly a linear plot with the impurity level falling by a factor of 100 in 5 years.

In applications such as semiconductor technology, where the important regions are confined to a depth of a few micrometres, a number of techniques have developed which produce these surface layers. Methods include vapour- or liquid-phase epitaxy and molecular beam epitaxy. The latter is a simplistic approach in which oven sources control deposition and composition of successive monolayers of crystal. The method is inherently slow and develops samples at rates as low as micrometres per hour but has the advantage that planes of impurity can be built in as atomically abrupt transitions from the substrate planes. To achieve a high purity, there are problems with source material and an ultrahigh vacuum chamber is required. The abrupt changes tend to produce a mismatch in the atomic structure and so a series of misfit dislocations develop. Nevertheless, from the viewpoint of semiconductor device production with GaAs, $Ga_{1-x}Al_xAs$, etc., the method is very valuable.

It is apparent that all methods of sample preparation are imperfect; so in both commercial and academic studies the defects which occur must be understood as these defects cannot be eliminated, as even for nominally "pure" material these are at least parts per million of impurities. The human defect of preferring to make measurements with available crystals rather than devoting more effort to preparing better crystals implies a diversity of defects; hence an understanding of the influence and structure of such imperfections guarantees the need for texts such as this for many years to come.

GENERAL REFERENCES

1 G. Arnold and P. Mazzoldi (eds), *Ion Beam Modification of Insulators*, Elsevier, Amsterdam (1986).
2 R. Bahan, M. Kleeman and J. P. Poirier (eds), *Physics of Defects*, North-Holland, Amsterdam (1981).
3 J. R. Beeler, Jr, *Radiation Effects. Computer Experiments*, North-Holland, Amsterdam (1983).
4 F. Beniere and C. R. A. Catlow (eds), *Mass Transport in Solids*, Plenum, New York (1983).
5 A. V. Chadwick (ed.), *Defects in Solids*, Plenum, New York (1987).
5a A. V. Chadwick and M. Terenzi (eds), *Defects in Solids—Modern Techniques*, NATO ASI series 147, Plenum, New York (1986).
6 S. Chandra, *Superionic Solids*, North-Holland, Amsterdam (1981).
7 Y. C. Chen, W. D. Kingery and R. J. Stokes (eds), *Defect Properties and Processing of High-technology Non-metallic Materials, Materials Research Society Symposium Proceedings*, Vol. 60, Materials Research Society, Pittsburgh, PA (1986).
8 J. H. Crawford, Y. Chen and W. A. Sibley (eds), *Defect Properties and Processing of High-technology Non-metallic Materials, Materials Research Society Symposium Proceedings*, Vol. 24, Elsevier, New York (1984).
9 P. H. Dederichs and R. Zeller, *Point Defects in Metals, II, Springer Tracts in Modern Physics*, Springer, Berlin (1980).
10 M. S. Dresselhaus (ed.), *Intercalation in Layered Materials*, Plenum, New York (1987).
11 F. L. Galeener, D. L. Griscom and M. J. Weber (eds), *Defects in Glasses, Materials Research Society Symposium Proceedings*, Vol. 61, Materials Research Society, Pittsburgh, PA (1986).
12 F. Gautier (ed.), *Electronic Structure of Crystal Defects and of Disordered Systems*, Les Editions de Physique, Les Ulis (1980).
13 W. Hayes and A. M. Stoneham, *Defects and Defect Processes in Non-metallic Solids*, Wiley, New York (1985).
14 R. A. Johnson and A. N. Orlov (eds), *Physics of Radiation Effects in Crystals*, North-Holland, Amsterdam (1986).
15 E. N. Kaufmann and G. K. Shenoy (eds), *Nuclear and Electron Resonance Spectroscopies Applied to Materials Science, Materials Research Society Symposium Proceedings*, Vol. 3, Elsevier, New York (1981).

16 W. Krakow, D. A. Smith and L. W. Hobbs (eds), *Electron Microscopy of Materials*, Materials Research Society Symposium Proceedings, Vol. 31, Elsevier, New York (1984).
17 M. Lannoo and J. C. Bourgoin, *Point Defects in Semiconductors*, Vols I and II, Springer, Berlin (1981).
18 S. Mahajan and J. W. Corbett (eds), *Defects in Semiconductors II*, Materials Research Society Symposium Proceedings, Vol. 14, Elsevier, New York (1983).
19 S. W. S. McKeever, *Thermoluminescence of Solids*, Cambridge University Press, Cambridge (1985).
20 F. R. N. Nabarro, *Dislocations in Solids*, Vols I−VII, North-Holland, Amsterdam (Vols I, II and IV, 1979; Vols III and V, 1980; Vol. VI, 1983; Vol. VII, 1986).
21 J. Narayan and T. Y. Tan (eds), *Defects in Semiconductors*, Materials Research Society Symposium Proceedings, Vol. 2, Elsevier, New York (1981).
22 M. Prutton, *Surface Physics*, Oxford University Press, Oxford (1983).
23 F. Rosenberg, *Fundamentals of Crystal Growth I*, Springer, Berlin (1981).
24 J. I. Takamuram, M. Doyama and M. Kiritani (eds), *Point Defects and Defect Interactions in Metals*, University of Tokyo Press, Tokyo (1982).
25 R. J. D. Tilley, *Defect Crystal Chemistry*, Blackie, London (1987).
26 P. A. Varotsos and K. D. Alexopoulos, *Thermodynamics of Point Defects and Their Relation with Bulk Properties*, North-Holland, Amsterdam (1987).
27 W. M. Yen and P. M. Selzer (eds), *Laser Spectroscopy of Solids*, Topics in Applied Physics, Vol. 49, Springer, Berlin (1981).

2
Thermodynamics of Point Defects

2.1 INTRODUCTION

Thermodynamic concepts and methods provide a very useful framework for discussing point defect properties and processes. Indeed, thermodynamics demands the occurrence of intrinsic point defects, such as vacancies or interstitials, at any temperature. The concentrations found in practice are commonly much higher than those corresponding to thermodynamic equilibrium. However, at very high temperatures (close to the melting point), the thermally induced defects frequently dominate over those coming from extrinsic sources. At low temperatures, an equilibrium situation cannot, in general, be established because of the slow rates associated with most point defect processes, and so we may be dealing with a metastable state. In spite of these cautions, thermodynamics is an appropriate tool for discussing defect processes, even for situations where equilibrium is not guaranteed.

This chapter is mostly concerned with solids containing small concentrations of point defects (i.e. dilute systems), so that interaction among them can be ignored. The situation becomes much more complicated for heavily defective materials where more appropriate methods have to be used. These are discussed in Chapter 13.

2.2 PARTITION FUNCTION AND THERMODYNAMIC POTENTIALS

The thermodynamic potentials[1,2] of any system are as follows: U, internal energy; S, entropy; F, free Helmholtz energy; H, enthalpy; G, free Gibbs energy. These are derived from the partition function Z, which contains all the thermodynamic information on the system. Z is defined by

$$Z = \sum_i g_i \exp\left(-\frac{E_i}{kT}\right) \tag{2.1}$$

where the index i stands for every energy level E_i of the system whose degeneracy is g_i.

From Z, the thermodynamic potentials are obtained through the following relations

$$F = -kT \ln Z$$

$$S = -\left(\frac{\partial F}{\partial T}\right)_V$$

$$U = F + TS$$
$$H = U + pV \tag{2.2}$$
$$G = F + pV$$

$$p = -\left(\frac{\partial F}{\partial V}\right)_T$$

It should always be kept in mind that Z and consequently all thermodynamic potentials depend on thermodynamic conditions, e.g. p and T.

For solid-state systems $pV \ll V$ so that $G \approx F$ and $H \approx U$. Also, TS is often much smaller than V and consequently $U \approx H \approx F \approx G$, although this should be checked for any particular situation.

2.3   THERMODYNAMIC POTENTIALS FOR THE PERFECT CRYSTAL

Let us consider a perfect crystal containing N atoms. In the harmonic approximation, any generic vibrational state is characterized by the vibrational quantum numbers $n_1, \ldots, n_{3N}$, where $3N$ is the total number of degrees of freedom. The energy E associated with such a state is

$$E_{n_1, \ldots, n_{3N}} = \Phi + \sum_{i=1}^{3N} (n_i + \tfrac{1}{2})h v_i \tag{2.3}$$

where Φ stands for the cohesive energy of the crystal lattice and v_i for the frequency of the normal mode i. Substituting equation (2.3) into equation (2.1) gives

$$Z = \exp\left(-\frac{\Phi}{kT}\right) \prod_{i=1}^{3N} \frac{\exp(-h v_i / kT)}{1 - \exp(-h v_i / kT)} \tag{2.4}$$

From equation (2.2), we finally obtain for the thermodynamic potentials

$$G \approx F = \Phi + \sum_i \left\{ \frac{hv_i}{2} + kT \ln\left[1 - \exp\left(-\frac{hv_i}{kT}\right)\right] \right\}$$

$$S = \sum_i \left\{ \frac{hv_i}{T} \frac{\exp(-hv_i/kT)}{1 - \exp(-hv_i/kT)} - k \ln\left[1 - \exp\left(-\frac{hv_i}{kT}\right)\right] \right\} \qquad (2.5)$$

$$H \approx U = \Phi + \sum_i hv_i \left(\frac{1}{2} + \frac{\exp(-hv_i/kT)}{1 - \exp(-hv_i/kT)} \right)$$

It should be noted that the expression for U is simply the sum of the cohesive energy Φ, the zero-point energies of the various modes and the excitation energy of these modes weighted according to the Boltzmann factor at the temperature T. Equations (2.4) and (2.5) are rather cumbersome but can be markedly simplified for high temperatures ($T \gg \theta_{Debye}$, i.e. $kT \gg hv_i$). Then

$$Z = \exp\left(-\frac{\Phi}{kT}\right) \prod_i \frac{kT}{v_i} \qquad (2.6)$$

and consequently

$$G \approx F = \Phi + kT \sum_i \ln\left(\frac{hv_i}{kT}\right)$$

$$S = k \sum_i \left[1 - \ln\left(\frac{hv_i}{kT}\right)\right] \qquad (2.7)$$

$$U = \Phi + 3NkT$$

As remarked above, Φ and v_i are assumed to correspond to well-defined thermodynamic conditions, e.g. p and T.

2.4 THERMODYNAMIC POTENTIALS FOR THE DEFECTIVE CRYSTAL

Let us now consider a crystal with N atoms and n point defects. The partition function Z can be factorized as

$$Z = Z_c Z_v \qquad (2.8)$$

where Z_c is a configurational factor associated with the geometrical degeneracy or multiplicity of energetically equivalent configurations that can be adopted by the n defects. Z_v is the vibrational factor associated with the vibrational energy levels of the defective crystal.

Z_c can be easily calculated by combinatorial methods. For example, for a simple cubic (SC) lattice, the values for vacancies and self-interstitials are listed in Table 2.1.

Table 2.1.
Configurational term Z_c for simple defects
in a simple cubic lattice

Defects	Z_c
n vacancies	$\binom{N}{n}$
n interstitials at cube centre	$\binom{N}{n}$
n interstitials at face centre	$\binom{N/3}{n}$
n interstitials at edge centre	$\binom{N/3}{n}$

In the high-temperature approximation the vibrational factor Z_v can be written as

$$Z_v(n) = \exp\left(-\frac{\Phi'}{kT}\right) \prod_i \frac{kT}{h\nu_i'} \tag{2.9}$$

where Φ' is the cohesive energy for the defective crystal and ν_i' the modified normal mode frequencies.

For low concentrations ($n \ll N$) so that interactions among defects can be ignored,

$$\Phi' = \Phi + n\epsilon_d \tag{2.10}$$

ϵ_d being the energy associated with each defect, i.e. the energy required to take an atom from its regular site to the surface (vacancy defect) or to take it from the surface to an interstitial site (interstitial defect). Equation (2.9) can be further simplified if it is assumed that the Einstein model applies and that the defect only modifies the vibrational frequency of a few neighbouring atoms, e.g. the p nearest neighbours. Then,

$$\begin{aligned} Z_v(n) &= \exp\left(-\frac{\Phi'}{kT}\right) \frac{(kT)^{3N}}{(h\nu')^{np}(h\nu)^{3N-np}} \\ &= Z_v(0)\left[\exp\left(-\frac{\epsilon_d}{kT}\right)\left(\frac{\nu}{\nu'}\right)^p\right]^n \end{aligned} \tag{2.11}$$

$Z_v(0)$ referring to the partition function in the absence of defects ($n = 0$).

From equation (2.11) the vibrational component in the thermodynamic potentials of a defective crystal can be written as

$$G_v(n) = G_0 + n\epsilon_d - kT \ln\left[\left(\frac{v}{v'}\right)^p\right]$$

$$S_v(n) = S_0 + nk \ln\left[\left(\frac{v}{v'}\right)^p\right] \tag{2.12}$$

$$H_v(n) = H_0 + n\epsilon_d$$

where G_0, S_0 and H_0 represent the thermodynamic potentials of perfect crystal.

The full thermodynamic potentials are obtained by adding the vibrational component (equations (2.12)) to that derived from the configurational partition function Z_c. This component has to be calculated for each particular case, by following the same procedure as for the examples in Table 2.1. For defects in a monatomic lattice and a low defect concentration $c = n/N \ll 1$, it can be shown that

$$G_c(c) = -kTn\left[1 + \ln\left(\frac{\alpha}{c}\right)\right]$$

$$\approx NkTc \ln\left(\frac{c}{\alpha}\right) \tag{2.13}$$

α being a factor depending on the type of defect (e.g. $\alpha = 1$ for vacancies, and $\alpha = 12$ for a nearest-neighbour bound pair of vacancies in a face-centred cubic (FCC) lattice).

We can now introduce the (vibrational) thermodynamic potentials associated with a single defect, which from equations (2.12) are written as

$$h = u = \frac{H_v - H_0}{n} = \epsilon_d$$

$$s = \frac{S_v - S_0}{n} = k \ln\left[\left(\frac{v}{v'}\right)^p\right] \tag{2.14}$$

$$g = \frac{G_v - G_0}{n} = \epsilon_d - kT \ln\left[\left(\frac{v}{v'}\right)^p\right]$$

The Gibbs free energy for a defective crystal is then

$$G(c) = G_0 + G_c(c) + G_v(c)$$

$$= G_0 + N\left[cg + kTc \ln\left(\frac{c}{\alpha}\right)\right] \tag{2.15}$$

which is the expression corresponding to the ideal solution model. We can use the molar concentration x instead of the atomic concentration c. $G(x)$ has the same equation (2.15) where x is substituted for c and N is the total number of moles.

As a rough estimate, $h \approx 10^{-2}$ eV for simple point defects in molecular crystals whereas, for covalent, ionic or metallic solids, h lies typically in the $1-10$ eV range. For the entropy s, we can predict from equations (2.14) that $s > 0$ for those defects (e.g. vacancies) inducing a dilatation of the lattice so that presumably $v' < v$. Defects, such as interstitials, inducing a lattice compression and consequently $v' > v$ should present a negative entropy $(s < 0)$. For inorganic materials, the contribution of the entropy term sT to the defect free energy g in equations (2.14) is generally smaller than u or h.

To avoid any ambiguity, the thermodynamic conditions at which the defect has been formed should be specified. This point may become relevant when comparing experimental values of the thermodynamic potentials with theoretical predictions[3]. Experimental data are generally taken at constant p and T, whereas defect calculations are often performed at constant lattice parameter (volume).

2.5 DEFECT CONCENTRATIONS AT THERMODYNAMIC EQUILIBRIUM

It is now possible to calculate the concentration of a given intrinsic defect, when the material is assumed to be in thermodynamic equilibrium at temperature T and pressure p (this variable is scarcely relevant). The equilibrium condition, i.e. minimum for the free energy, can be written as

$$\frac{\partial G(c)}{\partial c} = 0 \qquad (2.16)$$

whose solution yields the equilibrium concentration c_0.

In order to illustrate the calculation procedure, let us consider the simple case of vacancies in a monatomic lattice ($\alpha = 1$)

$$G = G_0 + N(cg + kTc \ln c) \qquad (2.17)$$

In equilibrium, we obtain from equation (2.16)

$$c_0 = \exp\left(-\frac{g}{kT}\right) \qquad (2.18)$$

The case of Schottky pairs in alkali halides or simple binary oxides is slightly more complicated. Considering the pair, irrespective of the location of the partner vacancies, as a single defect with formation free energy g_s, we obtain after simple algebra

$$c_0 = \exp\left(-\frac{g_s}{2kT}\right) \tag{2.19}$$

Deviations of the measured thermal concentrations of defects with regard to the predicted "Arrhenius" behaviour, are often found and attributed to changes in h and s with temperature.

2.6 REACTIONS AMONG DEFECTS—MASS ACTION LAW

Point defects in a solid can react among themselves and give rise to new defects. A simple example of this behaviour is the aggregation of point defects to form clusters. A general reaction involving initial defect species A_i and final species A_j' can be written as

$$a_1A_1 + \ldots + a_pA_p \rightleftarrows a_1'A_1' + \ldots a_g'A_g' \tag{2.20}$$

where the i and j are integer numbers which are prime among themselves.

In order to find the concentrations c_i and c_j' of the various defects that are in equilibrium at a temperature T, the total Gibbs free energy G of the system should again be minimized. This requires that $\delta G = 0$ for any deviation of the concentrations with regard to their equilibrium values. Deviations δc_i and $\delta c_j'$ consistent with reaction (2.20) can be written as

$$\delta c_i = -\lambda a_i$$
$$\delta c_j' = \lambda a_j' \tag{2.21}$$

λ being a constant. Now, the change in G induced by the concentration changes in equations (2.21) should be equated to zero, i.e.

$$\sum_j \frac{\partial G}{\partial c_j'} a_j' - \sum_i \frac{\partial G}{\partial c_i} a_i = 0 \tag{2.22}$$

By taking into account the fact that G is expressed as a sum of terms (equation (2.15)) with different parameters $g_{i \text{ (or } j)}$ and $\alpha_{i \text{ (or } j)}$ for each of the defect species, we obtain after some simple but tedious algebra

$$\prod_j c_j'^{\,a_j'} / \prod_i c_i^{\,a_i} = \frac{1}{\alpha}\exp\left(-\frac{g_R}{kT}\right) \tag{2.23}$$

where

$$\alpha = \prod_i \alpha_i^{\,a_i} / \prod_j \alpha_j'^{\,a_j'}$$

and

$$g_R = \sum_j g_j' a_j' - \sum_i g_i a_i$$

is the free energy of the reaction. Equation (2.23) is the mathematical expression for the mass action law.

A simple example of the application of the mass action law is provided by the aggregation of divalent cation impurities I and cation vacancies V to yield bound I−V dipoles D in alkali halides. The association reaction I + V ⇌ D ($\alpha = 12$) leads to the following expression for equation (2.23)

$$\frac{c_D}{c_I c_V} = \frac{1}{12} \exp\left(-\frac{g_A}{kT}\right) \tag{2.24}$$

where g_A stands for the binding free energy. As $c_I = c_V$, and $c_I + c_D = c$, the total impurity concentration, we can write, in terms of the degree of association given by $p = c_D/c$,

$$\frac{p}{(1-p)^2} = \frac{c}{12} \exp\left(-\frac{g_A}{kT}\right) \tag{2.25}$$

whose solution yields the dependence of p on temperature. This dependence[4] for different impurity concentrations and binding energies is illustrated in Figure 2.1.

2.7 THERMODYNAMICS OF ELECTRONIC STATES

In the previous sections, we have been exclusively concerned with the lattice structure of point defects and ignored electronic contributions. In other words, the partition function and thermodynamic potentials refer to the lattice configuration and do not include electron levels (except possibly for core or closed-shell electrons). This is a serious shortcoming when studying insulator or semiconductor solids where electron states can be markedly altered with increasing temperature or during defect reaction processes. Therefore a complete treatment of the thermodynamics of defects in these materials requires the electronic contribution to the partition function to be taken into account, and this is a rather complex problem for a general case. The situation can be very much simplified if the opposite view to that considered so far is adopted, i.e. the lattice structure is ignored. This applies to the problem of electron redistribution among electronic levels in a semiconductor, ignoring relaxation effects. Although this approach may have serious drawbacks, it is usually adopted in most treatments of the problem.

The partition function for the electron system can be calculated[2,3] within

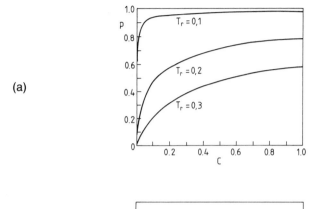

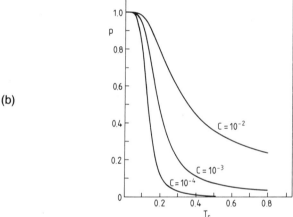

Figure 2.1.
Degree p of impurity–vacancy association predicted by the mass action law in an FCC ionic lattice: (a) as a function of impurity concentration; (b) as a function of temperature. The temperatures T_r are given in reduced units: $T_r = kT/E_A$, E_A being the binding energy of the dipole.

the framework of the Fermi–Dirac statistics and therefore the thermo-dynamic potentials. The relevant parameter is the Fermi level or chemical potential μ, defined as

$$\mu = \left(\frac{\partial F}{\partial N}\right)_{V,T} = \left(\frac{\partial G}{\partial N}\right)_{p,T}$$

where N is the total number of active electrons in the system. By standard methods, we calculate the distribution of electrons among the available energy levels, either forming bands or isolated levels in the interband gap region (donor and acceptor levels). In this way, the state of charge of impurities and lattice defects can, in principle, be determined.

2.8   INTERACTIONS AMONG DEFECTS: THE DEBYE−HÜCKEL METHOD

So far we have assumed that the concentration of point defects was sufficiently low that their mutual interactions can be neglected. In fact, elastic as well as electrical interactions do, generally, exist and are responsible for the formation of defect clusters. Even without the formation of aggregates, the interactions are relevant and give rise to changes in the space distribution of the defects and to changes in the thermodynamic potentials which lead to different equilibrium conditions from those predicted by the simple non-interacting model.

The situation has particular features for ionic solids in which many defects have an effective electrical charge. Then, the long-range Coulomb interaction among defects induces the formation, around a given defect, of an atmosphere or cloud of defects with opposite charge. The effect of such Coulomb interactions on thermodynamic potentials was originally discussed by Lidiard following the Debye−Hückel theory of electrolyte solutions. We are not giving the details of this and we only quote the result obtained for the chemical potential μ:

$$\mu = g + kT \ln a \qquad (2.26)$$

which is the same expression obtained for non-interacting defects, except for the substitution of the activity a instead of the concentration c. The activity a can be put in the form $a = \gamma c$, γ being the corresponding activity coefficient, which for point charged defects is given by

$$\ln \gamma = -\frac{q^2 \chi Z^2}{2\epsilon kT} \qquad (2.27)$$

where q is the electronic charge, Z the valence of the defect, ϵ the dielectric constant of the host material and χ the reciprocal of the screening length or size of the Debye−Hückel cloud. For finite-size defects, the activity coefficient has a slightly different form, but the formula is still valid. Since the expressions for the free energy and chemical potential are formally identical with those valid for defects in dilute solution, the mass action law (2.23) can also apply to Coulomb interacting defects as long as activities are used instead of concentrations.

As an example, the effect of Coulomb interactions on the defect association can be illustrated by considering the binding of divalent impurities to cation vacancies in alkali halides. The Debye–Hückel model predicts a lower degree of association than the non-interacting theory does; the differences are more marked, the higher the temperature or the greater the impurity concentration[5].

The Debye–Hückel model requires a clear distinction between the paired and free defects which is not often easy (more distant pairs can be also considered as bound "excited pairs"). Moreover, the charge and electrical potential are averaged over spheres around a particular charged species. Both difficulties can be removed if the thermodynamic potentials are calculated by summing the effect of all possible configurations of small clusters of defects. This cluster method is considered in more detail in Chapter 13.

2.9 REACTIONS INVOLVING THE SURROUNDING ATMOSPHERE

So far we have assumed that the material is surrounded by a neutral (passive) medium or atmosphere. In other words, reactions between the surrounding medium and the material (e.g. exchange of matter) have not been considered. However, in many practical cases, this is not so and exchange of atomic species takes place. One particular example is the additive colouring procedure in alkali halides, mentioned in Chapter 1, where alkali-metal atoms are incorporated from a gas phase into the crystal through a diffusion process.

A very simple reaction between the atmosphere and the solid material is the solution of hydrogen into a metal. If the hydrogen is dissolved in atomic form, the process can be described by the reaction $H + H \rightleftarrows H_2(g)$. Using the common chemical nomenclature of bracketed species to indicate concentration the mass action law can be written as

$$[H] \propto p_{H_2}^{1/2} \qquad (2.28)$$

i.e. the atomic concentration of hydrogen varies as the square root of the hydrogen partial pressure. In fact, the pairing of hydrogen atoms modifies such a relation into

$$[H] \propto \alpha p_{H_2}^{1/2} + \beta p_{H_2} \qquad (2.29)$$

where the coefficients α and β depend on the degree of pairing. More subtle complications can also arise but they are not considered here.

One of the most widely studied classes of materials involving equilibrium between the solid and gas phase is provided by the non-stoichiometric transition-metal oxides, e.g. $Mn_{1-x}O$, $Fe_{1-x}O$ and $Co_{1-x}O$. Owing to the variable valence of the cations, these oxides gain oxygen from the gas phase according to the reaction

$$\frac{1}{2}O_2 + 3M_M \rightarrow 2M_M^{\cdot} + V_M'' + (MO)$$

where (MO) indicates metal oxide created on the surface of the crystal, $M^{\cdot}$ represents an oxidized lattice cation and V_M'' a cation vacancy having a (double) negative effective charge. Application of the law of mass action to this equation gives

$$p_{O_2}^{1/2} = K_{ox}[M_M^{\cdot}]^2[V_M''] \tag{2.30}$$

If we note, however, that the concentration of vacancies equals the deviation from stoichiometry x and that the oxidation reaction implies that

$$[M_M^{\cdot}] = 2[V_M''] \tag{2.31}$$

then we have, assuming that activities may be equated to concentrations,

$$p_{O_2}^{1/2} = 4K_{ox}x^3$$

i.e.

$$x \propto p_{O_2}^{1/6} \tag{2.32}$$

Such behaviour is indeed observed at low oxygen partial pressures and consequently low deviations from stoichiometry. As x increases, the exponent n of p_{O_2} in the above proportionality deviates from $\frac{1}{6}$, which is commonly attributed to defect clustering. Indeed, if we propose that vacancies trap a single oxidized cation on an adjacent site, then it can be readily shown that we would have $x \propto p_{O_2}^{1/4}$. A complex of one vacancy and two holes would give $x \propto p_{O_2}^{1/2}$.

Measurement of the oxygen exponent can therefore be used to study the defect structure of the oxide. A note of caution is necessary, however. The argument that we have advanced above, rests on ideal solution theory, and deviations from for example the predicted slope of $\frac{1}{6}$ may be due in part to deviations from ideality rather than to defect clustering. Nevertheless, this type of measurement has been made extensively; detailed discussions are given by Kofstad[6], for example.

Before leaving this topic, we should note that there is now good evidence that, for the more grossly non-stoichiometric oxides (e.g. $Fe_{1-x}O$ where x may attain 0.15), there is very extensive clustering involving the formation of complex vacancy aggregates. A detailed discussion is given by Catlow[7].

A similar analysis has been applied to understand the defects induced by oxygen emission during reduction of ternary oxides of the perovskite and related families[8]. In that process, oxygen vacancy centres with different charge states are presumably formed (e.g. F^+ or F centres respectively containing one or two trapped electrons). Depending on the relative stability of those states, different dependences of the electrical conductivity on the oxygen partial pressure are obtained from the mass action law. Experimental data for $SrTiO_3$, $BaTiO_3$ and $LiNbO_3$ are plotted in Figure 2.2. For $LiNbO_3$, the data fit a $\sigma \propto p_{O_2}^{-1/4}$ dependence, suggesting that the F^+ centre is stable but not the F centre. In contrast, the data for $SrTiO_3$ and $BaTiO_3$ are in accordance with the law $\sigma \propto p_{O_2}^{-1/6}$, so that oxygen vacancies are not stable traps for electrons.

2.10 DIFFUSION OF POINT DEFECTS

A point defect located at a given lattice site is separated from a neighbour equivalent site by an energy barrier E_m. This barrier can be overcome with the assistance of thermal energy and therefore the defect can migrate through the material by successive equivalent jumps. The same behaviour applies to a disordered material although in that case it is not appropriate to speak of equivalent sites or jumps. Here, we mostly refer to crystalline materials in which the diffusion processes can be more easily analysed.

The energy barrier E_m is the difference between the energy associated with the defect in an appropriate intermediate (saddle-point) configuration

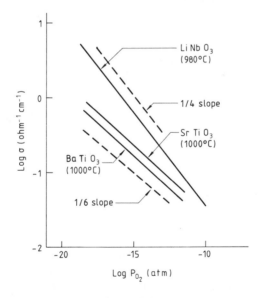

Figure 2.2.
Dependence of electrical conductivity on oxygen partial pressure for ternary oxides.

during jumping and that corresponding to the initial location. In fact, the thermodynamic potentials h_m, f_m and g_m for defect migration can be defined in a similar way to that already discussed for defect formation.

For the theoretical treatment of the diffusion process an appropriate framework is found in the thermodynamics of irreversible processes as well as in general kinetic or statistical theories. However, we are going to adopt a simple phenomenological model that is quite adequate for dealing with most practical situations. More detailed accounts of diffusion can be found in some excellent books (e.g. those by Flynn[9] and Murch and Nowick[10]).

Let us consider a solid at constant temperature and having a non-uniform concentration of point defects. The diffusion process will tend to equalize concentrations and will be governed to first order by the well-known Fick law

$$j = -D \ \text{grad} \ c \tag{2.33}$$

where j is the net current density of defects, $c(x, y, z)$ the defect concentration and D the so-called diffusion coefficient, which is a function of the thermodynamic conditions. For an isotropic or cubic material, D is a scalar. However, for an anisotropic solid, D is a tensor whose structure is determined by crystal symmetry[11]. We restrict the discussion to the isotropic situation since it contains all the physics of the process.

As a general rule, the coefficients D for self-diffusion (host atoms) or diffusion of substitutional impurities in metallic, ionic or covalent solids are, typically, in the approximate range $10^{-10}-10^{-8}$ cm^2 s^{-1} for temperatures close to the melting point. The diffusion of small interstitial impurities such as hydrogen or carbon in metals is generally much faster and D values of about 10^{-4} cm^2 s^{-1} are easily reached in the same conditions mentioned above. For comparison purposes, it should be recalled that, for liquids at room temperature, $D \approx 10^{-6}-10^{-5}$ cm^2 s^{-1} and, for gases under normal conditions, $D \approx 10^{-1}-1$ cm^2 s^{-1}.

2.11   RANDOM-WALK APPROACH—CORRELATION FACTORS

The theoretical calculation of D in terms of the microscopic jump processes is complicated. By using the random-walk approach a simple general expression for D, under the assumption that diffusion proceeds through random discrete jumps of equal length, can be obtained; this reads

$$D = \tfrac{1}{6}\Gamma r^2 \tag{2.34}$$

Γ being the jump frequency and r the jump length.

For the particular case of vacancy diffusion in a monatomic FCC lattice (e.g. copper), we immediately obtain

$$D = \tfrac{1}{12}\Gamma a^2 \tag{2.35}$$

where a is the lattice parameter. In contrast, the coefficient for substitutional impurity diffusion (or self-diffusion) in the same lattice would be

$$D = \tfrac{1}{12}\Gamma c a^2 \tag{2.36}$$

under the assumption that diffusion proceeds through impurity (or host) exchange with neighbouring vacancies (see below); c is the atomic concentration of vacancies. Note that the diffusion mechanism is the same in both cases but that D is much smaller for impurity than for vacancy diffusion.

It is generally assumed that the jump frequency Γ depends on temperature according to the law $\Gamma = \Gamma_0 \exp(-E_m/kT)$, where Γ_0 ($\approx 10^{13}$ s^{-1}) is an effective frequency and E_m is the height of the energy barrier to be overcome during the jump. Consequently the diffusion coefficient D is written as

$$D = D_0 \exp\left(-\frac{E_m}{kT}\right) \qquad (D_0 = \tfrac{1}{6}\Gamma_0 r^2) \tag{2.37}$$

In a first approximation, the effective frequency Γ_0 should depend on the isotopic mass M of the active chemical species as $\Gamma_0 \propto M^{1/2}$. A more rigorous calculation of Γ which takes into account the "many-body" nature of the problem was carried out by Vineyard[12]. For a solid containing N atoms, he obtained

$$\Gamma_0 = \prod_i v_i / \prod_j v_j' \tag{2.38}$$

where v_i and v_j' are the normal mode frequencies when the defect is in the stable and saddle-point configuration, respectively.

In many cases the assumptions made for the derivation of equation (2.34) are not obeyed. For example, the jump probability of the defect may not be the same for all neighbouring equivalent sites but it is conditioned by the previous jump (memory effect). For such a case, we can define a correlation factor f by $f = D/D_R$, D being the true diffusion coefficient and D_R that corresponding to the random case. For vacancy diffusion it is clear that no correlation exists among successive jumps and therefore $f = 1$. However, for substitutional impurity diffusion (or self-diffusion), $f \neq 1$. For FCC, body-centred cubic (BCC) and SC the f values are shown in Table 2.2.

Finally, we should comment that according to chemical rate theory the correct expression for Γ should involve the appropriate thermodynamic potential instead of the barrier energy E_m. In particular, for diffusion at

Table 2.2.
Correlation factor for self-diffusion
in cubic lattices

Lattice	f
FCC	0.781
BCC	0.721
SC	0.655

constant T and p, equation (2.37) should be written as

$$D = D_0\left(-\frac{g_m}{kT}\right)$$

$$= D_0 \exp\left(-\frac{S_m}{k}\right) \exp\left(-\frac{h_m}{kT}\right)$$

$$(2.39)$$

S_m and h_m being the entropy and enthalpy respectively for migration.

There are a number of complicating factors that may induce deviations from typical Arrhenius behaviour (equation (2.37)) for the diffusion coefficient. For example, the occurrence of several competing mechanisms for diffusion or the role of other defects which produce temporary trapping of the diffusing species. However, even for a single mechanism, deviations can arise from anharmonic and quantum effects. It can be shown that anharmonic effects should mainly affect the pre-exponential factor D_0, although it is commonly observed that even for highly anharmonic materials the Arrhenius plot is obeyed over a rather extended temperature range. The role of quantum effects is briefly commented on in section 2.17.

2.12   ISOTOPE EFFECT

This effect refers to the dependence of the diffusion coefficient D on the mass of the diffusing atom. The diffusion coefficient D_α of a given isotope α can be expressed as $D_\alpha = A_\alpha \Gamma_\alpha f_\alpha$, where A_α is a geometrical factor, Γ_α the jump frequency of the isotope and f_α the correlation factor; so, in principle, it is expected that $\Gamma_\alpha \propto m_\alpha^{-1/2}$ and $D_\alpha \propto m_\alpha^{-1/2}$. However, the situation is more complicated, since the correlation factor f_α is also dependent on isotopic mass. A simple approach leads to the formula

$$f_\alpha = \frac{u}{u + \Gamma_\alpha} \qquad (2.40)$$

u being the competing jump frequency for other (host) atoms which is, in general, different from Γ_α. By using this formula, we obtain

$$f_\alpha = \frac{D_\alpha/D_\beta - 1}{(m_\beta/m_\alpha)^{1/2} - 1} \tag{2.41}$$

which can be considered as the basic formula describing isotope diffusion effects. It does not take into account many-body effects illustrated by Vineyard's treatment as well as other corrections which lead to the introduction of a correction factor. Anyhow, the measurement of the isotope effect has often been used to obtain f and so to confirm a given diffusion mechanism.

2.13 THE NERNST−EINSTEIN RELATIONSHIP

Since electrical transport and mass transport (diffusion) in ionic crystals are both related to elementary jump events of the mobile species, it might be expected that some general relationship between the electrical mobility μ and the diffusion coefficient D exists. A simple derivation, using general statistical arguments has been given by Mott and Gurney[13] for a single diffusing species. Consider that the diffusing ions are in equilibrium under an applied electric field. The total electric field E (including the space charge field) at a given point is given by $E = -\text{grad } \mathcal{V}$, $\mathcal{V}$ being the electrical potential. The distribution of ions should follow the Maxwell−Boltzmann statistics, i.e.

$$n(r) \propto \exp\left(-\frac{q\mathcal{V}(r)}{kT}\right) \tag{2.42}$$

q being the ion charge. By taking gradients in both members,

$$\text{grad } n = -\frac{nq}{kT}\text{grad } \mathcal{V}$$
$$= \frac{nqE}{kT} \tag{2.43}$$

However, the equilibrium condition of the system should be written as

$$nq\mu E = qD \text{ grad } n \tag{2.44}$$

By comparing this with equation (2.43), we immediately obtain

$$\frac{\mu}{D} = \frac{q}{kT} \tag{2.45}$$

or

$$\frac{\sigma}{D} = \frac{nq^2}{kT} \tag{2.46}$$

which constitute the Nernst–Einstein equations. So, in situations where these relationships are assumed, measurements of ionic conductivity could provide a convenient method for determining diffusion coefficients.

2.14   DIFFUSION MECHANISMS

Some well-documented mechanisms for atomic diffusion via point defects have been shown to operate in a large variety of materials.

(1)   The first mechanism which has been already mentioned as an illustrative example is vacancy diffusion. The atoms migrate through successive jumps into neighbouring vacancies. This mechanism implies a correlation factor which depends on lattice type (see Table 2.2). However, the effective activation enthalpy h for this mechanism in a monatomic lattice is the sum of the proper migration enthalpy h_m plus the formation enthalpy h_f of the vacancy. The latter component derives from the thermal concentration of vacancies that is available for jumping. This is the operative mechanism for self-diffusion and substitutional impurity diffusion in most metals, ionic and covalent solids. However, the situation becomes more complicated in many cases because of the role played by vacancy clusters (divacancy, trivacancy, etc.) or vacancy–impurity complexes. Therefore, a detailed analysis of diffusion in a given material requires careful consideration of all these possibilities. Anyhow, these contributions are often responsible for deviations of the coefficient D from a simple Arrhenius behaviour.

(2)   A mechanism applicable to interstitial atoms is the so-called interstitial mechanism involving the jump of the atom from an interstitial site to a neighbouring site. For this case, the correlation factor f is 1 and the activation enthalpy is simply that corresponding to migration. Interstitials generally diffuse through this mechanism in most materials.

(3)   For some interstitial systems, another mechanism (called interstitialcy) operates. It involves the exchange of the interstitial atom with a regular host atom which is ejected from its site and then occupies another interstitial position. It usually requires rather low activation energies and provides an interesting channel for fast diffusion. It has been shown to apply to some interstitial halogen centres in alkali halides and silver interstitials in silver halides.

Examples of diffusion mechanisms are mentioned in Chapter 7.

Some other mechanisms have been invoked (e.g. the Zener mechanism) but they lack adequate experimental support or only apply to specific situations. In contrast, for diffusion in dielectric solids, electronic effects may be important as reported[14] for transition-metal ions in AgCl. Apparently, crystal field effects are responsible for relevant contributions to the activation enthalpy, which has been found to vary systematically over the first transition-metal row.

2.15   ENHANCED DIFFUSION

The above discussion of diffusion processes implicitly assumes that bulk processes, typical of the perfect material are being dealt with. In many cases, D values which are much higher than expected for such perfect bulk processes[10] are measured experimentally. The generic reason for this behaviour lies in the existence of extended defects or defective regions in the material which can act as easy paths for atom migration. In fact, there is a large volume of experimental work documenting diffusion along grain boundaries and dislocations in many materials, especially metals. The situation is very clearly illustrated by the diagram prepared by Gjostein[15] for FCC metals. Indeed, for polycrystalline materials, grain boundary effects almost certainly predominate in most diffusions.

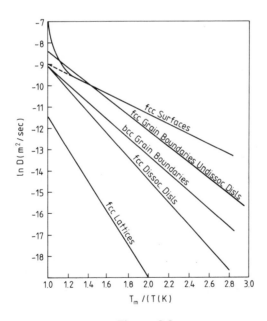

Figure 2.3.
Diagram showing the diffusion coefficients in FCC metals contributed by various diffusion-enhancing processes. Temperatures are normalized to the melting point T_m.

Theoretical models have been developed to take into account the role of dislocations. In a simple picture the effective diffusion coefficient can be written as $D_{eff} = D_l + \beta(c_d/c_l)D_d$ where the subscripts d and l refer to dislocation and lattice respectively. c_d/c_l is the corresponding ratio of the concentrations, and β is the atomic fraction at the dislocations.

However, the presence of other point defects may also contribute to an enhanced diffusion coefficient. One recent example is provided by the enhanced deuterium diffusivity in lithium-doped MgO. The diffusion coefficient D at 1900 K has been shown[16] to be three orders of magnitude larger in MgO:Li than that in pure MgO. The effect has been attributed to the presence of Li_2O precipitates.

Enhanced diffusion has also been observed during irradiation, but this is commented on in Chapter 3 and Chapter 7.

2.16   VERY FAST TRANSPORT—SUPERIONIC CONDUCTORS

Fast transport occurs in metallic as well as in covalent and ionic materials. For all solids, light interstitial atoms, such as hydrogen (also muons and pions) may reach very high diffusion coefficients. For these atoms, anharmonic and quantum effects may be important for a correct description, particularly at low temperatures.

Much interest has arisen in rapid transport in ionic solids because of technological implications (e.g. solid-state batteries). There are some books or general reviews that the reader may consult for a detailed account (e.g. those by Hagenmuller and Van Gool[17] and Collongues et al.[18]). One important characteristic of these ionic materials is that they are not close packed so that they contain tunnels or passage-ways to allow for a rapid transit of ions. These tunnels are formed by connected void sites where ions of a given sublattice reside and move. Typically, there are more sites than atoms in these tunnels, and the sites are often randomly occupied at the appropriate temperatures. Therefore, ionic transport is quite selective, i.e. it applies to only one of the sublattices. As an example, β-alumina compounds present cationic conduction, whereas fluorite-like oxides are anionic conductors.

With regard to the diffusion or conduction behaviour, superionic conductors have some main properties.

(1)   Very high values and small temperature dependence of D and σ.

(2)   Very small values of D_0 and σ_0. This implies a small value of the attempt frequency ν_0 or of the entropy s_m of migration.

There have been a variety of theoretical approaches adopted to understand this fast conduction, some of them stressing the connection of this behaviour with an order–disorder transformation of the conducting sublattice.

2.17   QUANTUM DIFFUSION

The above picture of diffusion has assumed that the jump of the migrating atom from a given lattice site to the next site takes place by thermally overcoming the energy barrier separating the two sites but this is not always the case. There are situations where diffusion can occur by "under-barrier" jumps, i.e. quantum tunnelling through the barrier. This effect can give rise to coherent (band-type) motion of the atoms at low temperatures. At high temperatures the coherent motion turns into an incoherent thermally activated tunnelling, i.e. diffusion enters into the so-called hopping regime. This type of behaviour also yields an Arrhenius-type dependence on temperature and cannot be easily distinguished from ordinary "above-barrier" diffusion.

A model for quantum diffusion was put forward by Flynn and Stoneham[19] (also discussed by Kagan and Klingerr[20]) which is based on the theory of small-polaron motion, initially developed by Holstein[21]. Quantum diffusion has been revealed when studying muon diffusion in metals, probably because of the light mass of that interstitial particle, together with the perfection of the crystals and the low concentration involved. Quantum effects are also apparently important in order to understand some features of the low-temperature diffusion of hydrogen in metals, particularly an anomalous isotope effect.

2.18   PROCESSES RELATED TO DIFFUSION

There are a number of relevant solid-state processes which are directly related to diffusion[22]. Some of them are of great technological importance, particularly in the fields of metallurgy and ceramics, e.g. sintering, oxidation of metals, ageing of alloys and creep. We are going to briefly comment on two of them, as illustrative examples.

(a)   Oxidation of metals.   Although this is a very complex problem, it is possible to bring out the role of diffusion by recourse to a simplified and idealized simple model. It may be assumed that the oxide film developing onto a metal surface increases its thickness by diffusion of the metal atoms through the oxide. If the metal concentration in the oxide film is taken to be inversely proportional to its thickness, then the Fick equation can be written

$$\frac{\mathrm{d}z}{\mathrm{d}t} = \frac{\chi}{z} \tag{2.47}$$

where the first term represents the rate of increase in the oxide film thickness and therefore the flow of metal atoms contributing to film growth. The solution of equation (2.47) is

$$z^2 = 2\chi t \tag{2.48}$$

which expresses the well-known parabolic law for the kinetics of oxidation. Of course, a number of complicated features have to be included in a more realistic model[23] but the basic idea is still correct.

(b) Creep. This is defined as the straining of a material at a constant stress and temperature[24]. As a general rule, there is an initial stage of rapid creep, followed by another slower stage in which the strain rate is constant. This latter behaviour is associated with the climb of dislocations to overcome the strain-induced obstacles and therefore it is governed by point defect diffusion. The constant creep rate is usually written as

$$\frac{d\epsilon}{dt} = C \frac{\sigma^m}{kT} \exp\left(-\frac{h}{kT}\right)$$

where C and m are constants. h is the activation energy for the creep process; this has been found to be equal to the self-diffusion activation enthalpy for a number of metals. In other words, $h = h_f + h_m$, h_f and h_m being the formation and migration enthalpies respectively for vacancies, which are responsible for the diffusion mechanism and consequently dislocation climb.

REFERENCES

1 T. L. Hill, *An Introduction to Statistical Thermodynamics*, Addison-Wesley, Reading, MA (1960).
2 C. Kittel, *Thermal Physics*, Wiley, New York (1969).
3 C. R. A. Catlow, J. Corish, P. W. M. Jacobs and A. B. Lidiard, *Report* No. AERE-TP-873, Atomic Energy Research Establishment (1980).
4 A. B. Lidiard, Ionic conductivity, *Encyclopedia of Physics*, Vol. XX, Springer, Berlin (1957).
5 C. Wagner, *Annu. Rev. Mater. Sci.* 7, 1 (1977).
6 P. Kofstad, *Non-stoichiometry, Diffusion and Electrical Conductivity in Binary Metal Oxides*, Wiley, New York (1972).
7 C. R. A. Catlow, in O. T. Sorensen (ed.), *Non-stoichiometric Oxides*, Academic Press, New York (1981).
8 L. Arizmendi, J. M. Cabrera and F. Agulló-López, in A. Domínguez-Rodríguez, J. Castaing and R. Márquez (eds), *Basic Properties of Binary Oxides*, Universidad de Sevilla, Seville (1983).
9 C. P. Flynn, *Point Defects and Diffusion*, Oxford University Press, Oxford (1972).
10 G. E. Murch and A. S. Nowick (eds), *Diffusion in Crystalline Solids*, Academic Press, New York (1984).

11 J. F. Nye, *Physical Properties of Crystals*, Clarendon, Oxford (1957).
12 G. H. Vineyard, *J. Phys. Chem. Solids* **3**, 121 (1957).
13 N. F. Mott and R. W. Gurney, *Electronic Processes in Ionic Crystals*, Oxford University Press, Oxford (1948).
14 A. P. Batra, J. P. Hernández and L. M. Slifkin, *Phys. Rev. Lett.* **36**, 876 (1976).
15 N. A. Gjostein, *Diffusion*, American Society of Metals, Metals Park, OH (1973).
16 R. González, Y. Chen and K. L. Tsang, *Phys. Rev. B* **26**, 4637 (1982).
17 P. Hagenmuller and W. Van Gool, *Solid Electrolytes*, Academic Press, London (1978).
18 R. Collongues, A. Kahn and D. Michel, *Annu. Rev. Mater. Sci.* **9**, 123 (1979).
19 C. P. Flynn and A. M. Stoneham, *Phys. Rev. B* **1**, 3966 (1970).
20 Yu. Kagan and M. I. Klingerr, *J. Phys. C: Solid State Phys.* **7**, 2791 (1974).
21 T. Holstein, *Ann. Phys.* (*NY*) **8**, 343 (1959).
22 A. M. Stoneham, *Adv. Phys.* **28**, 457 (1979).
23 K. R. Lawless, *Rep. Prog. Phys.* **37**, 231 (1974).
24 P. Haasen, *Physical Metallurgy*, Cambridge University Press, Cambridge (1978).

3
Energetic Methods of Defect Production

3.1 INTRODUCTION

In this chapter, we consider defect formation as a consequence of displacements of lattice atoms by energetic particles[1]. The most direct approach in a study of point defects is to use electron irradiation as the electron energy may be adjusted so that only the simplest of vacancy and interstitial structures are formed. When the irradiations and measurements are made at a low temperature (about 4 K), then these defects can be studied with minimal problems from subsequent defect aggregation, motion or annealing.

More complex defect structures develop during ion beam implantation or neutron irradiation because of the localized high-energy transfer which can produce a cascade of atomic displacements. Despite the complexity of the problem from a viewpoint of defect studies, ion implantation is central to a wide range of materials technology, most obviously in semiconductor device production.

In this chapter, emphasis is on the mechanisms of energy transfer to the material and the depth of penetration into the solid, and an estimate will be given of the number of atomic displacements which ensue.

3.2 ELECTRON IRRADIATION OF SOLIDS

3.2.1 Energy transfer to the lattice

As for all charged particles, electrons moving through a material lose energy by electronic excitation of the lattice. A second possibility is direct momentum transfer to the atoms by collision. At high energies (1 MeV or higher)

an additional loss term from the production of bremsstrahlung radiation
becomes important.

Electron irradiation can produce defects either by direct momentum
transfer or by causing more subtle lattice relaxations following electronic
excitation (see Chapter 4). For metallic and most covalent and ionic materials, energy transfer by collision is the dominant mechanism for inducing
damage by atomic displacements. The use of relativistic electrons to form
simple vacancy and interstitial defects is well documented for all types of
material from metals to insulators and offers a direct route to the identification of such defects. The choice of electron energy is determined by the
need to knock atoms adiabatically from the lattice sites. For simplicity this
should be done within the framework that there are no impurity or stoichiometry changes, that only one type of atom is displaced in a compound
material, that the energy is only sufficient to move one atom and does not
cause divacancies or a disordered region and finally that no diffusion is
allowed. These requirements can be fulfilled when the irradiation damage is
made by relativistic electrons to pure targets at a low temperature.

At energies in the range 0.5–2.0 MeV the electron has sufficient momentum to displace lattice atoms directly but insufficient momentum for multiple damage events. Even with a compound system there will exist an energy
region in which the energy transferred during a collision is above the
threshold for displacement of the lighter atom but is insufficient to displace
the heavier atom.

Energetic electrons have a relativistic mass, and their kinetic energy is

$$E_1 = (m - m_0)c^2 \qquad (3.1)$$

where $m = m_0(1 - v^2/c^2)^{-1/2}$, v being the electron velocity. In a head-on
collision the maximum energy E_m is transferred to an atom of mass M_2
where

$$E_m = \frac{2E_1(E_1 + 2m_0c^2)}{M_2c^2} \qquad (3.2)$$

Most collisions are not head on; so, to compute the total probability of
displacing an atom, we use the differential cross-section $d\sigma$ for Rutherford
scattering of fast electrons through an angle θ given by

$$d\sigma(\theta) = \left(\frac{Ze^2}{2mv^2}\right)^2 \operatorname{cosec}^4\left(\frac{\theta}{2}\right) d\Omega \qquad (3.3)$$

The solid angle of this annular cone is

$$d\Omega = 2\pi \sin\theta \, d\theta = 4\pi \sin\left(\frac{\theta}{2}\right)\cos\left(\frac{\theta}{2}\right) d\theta \qquad (3.4)$$

We are dealing with relativistic electrons and so

$$d\sigma(\theta) = \left(\frac{Ze^2}{m_0c^2}\right)^2 \frac{\pi}{\beta^4\gamma^2} \cos\left(\frac{\theta}{2}\right) \operatorname{cosec}^3\left(\frac{\theta}{2}\right) d\theta \qquad (3.5)$$

where $\beta = v/c$ and $\gamma = 1/(1 - \beta^2)^{1/2}$.

Unfortunately we are treating the electron as a classical particle, which is only true for a de Broglie wavelength which is short compared with the distance of closest approach of the electron to the atom. To satisfy this condition we require that $Ze^2/\hbar c > 1$. Evaluating this term gives $Z/137$ ($=\alpha$), which is much less than unity. A more formal treatment for electron scattering has been made by Mott[2], and the quantum mechanical cross-section is related to the classical Rutherford scattering by $d\sigma_{QM} = R_s$ $d\sigma_{classical}$. The factor R_s has an approximate analytical form given by McKinley and Feshbach[3]:

$$R_s = 1 - \beta^2 \sin^2\left(\frac{\theta}{2}\right) + \pi\alpha\beta \sin\left(\frac{\theta}{2}\right)\left[1 - \sin\left(\frac{\theta}{2}\right)\right] \qquad (3.6)$$

Hence the scattering cross-section becomes

$$d\sigma(\theta) = \frac{\pi Z^2 e^4}{m_0^2 v^4}(1 - \beta^2)\left\{1 - \beta^2 \sin^2\left(\frac{\theta}{2}\right)\right.$$

$$\left. + \pi\alpha\beta \sin\left(\frac{\theta}{2}\right)\left[1 - \sin\left(\frac{\theta}{2}\right)\right]\right\} \cos\left(\frac{\theta}{2}\right) \operatorname{cosec}^3\left(\frac{\theta}{2}\right) d\theta \qquad (3.7)$$

To relate this angular dependence to energy, we noted that the energy transferred at an angle θ is related to that for a head-on collision by

$$E = E_m \sin^2\left(\frac{\theta}{2}\right) \qquad (3.8)$$

Thus the differential cross-section for energy transfer becomes

$$d\sigma(E) = \frac{\pi b^2(1 - \beta^2)E_m}{4}\frac{1}{E^2}\left\{1 - \beta^2\frac{E}{E_m} + \pi\alpha\beta\left[\left(\frac{E}{E_m}\right)^{1/2} - \frac{E}{E_m}\right]\right\} dE \qquad (3.9)$$

where

$$b = 2Ze^2/m_0v^2. \qquad (3.10)$$

3.2.2 Production of displacements—threshold energy

In order to determine whether an atom will be displaced from its lattice position, it is usually assumed that a minimum well-defined energy or threshold is required. If the threshold energy for displacements is E_D, we obtain the total cross-section for displacement by integration from E_D to E_m, i.e.

$$\sigma = \frac{\pi b^2 (1 - \beta^2)}{4} \left(\frac{E_m}{E_D} - 1 - \beta^2 \ln\left(\frac{E_m}{E_D}\right) \right.$$
$$\left. + \pi\alpha\beta \left\{ 2\left[\left(\frac{E_m}{E_D}\right)^{1/2} - 1 \right] - \ln\left(\frac{E_m}{E_D}\right) \right\} \right) \qquad (3.11)$$

If the maximum transferred electron energy is close to the threshold energy for displacement, then only the primary event produces damage, and the above formula approximates to

$$\sigma = \frac{\pi Z^2 e^4 (1 - \beta^2)}{\beta^4 m_0^2 c^4} \left(\frac{E_m}{E_D} - 1 \right) \qquad (3.12)$$

Typical cross-section curves are shown in Figure 3.1 for a variety of displacement energies of carbon atoms as a function of the primary-electron energy. The cross-section increases rapidly from a threshold value of the incident electron energy (depending on displacement energy) and then becomes effectively constant. The subject was developed extensively in the book by Corbett[1], including effects of secondary defect formation. To compute the total defect concentration and its depth distribution, we require the range−energy relation for relativistic electrons. Various empirical expressions have been used; for example, Glendennin[4] gives a range R for

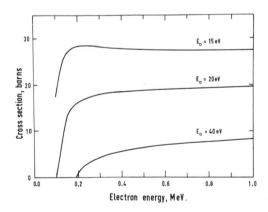

Figure 3.1.
Examples of the calculated cross-sections for atomic displacements in carbon for fast electron bombardment. Curves are shown for different assumed displacement energies E_D.

unit-density material in units of milligrams per square centimetre for mega-electronvolt electrons as

$$R = 407E^{1.38}, \qquad 0.15 \text{ MeV} < E < 0.8 \text{ MeV}$$

$$R = 542E - 133, \qquad E > 0.8 \text{ MeV}$$

Alternative relations by Katz and Penfold[5] and Kobetich and Katz[6] take the forms

$$R = 412E^{1.265 - 0.0954 \ln E}, \qquad E < 2.5 \text{ MeV}$$

$$R = 530E - 106 \qquad E > 2.5 \text{ MeV}$$

and

$$R = 537E\left(1 - \frac{0.9815}{1 + 3.123E}\right), \qquad 0.3 \text{ keV} < E < 20 \text{ MeV}$$

As an example of the range, a 2 MeV electron is stopped in 2.38 mm in Al_2O_3, which has a density of 3.97 g cm^{-3}. There is some uncertainty in the range of very-low-energy electrons, but a typical energy loss is 10 eV nm^{-1} in unit-density material.

A crucial factor for the calculation of the number of displaced atoms is the displacement energy. Experimentally this energy can be found by comparing the computed displacement curve as a function of primary-electron energy with an observed parameter change, such as resistivity or the strength of an optical absorption band. The example shown in Figure 3.2 gives the

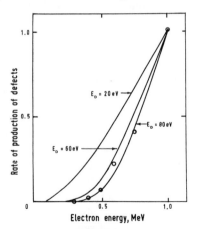

Figure 3.2.
A comparison of the normalized displacement cross-section curves with experimental results for diamond. The optical absorption band at 2.0 eV was used as a measure of the damage.

curves for the rate of production of defects in diamond for threshold energies of 20, 60 and 80 eV. These are compared with experimental results for the growth of the optical absorption band at 2.0 eV. The curves, when normalized at an electron energy of 1.0 MeV, show that the high value of displacement energy is appropriate in this material. Similar results have been obtained with the conductivity changes produced in semiconducting diamonds. It is important to note that the curve shape must be used rather than the threshold energy at which damage becomes detectable. The latter approach may merely reflect the sensitivity of the measuring technique, particularly if minor processes exist which can cause some subthreshold events.

One minor process which can occur in compound materials is the transfer of energy from the electron to a heavy atom via collisions with a light atom. The closer the masses m, M_1 and M_2, the more efficient is the transfer process. As an example consider Al_2O_3, which has ions of mass 27 and 16. If the displacement energy for aluminium is 40 eV, then a direct collision requires a primary electron energy of 0.385 MeV. However, the more efficient two-stage collision in which an electron strikes an oxygen ion which in turn strikes an aluminium ion will, in the optimum case, deliver 40 eV to the aluminium starting from a 0.29 MeV electron. Work on colour centre formation in sapphire (Al_2O_3) has shown that there is a finite rate of production of the 6.1 eV optical absorption band (corresponding to the oxygen vacancy) even at 0.3 MeV, although the threshold energy predicted from curves such as those in Figure 3.1 is above 0.38 MeV. It should be noted that indirect collision routes involving lighter ions such as hydrogen (e.g. as OH contaminants) are even more efficient; so an $e^- \rightarrow H \rightarrow 0$ process would cause the apparent threshold to drop towards 0.13 MeV.

3.2.3   Displacement energies in real lattices

A lattice is not isotropic, and so the energy required to displace an atom from the lattice site will depend on the direction in which the atom moves. The effect will be most pronounced in an anisotropic lattice, which has open channels as well as close-packed directions.

Detailed calculations of the displacement energy as a function of direction were shown rather concisely in the work of Erginsoy et al.[7] for iron. The contours of equal displacement energy about three crystalline axes are plotted in Figure 3.3. The calculation also emphasized the role of replacement collision sequences as a means of separating vacancies and interstitials (see section 3.2.4).

Because electron irradiation transfers most energy in a direct forward collision, the number of defects formed in thin samples directly reflects this

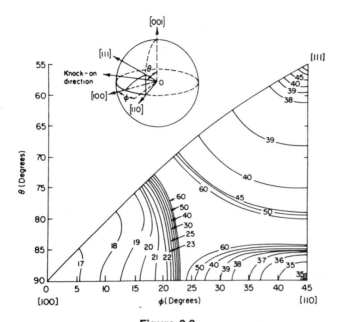

Figure 3.3.
Calculated contours of constant-displacement threshold in the triangle bounded by $\langle 100 \rangle$, $\langle 110 \rangle$ and $\langle 111 \rangle$ directions in iron.

anisotropy of E_D. In thin samples the electron does not diverge significantly from the incident direction and it remains essentially monoenergetic; hence we can sense the directionality of the threshold energy. Note, however, that for most calculations of the number of displaced atoms this directionality feature is ignored and an average value appropriate for thick targets is used.

3.2.4 Secondary effects

The primary struck atoms can also induce additional displacements and damage if they possess sufficient energy. Typically for incident electrons in the energy range less than 0.75 MeV the number of these secondary displacements is negligible. At higher incident electron energies the struck particle may in turn displace a second or third lattice atom. For electron irradiation, where such events are fairly minor, allowance could be made for them by introducing a term with an energy $2E_D$. With higher mass incident particles, as used in neutron or ion bombardment, a more complete description of these damage cascades has been developed[8,9] and this is mentioned in section 3.4.2.

One important feature associated with the secondary processes is the occurrence of replacement collisions in which the impinging atom moves to a neighbouring lattice site after having transferred its kinetic energy to it. This may initiate a sequence of replacements leading to the formation of an interstitial atom far away from the vacancy, which is left at the initial collision site. A well-separated Frenkel pair is thus formed and is more stable than a close vacancy–interstitial pair. These replacement collisions may induce appreciable compositional disorder in ordered alloys or compounds.

The crystalline structure of a material also has an effect on the secondary collision cascade. It may happen that the knocked-on atom has received momentum along a direction close to a close-packed row of the lattice. Then a focusing effect of the collision axis towards the close-packed direction takes place, leading to the propagation of the incoming energy several lattice distances away from the starting collison event. This preferential energy propagation is also required to account for sputtering patterns. The focusing sequence is illustrated in Figure 3.4 where the successive collision angles θ_n fulfil the condition $\theta_n < \theta_{n-1}$. For a given geometry, the critical condition for focusing is controlled by the space D between the ions and their diameter $2R$. Successive deflection angles are θ_1 and θ_2 and, in a hard-spheres model, are related by

$$\frac{\theta_2}{\theta_1} = \frac{D}{2R} \tag{3.13}$$

Hence focusing occurs if $\theta_2 < \theta_1$, i.e. $D < 4R$. If the inter-ion gap is too large, the collision sequence will defocus.

3.3   ION IMPLANTATION AND DAMAGE

In modern technology, ion implantation is a powerful tool for control of surface properties, as has been shown by Townsend[10], Ziegler[11] and Hubler et al.[12], for example; nonetheless the process is highly destructive and an understanding of the defect processes is essential. In this section a brief discussion is given of the energy transfer to the lattice which leads to the implant and damage distributions in stable amorphous materials. This is clearly a convenient theoretical simplification but in practice the results for crystalline materials are not significantly different unless there are additional considerations such as defect generation by electronic excitation or radiation-enhanced motion. Ion beam channelling is apparent in crystals for small ion doses (less than 10^{13} ions cm^{-2}) but at higher ion doses the radiation damage reduces channelling. In Chapter 14 the economics of industrial ion beam implantation will be considered for various materials.

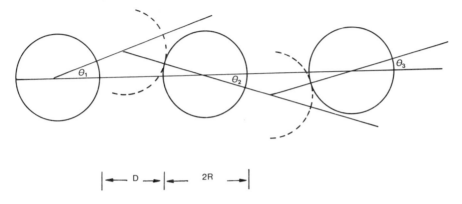

Figure 3.4.
A collision sequence for ions is focused (i.e. $\theta_1 > \theta_2 > \theta_3$) if $D < 4R$.

3.3.1 Mechanisms of energy transfer

When an energetic ion is incident on a solid target, there is a very rapid transfer of energy from the ion to the target atoms. The total rate of energy transfer can be many kiloelectronvolts per micrometre of ion path. Obvious consequences are that the original lattice structure becomes disordered (radiation damage), atoms leave the surface (sputtering) and the implanted ions remain buried at the end of their damage track (doping). The energy transfer occurs via several mechanisms with the dominant processes being excitation of electronic states or nuclear collisions between the incident ions (of mass M_1) and the target ions (of mass M_2). In studies of metals and semiconductors the radiation damage and sputtering are almost entirely governed by the energy deposited in nuclear collisions.

The mechanisms have an energy-dependent cross-section which varies with M_1 and M_2, but the damage or sputtering effects are only significant for energies below about 500 keV and the peak cross-section typically occurs in the energy range 10–100 keV. By defining scaling functions involving mass M and atomic number Z, we can generate a "universal" curve for the energy E dependence of the nuclear stopping cross-section (i.e. dE/dx) for all ion–target combinations. In the original description by Lindhard et al.[13] the universal energy parameter ϵ_u appears as

$$\epsilon_u = \frac{EM_2[0.885a_0(Z_1^{2/3} + Z_2^{2/3})^{-1/2}]}{Z_1Z_2e^2(M_1 + M_2)} \qquad (3.14)$$

where a_0 is the Bohr radius. The form of the nuclear collision cross-section curve as a function of energy is given by Figure 3.5. It reaches a maximum value in the kiloelectronvolt energy range and approaches the classical Rutherford cross-section at higher energies.

The rate of electronic energy deposition into the target is proportional to ion velocity for energies below about 1 MeV. At higher velocities the cross-section decreases with increasing energy and the target becomes progressively more "transparent". This occurs when the ion velocity is greater than the orbital velocity of the electrons.

In the current applications of ion implantation for semiconductor or metal processing, the ion energies are typically below 400 keV; so only nuclear collision damage is a major consideration. The dose scale for semiconductors is characteristically some $10^{12}-10^{15}$ ions cm^{-2} (e.g. As$^+$ or P$^+$ into silicon); this leads to radiation damage but negligible sputtering, as is discussed later. Metal processing is less simple and in excess of 10^{17} ions cm^{-2} may be required to achieve a property modification (e.g. for nitriding). At these high doses all the atoms in the implanted layer will have been displaced many times and the surface will have receded by sputtering. The energy deposited in electronic excitation is important in the implantation of insulators as lattice relaxations caused by electronic rearrangement may lead to point defect formation and/or defect annealing. The displacement energy for an electronically excited lattice may be only a few electronvolts as the lattice relaxes, rather than the 25 eV needed for dynamic events; so there may also be synergistic effects between the electronic and nuclear collision processes of energy transfer. In some insulators (e.g. the alkali halides),

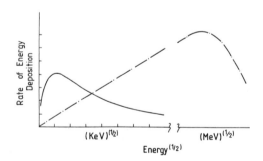

Figure 3.5.
The rate of energy deposition is shown for electronic excitation ($-\cdot-$) and nuclear collisions ($——$). The maxima occur near a few megaelectronvolts or tens of kiloelectronvolts for these processes for most ion-target combinations. The nuclear stopping curve can be scaled to fit all cases.

displacements may be almost entirely by electronic excitation; this has been observed during energetic H^+ or He^+ implantation[14,15].

The literature on ion implantation effects in insulators contains examples where light ions are used at a few megaelectronvolts to penetrate several micrometre into the solid[14]. In these cases the nuclear-collision-induced damage is mainly buried at the end of the ion range and sputtering is negligible.

3.3.2   Ion ranges and damage distributions

Considerable effort has been made to understand the ion−target collision processes and hence to predict the depth distribution of the implanted ions, the distribution of the radiation damage and the lateral spreading of the ion beam. In earlier theoretical models the interaction potentials were chosen with an analytic form such as those proposed by Thomas and Fermi, by Molière or by Lenz and Jensen. More recently, greater computing power has enabled the use of more complex numerical expressions for the inter-action potentials. Many tabulations of the ion range, nuclear and electronic energy loss rates now exist[11,16,17] and, as the data base of experimentally determined values has increased, more recent tables include empirical cor-rection factors so that ion ranges may be confidently predicted to better than $\pm 10\%$.

For lower-energy implants where nuclear collisions are important, the implanted ion distribution is approximately Gaussian in form and the con-centration with depth is characterized by moments termed the projected range R_p and the range spread ΔR_p. The concentration at a depth x for an incident ion flux of ϕ ions cm^{-2} in a target of atomic density N is

$$C(x) = \frac{\phi}{(2\pi)^{1/2}N\,\Delta R_p} \exp\left(\frac{-(x-R_p)^2}{2\pi\,\Delta R_p^{\,2}}\right) \qquad (3.15)$$

Hence R_p corresponds to the maximum concentration of implanted ions. Because of ion scattering, the total path R_{tot} is greater than the projected range R_p, and an approximation of the total range is given for low-energy ions as

$$R_{tot} \approx \left(1 + \frac{M_2}{3M_1}\right)R_p \qquad (3.16)$$

Experimental values of the ion penetration as a function of energy are shown in Figure 3.6.

The damage distributions are of a similar form but peak closer to the surface as atomic displacements occur not only from collisions with the incident ion but also as a consequence of energetic displaced target ions.

The Gaussian form of these distributions reflects the statistical nature of the scattering events. Figure 3.7 shows a comparison between the experimentally[18] determined ion range and damage distributions for iron bombarded with 100 keV N^+ ions and a theoretical calculation for this case[19] which gives $R_p \approx 105$ nm, $\Delta R_p \approx 45$ nm and the damage peak at about $0.7R_p$. There is thus good agreement between theory and experiment.

3.3.3 Lateral deviations

As a result of multiple collisions the ions will be deviated from their original direction and there will be a lateral spreading R_{lat} of the ion beam in the solid. The degree of spreading will increase with increasing depth into the target and it is a function of the mass ratio M_2/M_1. There is least deviation for heavy ions incident into a light target; for $M_2/M_1 \approx 0.1$ the spread R_{lat} is approximately ΔR_p and is some 20% of R_p. The spreading increases with increasing M_2/M_1 ratio and is about 50% of R_p for $M_2 = M_1$; it rises to $2R_p$ for $M_2/M_1 = 10^{20}$.

In practical terms the low values of R_{lat} are of considerable importance as they imply that, if a target is doped by ion implantation in a region defined

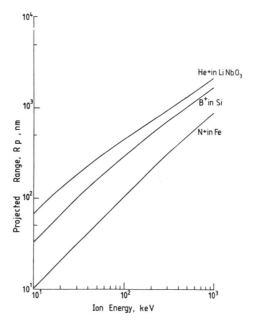

Figure 3.6.
Examples of projected ion ranges for several ion–target combinations.

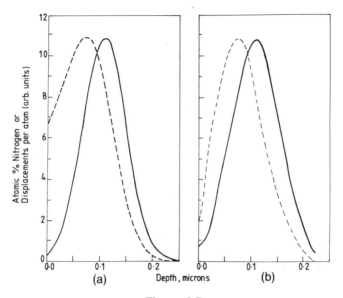

Figure 3.7.
A comparison of (a) experimental data and (b) theory for the impurity profile
(——) and damage distributions (– – –) produced by a 100 keV N^+ ion implantation in iron. The theoretical calculation assumes an ion dose of 10^{16} ions cm^{-2}
whereas the experimental data were for 10^{17} ions cm^{-2}.

by a mask window, then the surface pattern is only slightly degraded with
increasing depth into the solid. For semiconductor manufacture it is possible
to preserve the high definition of lithographic masks and hence to form
abrupt junctions or close positioning of differently doped regions; because
this can minimize capacitance effects, the devices can be designed for high-
speed operation. This feature is a major advantage of ion implantation and
may be contrasted with earlier techniques of diffusion doping at high
temperatures in which the dopants spread laterally beneath the mask as well
as entering the crystal. It is also evident that thermal diffusion gives a
concentration profile which is peaked at the surface whereas by varying the
ion energy we can, in principle, tailor the depth profile of the ion-implanted
dopants.

3.3.4 Channelled ion ranges

In the preceding discussion and in range tabulations the target is assumed to
be amorphous and isotropic. Computationally this is a simpler problem than
a calculation of ion ranges in a perfect crystalline solid, as in a crystal the

stopping cross-sections are not isotropic and after each collision event the new positions of all the lattice ions must be computed. In particular, the stopping cross-sections are a function of the angle at which the ion approaches crystalline directions. If the beam is directed parallel to a major crystal axis, there is a reduced interaction and only glancing-angle collisions which lose little energy. Further, the planar potential tends to direct the ions down the open channels. Channelling may be either axial or planar. Intentional use of channelling directions can steer ions deep into the crystal with some ions travelling more than $10R_p$. However, crystal perfection and channel alignment are critical so that, as radiation damage develops, there is more dechannelling. Therefore, for reproducibility of implants at dose levels above some 10^{13} ions cm^{-2}, it is preferable to avoid channelling effects. An "amorphous" direction may be selected by moving a few degrees away from any major crystallographic direction. Alternatively the crystal may be amorphized before impurity doping and then the sample subsequently heated to anneal the damage and to re-form a perfect crystalline layer of doped material.

3.4 SIMPLE MODEL OF ION BEAM DAMAGE

3.4.1 Primary displacements—cross-sections

The problem of ion–atom collision is complex because of several factors such as the screening of interacting nuclei by the surrounding electronic clouds and the dependence of the effective ion charge on kinetic energy. These are formidable problems not yet solved in a general and satisfactory form although empirical approaches are often adequate.

The kinematics of the two-atom collision is shown in Figures 3.8(a) and 3.8(b), depicting the situation in the laboratory reference system and in the centre-of-mass system, respectively. In the laboratory system the bombarding atom is assumed to have velocity u_1 and kinetic energy $E_1 = \frac{1}{2}M_1u_1^2$, whereas the target atom is at rest ($u_2 = 0$). After the collision, the final velocities are v_1 and v_2 in the laboratory system. It can be easily shown that

$$v_2 = 2V_{CM}\sin\left(\frac{\phi}{2}\right) = 2\frac{\mu u_1}{M_2}\sin\left(\frac{\phi}{2}\right) \tag{3.17}$$

where ϕ is the collision angle in the centre-of-mass system and $\mu = M_1M/(M_1 + M_2)$ the reduced mass. V_{CM} is the velocity of the centre of mass in the laboratory system. The kinetic energy transferred to the target atom is

$$T = \frac{1}{2}M_2v_2^2 = \frac{4M_1M_2}{(M_1 + M_2)^2}E_1\sin^2\left(\frac{\phi}{2}\right) \tag{3.18}$$

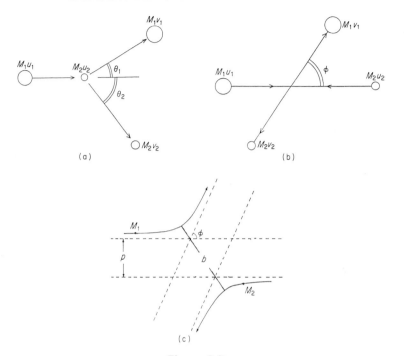

Figure 3.8.
Examples of a two-body collision in (a) the laboratory coordinate system and (b) a centre-of-mass frame. The quantity p, termed the impact parameter, and the closest-approach distance b are shown in (c).

The maximum transferred energy T_{m} occurs for $\phi = \pi$:

$$T_{\mathrm{m}} = \frac{4M_1M_2}{(M_1 + M_2)^2}E_1 \tag{3.19}$$

so that

$$T = T_{\mathrm{m}} \sin^2\left(\frac{\phi}{2}\right) \tag{3.20}$$

Important kinematic parameters in the collision are the distance b of closest approach, which is given by

$$b = \frac{2Z_1Z_2e^2}{\mu u_1^2} \tag{3.21}$$

where Z_1e and Z_2e are the net atom charges, and the impact parameter p (= $b \cos \phi/2$) which are shown in Figure 3.8(c).

The differential cross-section for transfer of an energy T in the collision depends on the interaction potential between colliding ions. That potential is usually written as a screened Coulomb potential

$$V(r) = \frac{Z_1 Z_2 e^2}{r} \exp\left(-\frac{r}{a}\right) \qquad (3.22)$$

where r is the interatomic separation. The screening correction to the purely Coulomb interaction is provided by the factor $\exp(-r/a)$, a being a screening radius of the order of the Bohr radius a_0. Lindhard et al.[13,21] and Firsov[22] have suggested on the basis of the Thomas−Fermi model of an atom that $a = 0.8853 a_0 (Z_1^{2/3} + Z_2^{2/3})^{-1/2}$.

Two limiting cases which lend themselves to a simpler analysis can be considered. They refer to the strength of the electronic screening factor characterized by the parameter $\xi = b/a$.

For weak screening ($\xi \ll 1$), $V(r)$ can be approximated by a Coulomb potential and the differential cross-section follows the Rutherford formula

$$d\sigma = \frac{\pi b^2}{4} \cos\left(\frac{\phi}{2}\right) \operatorname{cosec}^3\left(\frac{\phi}{2}\right) d\phi = \frac{\pi b^2}{4} T_m \frac{dT}{T^2} \qquad (3.23)$$

T_m being the maximum transferred energy.

This cross-section favours small collision angles θ and consequently small energy transfers. It applies to light bombarding ions with a high kinetic energy. As an example, the weak-screening approximation is reasonable for protons with $E_1 \geqslant 100$ keV, in all target materials.

For strong screening ($\xi \gg 1$), the overlapping of electronic clouds is strongly forbidden and the situation can be assimilated to a hard-sphere collision. The cross-section then appears as

$$d\sigma = \sigma_0 \frac{dT}{T_m} \qquad (3.24)$$

σ_0 being the total scattering cross-section.

The general case (with arbitrary ξ) is quite complicated and is discussed in various references[21−24].

By assuming a well-defined energy threshold E_D for atomic displacement, the displacement cross-section is

$$\sigma_d = \int_{E_D}^{\infty} \sigma(T) \, dT \qquad (3.25)$$

For Rutherford collision (weak screening),

$$\sigma_d = \frac{\pi b^2}{4}\left(\frac{T_m}{E_D} - 1\right) \approx \frac{\pi b^2}{4} \frac{T_m}{E_D} \qquad (3.26)$$

For hard-sphere collisions (strong screening),

$$\sigma_d = \frac{\sigma_0}{T_m}(T_m - E_D) \approx \sigma_0 \qquad (3.27)$$

3.4.2 Secondary displacements—overall damage

Each primary displaced atom can have enough energy to eject other ions from their lattice sites so that a cascade of secondary displacements is produced. The calculation of the number of displaced atoms is usually carried out under some simplifying assumptions.

(1) Sequence of two-body collisions.
(2) Well-defined thresholds.
(3) Random (glassy material).

Several approaches have been developed to obtain the average concentration of secondary displacements.

Here, we follow one first put forward by Kinchin and Pease[8], which takes into account the possibility of replacement collisions. As a result of the secondary collisions, the following cases are considered.

(1) If $T_1 > E_D$ and $T_2 > E_D$, the two colliding atoms are displaced and the number of displaced atoms has increased by one.

(2) If $T_1 \lesssim E_D$ and $T_2 < E_D$, only the incoming knocked-on atom will remain displaced, without any increase in the number of displaced atoms.

(3) If $T_1 < E_D$ and $T_2 > E_D$, the initial knocked-on atom will replace the target atom, which will become displaced. No net increase in the number of displaced atoms will occur.

Let us consider a primary knocked-on atom with kinetic energy E, which induces outgoing atoms with energies T_1 and T_2, after the first secondary collision. The number v of secondary displacements as a function of the primary knocked-on energy will obey the equation

$$v(E) = \int_{E_D}^{E} \frac{\sigma(T_1)}{\sigma_0} v(T_1)\, dT_1 + \int_{E_D}^{E} \frac{\sigma(T_2)}{\sigma_0} v(T_2)\, dT_2$$

$$= 2 \int_{E_D}^{E} \frac{\sigma(T)}{\sigma_0} v(T)\, dT \qquad (3.28)$$

For secondary ions, the hard-sphere situation, considered above, generally applies so that

$$v(E) = \frac{2}{E} \int_{E_D}^{E} v(T)\, dT \qquad (3.29)$$

After differentiation one notes $E(dv/dE) = v$ which has the linear solution

$$v(E) = cE \tag{3.30}$$

In the Kinchin–Pease model, the initial condition is $v(2E_D) = 1$, so that

$$v(E) = \frac{E}{2E_D} \tag{3.31}$$

The function $v(E)$ is illustrated in Figure 3.9.

All other approaches lead to the same linear dependence for $E \gg E_D$ but differ[9] for $E \leq E_D$.

The average number $\langle v \rangle$ of secondary displacements produced by a primary ion of energy E is then given by

$$\langle v \rangle = \int_{E_D}^{T_m} v(T_1)\sigma(T_1)\,dT_1 \bigg/ \int_{E_D}^{T_m} \sigma(T_1)\,dT_1$$

$$= \frac{\langle T_1 \rangle}{2E_D} \tag{3.32}$$

For a primary collision governed by the Rutherford scattering,

$$\langle T_1 \rangle \approx E_D \ln\left(\frac{T_m}{E_D}\right) \tag{3.33}$$

and

$$\langle v \rangle = \frac{1}{2} \ln\left(\frac{T_m}{E_D}\right)$$

As an example, $\langle v \rangle = 5$ for a deuteron of energy $E = 10$ MeV.

The calculation of the total number of displacements in the sample is extremely complicated since the energy losses of the primary ion in the solid should be taken into account and a summation of $\langle v \rangle$ performed over all primary collision events in the whole ion range. The strong and even dominant contribution of electronic losses to the stopping power adds to the complexity of such a procedure. In contrast, the separation between primary knocked-on atoms is quite small, indicating that a heavy damage pattern will appear along the ion track.

3.4.3   Highly disordered regions—many-body effects

Nuclear collisions generate a cascade of damage during the stopping of each ion. Additionally the accumulation of the disorder leads to a variety of complex defects or even amorphization of a lattice. For example, for iron nitriding (e.g. see Figure 3.7), doses of 10^{17} N$^+$ ions cm^{-2} are used at 100

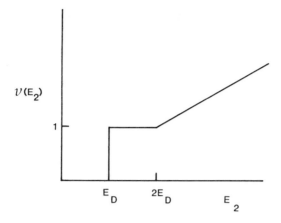

Figure 3.9.
The relationship between the number $v(E_2)$ of displaced atoms and the recoil energy E_2 in the simple model of Kinchin and Pease.

keV and hence each lattice atom may be displaced as many as 50 times. However, the number of defects retained in the target is not easily estimated. Defects may be generated in isolation along the ion track where the rate of energy transfer is low or they may be in close proximity if the deposition rate is high. The defects formed by the struck atoms may, in turn, lead to a defect cascade which occupies some tens of nanometres. A region so highly disordered is not in thermal equilibrium and, in extreme cases, we may attempt to describe the disorder by a "thermal spike" in which the average ion energy would correspond to an equivalent "temperature" of, say, 5000 K. Depending on the method of energy dissipation from the cascade the disorder may survive for times ranging from 10^{-14} to 10^{-9} s. The final state of the region is critically dependent on these relaxation processes and it can vary from a glassy amorphous region to a region in which there is perfect epitaxial regrowth and no defects survive.

The problem may be further confused by the presence of implanted impurities or radiation-enhanced diffusion of the defects. In practice, it is simpler to take a pragmatic approach and to choose the beam conditions which provide the desired property change. For the most heavily studied systems, conditions may be selected which generate an amorphous layer; alternatively, damage may be removed by subsequent furnace heat treatment or by laser or electron beam pulse annealing. Despite the difficulties in predicting absolute defect concentrations, computer modelling has been successful in predicting the form of the defect distribution with depth[25].

It might be imagined that defect generation leading to a random atomic arrangement is limited to crystalline targets. Surprisingly this is not so, in part because point defects only involve short-range order and in part because amorphous materials can exist at various atomic packing densities. Amorphous quartz (i.e. silica) is in this category and during ion implantation relaxes to a higher-density glass. The material is of interest because of its use in the semiconductor industry and it is potentially useful for integrated optics as the higher density increases the refractive index and so forms an optical waveguide. More immediately it provides a convenient example for a number of the preceding comments. That is, it can be compacted purely by electronic excitation; electronic excitation following ion implantation causes some recovery of the ion-generated compaction; the defects associated with the electronic damage are less stable than the more extensive defects formed by nuclear collisions. Figure 3.10 describes the enhancement in refractive index (from compaction) of nitrogen-implanted silica. In curve a it is apparent that the compaction saturates from the surface to a depth close to R_p. However, after annealing at 450°C, the electronically induced damage is removed, and curve b describes the effects of nuclear collision damage. A comparison with experimental data[26] has been made using the TRIM algorithm of Biersack and Eckstein[25] and the computer simulation of the vacancies produced by the ions is shown in curve c. The agreement is good. Curve d gives the TRIM estimate of the impurity distribution. This has been confirmed at higher ion doses (greater than 10^{17} ions cm^{-2}) where a chemical effect of the implant is seen as the production of an oxynitride of silicon.

3.4.4 Computer simulations

The analysis of the collision damage process can be adequately investigated by computer simulation techniques. Monte Carlo methods, as well as detailed solution of the equations of motion for the atoms in a small crystallite stuck by an energetic particle, have been used. These simulations have provided "evidence" for the occurrence of a number of features of damage such as replacement collisions, channelling and focusing. Moreover they have given support to some proposals on the stable defect configurations which are reached at the end of the collision process. Early examples included the split interstitial in face-centred cubic (and body-centred cubic) irradiated metals. More recently the stability of defect clusters (e.g. discussed in section 3.7) has been considered. However, a word of caution is often appropriate when the defect "stability" is sensitive to the choice of interaction potential. Commonly used algorithms for defect production are based on the previously mentioned TRIM program of Biersack and Eckstein[25].

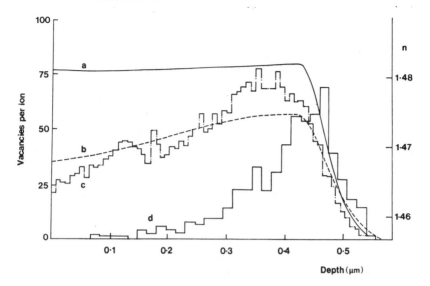

Figure 3.10.
An example of the damage induced by electronic excitation and nuclear collisions in silica by implantation of 0.18 MeV N^+ ions. The defects are sensed by changes in refractive index. Curve a shows a saturated index change in the region of electronic stopping. This "electronic" damage is annealed at 450°C (curve b), whereas the nuclear collision damage remains. Curve c shows a computer prediction of the nuclear damage profile and curve d the nitrogen impurity distribution.

3.5   SPUTTERING

When the collision cascades intersect the surface of the material, atoms will be ejected. In cases where this proceeds very rapidly, it sets a limit to the number of ions that can be implanted in the target as an equilibrium between doping and surface erosion is reached. Fortunately this problem only occurs for very-high-dose implants of low-energy heavy ions. The sputtering yield, i.e. the number of atoms ejected per incident ion, is a function of M_1, M_2, E, the angle of beam incidence and the target temperature. Theoretical descriptions of the sputtering yield as a function of these parameters are well formulated. In the most commonly used energy range the actual values of the yield vary from numbers near unity for the example of 100 keV N^+ ions incident on iron to about 35 for 45 keV Bi^+ ions incident on copper. In practical terms this means that, even for the high doses (10^{17} ions cm^{-2}) used in the metal nitriding example, the surface will recede by not more than some 100 monolayers. For semiconductor applications at small doses (10^{12} ions cm^{-2}) the problem is negligible. However, it should be noted that a limitation would occur if the intended treatment were to involve say 10^{17} Bi$^+$ ions in copper as the R_p value is a mere 10 nm and the predicted erosion of the surface is 1260 nm. In practice, a 10 nm layer of a Bi–Cu alloy would be generated on the final surface.

In the optical applications of ion implantation (e.g. with megaelectronvolt He$^+$ ions in quartz), sputtering is not an important consideration as cascade effects are buried deep below the surface. More caution is required when considering other insulators as, for many halides and oxides, electronic energy deposition can lead to sputtering (see Chapter 4). This is particularly apparent in the weakly bonded van der Waals solids such as the ices of H_2O, argon and neon which dissociate with input energies of 1 or 2 eV per atom. There has also been speculation that the differences in sputtering yields of UO_2 and UF_4 taken under the same conditions (i.e. 0.06 and 200) result from an electronically excited reconstruction into an unstable van der Waals layer. A related problem of changes noted in the refractive indices of $LiNbO_3$ and $Bi_4Ge_3O_{12}$ have been discussed in terms of a lattice relaxation that extends throughout the region of electronic excitation.

3.6 NEUTRON IRRADIATION DAMAGE

Materials exposed to neutron irradiation suffer defect production via two routes: direct collisions with atoms and nuclear reactions after capture of neutrons by nuclei. In the first, elastic scattering of the neutrons transfer energy to displace atoms in the same fashion as ion bombardment does. However, the small collision cross-section implies long particle ranges and hence, for samples with dimensions of a few centimetres, the neutron irradiation may be homogeneous. The most commonly used neutron source is a nuclear reactor whose energy spectrum extends from subthermal and thermal values to an upper limit of about 10 MeV. Although the precise energy distribution is initially dependent on reactor design, the peak energy is typically located at about 1.5 MeV. Monoenergetic neutron beams can be selected by suitable filtering. Neutrons produced in other sources may have a more peaked energy spectrum.

The initial interest in neutron damage was in the construction of fission reactor materials whereas the current incentive is to understand the problems of modern fusion reactor materials. In the tokamak-type fusion reactor, energy is obtained from the 14.1 MeV released by the reaction

$$d + t \rightarrow \alpha + n \qquad (3.34)$$

The regeneration of t (tritium) is obtained through another reaction

$$n + {}^7Li \rightarrow {}^3H + \alpha + n \qquad (3.35)$$

which uses lithium as fuel.

Alternative sources may be driven by laser compression pulses. As single pulse demonstration systems the use of terawatt powers is technologically routine but it must be realized that in a power station generator fusion system the aim is to release megawatts or gigawatts of power and the neutron flux will be considerable. There is thus a major potential problem of neutron damage in the optical components or laser amplifiers that are needed for terawatt lasers. In this example, simple "colour centres" in a neodymium glass laser amplifier could act as localized points of power absorption that could destroy the amplifier. Hence the transition from single pulses to power systems may not be trivial.

3.6.1 Concentration of displaced atoms

At variance with ions, the main mechanism for energy loss in neutron irradiation is provided by elastic collisions. The angular distribution of outgoing neutrons is isotropic in the centre-of-mass system, and the differential cross-section for energy transfer is, as for hard-sphere collisions,

$$\sigma(T) = \frac{\sigma_0}{T_m} \qquad (3.36)$$

where σ_0, the total scattering cross-section, is approximately independent of neutron energy ($\sigma_0 \approx 10^{-24}$ cm^2).

From the maximum transferred energy, T_m can be expressed as

$$T_m \approx \frac{4E_1}{M} \qquad (3.37)$$

and the average transferred energy is given by

$$\langle T \rangle = \int^{T_m} \frac{T\sigma(T)}{\sigma_0} = \frac{T_m}{2} \qquad (3.38)$$

The displacement cross-section is

$$\sigma_d = \int_E^{T_m} \frac{\sigma_0}{T_m} \, dT \approx \sigma_0 \qquad (3.39)$$

If the atom density is N, the mean free path between displacements is $L_d = 1/N\sigma_d$, i.e. several centimetres; the pattern of primary damage is much simpler than for energetic ions. Because of this and as a result of the simplicity of the displacement cross-section, the total number N_d of displaced atoms per unit time and volume can be easily calculated by using the Kinchin–Pease approach described in section 3.4.2. We first write:

$$N_d = \phi\sigma_d N\nu \qquad (3.40)$$

ϕ being the neutron flux and N the atom density of the material. For neutron irradiation

$$\langle\nu\rangle = \frac{\langle T_1\rangle}{2E_d} \approx \frac{1}{4}\frac{T_m}{E_d}\frac{E_1}{ME_d} \qquad (3.41)$$

assuming monoenergetic neutrons with energy E_1. As an example, about 1000 secondary displacements are produced by each primary collision with a reactor neutron of $E_1 \approx 1.5$ MeV.

Equation (3.40) can then be written as

$$N_d = \frac{1}{4}\phi\sigma_0\frac{NE_1}{ME_d} \qquad (3.42)$$

In principle, this is the concentration of Frenkel pairs produced per unit time if the recovery processes acting to heal the damage are ignored. Moreover, it should be noted that other processes such as correlated collision chains and other crystallographic or many-body effects already mentioned are not included in the simple theory. These processes may markedly modify the above estimate. Indeed, theoretical estimates based on equation (3.42) predict a much higher rate of damage than that experimentally measured.

3.6.2   Damage by nuclear reactions

In addition to the elastic collision damage a second source of radiation damage results from nuclear reaction after capture of a neutron by a nucleus; so many possibilities from (n, γ), (n, β), (n, p), (n, α) to (n, fission) ensue.

The (n, γ) reaction changes the nuclear mass, and damage results from ionization (in insulators) or by direct γ−nuclei recoils. If the displacement energy is E_d for an atom mass M, the number of such displacements is of the order of $(1/4ME_d)(E_\gamma/c)^2$.

Reaction of the (n, β) type may provide electron collision damage and again the recoil energy will produce defects. The recoil energy is about $(1/2M)(E_\beta/c^2)(E_\beta + 2m_ec^2)$; so, typically, recoiling ions have energies of tens of electronvolts and will produce a few point defects. In exceptional cases (e.g. for lithium) the recoil can be as much as 3600 eV; so in a light target (LiF) the combined γ and recoil event will displace 80−100 atoms per nuclear reaction.

If the neutron capture results in a fission product, as for uranium, the energetic heavy-mass fragments displace atoms in high-energy cascades. From the viewpoint of point defect studies this is an extreme case of ion implantation. However, the scale of the ion range is so great that the tracks around a fission site in natural minerals (e.g. mica) can be seen directly (Section 4.16). Fission tracks can be revealed by chemical etching of the disordered material where the track density is high. Fission tracks are often discussed as a separate subject and indeed the high-density collision cascades amorphize crystalline material so that in old minerals the entire specimen may change state. Such amorphized crystals are referred to as metamict.

3.7 EXTENDED DEFECTS—VOIDS AND BUBBLES

A feature commonly encountered in heavily irradiated material is the growth of very large aggregates of vacancies. This is a consequence of the existence of very efficient traps for interstitials in the materials (e.g. edge dislocations), leading to an excess population of vacancies. The aggregates can exist in isolation and are termed voids, or alternatively the vacancy cluster can be stabilized by the inclusion of gas atoms. If the gas atoms dominate, then the defects are termed bubbles. Such defects are most readily generated in nuclear reactor materials because of the high radiation flux and the abundance of helium (from α particles) or oxygen liberated from UO_2 or steel alloys. The presence of the voids, is structurally undesirable (e.g. embrittlement) and expansion of the material can be considerable.

In the normal working life of a reactor structure the expansion has been estimated to be between 10 and 100%. In crystalline or oriented polycrystals the effect may be more pronounced if the voids align with crystal axes, and linear changes in dimensions of factors of ten can be induced by prolonged irradiation. Voids frequently show signs of an ordered structure within a crystal. Indeed, in some cases a superlattice structure may develop. Lines of voids are recorded in alumina crystals. Faceted void boundaries may produce square or octahedral sections as observed by transmission electron microscopy.

Void formation in compound materials is more complex than for simple elemental systems as an equal number of each type of vacancy must be generated. Thus in alkali halides or systems which preferentially generate vacancies on only one sublattice the formation of voids in the halogen sublattice by condensation of halogen vacancies may be accompanied by a parallel collapse of the metal ions into a colloidal aggregate.

In general the rate of void growth is a function of the rate K of atomic displacement, the irradiation time t, the defect sink concentration S and a term F which allows for changes in defect migration speeds with temperature. Bullough et al.[27] and Bullough and Nelson[28] wrote the fractional volume change as

$$\frac{\Delta V}{V} = KtSF \tag{3.43}$$

At very high temperatures, vacancies evaporate from the voids; so the critical temperature regime where void generation is a problem is dependent on displacement rate. Figure 3.11 shows that with typical damage rates the maximum rate of void growth occurs in precisely the temperature range at which the reactors operate. Inhibition of void growth can be controlled by the trapping of gas atoms at localized defects and hence prevent a rise in "gas" pressure which could stabilize void walls. For this reason, some titanium-rich steels have been favoured as there is a strong chemical trapping of liberated oxygen by the titanium.

The problems of void and bubble formation cannot be separated, as in many cases the walls of the void are stabilized by the pressure of a small number of gas atoms trapped within the cavity. This stabilization is particularly relevant in favouring void formation rather than collapse of vacancy aggregates into prismatic loops. A simplistic estimate of the pressure within a bubble, stabilized by gas pressure, gives

$$p = \frac{2\gamma}{r}$$

where γ is the surface energy per unit area and r is the bubble radius. On the assumption that the system acts as a "perfect" gas, $pV = \frac{3}{2}nkT$ and $V = \frac{4}{3}\pi r^3$; hence the number n of gas atoms is $[16n\pi/9kT]r^2$. Therefore, if

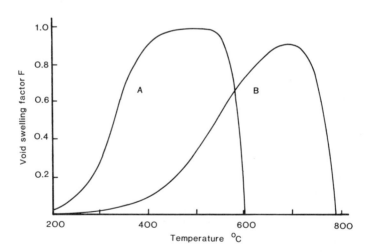

Figure 3.11.
Effects of change in particle flux on the temperature dependence of void swelling. The examples are for two dose rates of A $= 10^{-6}$ displacements atom^{-1} s^{-1}; curve B, 10^{-3} displacements atom^{-1} s^{-1}.

bubbles coalesce into a larger volume governed by the pressure term, the final bubble radius R should be given by $R^2 = \Sigma_r r_n^2$. This is observed in practice although the "perfect" gas model is not truly appropriate. In fact, bubble migration and coalescence provide a bubble growth mechanism alternative to that caused by vacancy incorporation. The binding energy between gas atoms and vacancies is a function of the ratio of gas atoms to vacancies. For helium atoms the binding energy is reduced as the ratio is increased but conversely the binding energy of a vacancy to a bubble or void increases with increasing ratio. Simple capture and loss processes of vacancies or gas atoms do not necessarily proceed by random thermal diffusion as the stress field around the defect can provide a directional term in the diffusion. During irradiation there is the further possibility of radiation-enhanced diffusion of a strongly bound helium−vacancy pair.

Near a surface, bubbles distort the surface and may burst, releasing gas and ejecting atoms from the surface. This surface destruction is termed blistering or exfoliation. Numerous examples are given in reviews of the subject, e.g. those by Clinard and Hobbs[29] and Guseva and Martynenko[30].

REFERENCES

1 J. W. Corbett, *Solid State Phys.*, *Suppl.* 7, (1966).
2 N. F. Mott, *Proc. R. Soc. London, Ser. A* **124**, 426 (1929); **135**, 429 (1932).
3 W. A. McKinley and H. Feshbach, *Phys. Rev.* 74, 1759 (1948).
4 L. E. Glendennin, *Nucleonics* 2, 12 (1948).
5 R. Katz and A. S. Penfold, *Rev. Mod. Phys.* 24, 28 (1952).
6 E. J. Kobetich and R. Katz, *Phys. Rev.* 170, 391 (1968).
7 C. Erginsoy, G. H. Vineyard and A. Englert, *Phys. Rev.* 133, A595 (1964).
8 G. H. Kinchin and R. S. Pease, *Rep. Prog. Phys.* 18, 1 (1955).
9 M. W. Thompson, *Defects and Radiation Damage in Metals*, Cambridge University Press, Cambridge (1969).
10 P. D. Townsend, *Contemp. Phys.* 27, 241 (1986).
11 J. F. Ziegler, *Ion Implantation Science and Technology*, Academic Press, New York (1984).
12 G. K. Hubler, O. W. Holland, C. R. Clayton and C. W. White (eds), *Ion Implantation and Ion Beam Processing of Materials*, *Materials Research Society Symposium Proceedings*, Vol. 27, Elsevier, New York (1984).
13 J. Lindhard, M. Scharff and H. E. Schiøtt, *K. Dan. Vidensk. Selsk., Mat.-Fys. Medd.* 33, No. 14 (1963).
14 P. D. Townsend, *Rep. Prog. Phys.* 50, 501 (1987).
15 L. H. Abu-Hassan and P. D. Townsend, *J. Phys. C: Solid State Phys.* 19, 99 (1986).
16 J. Ziegler, J. P. Biersack and U. Littmark, *The Stopping Range of Ions in Solids*, Pergamon, New York (1985).
17 D. K. Brice, *Ion Implantation Range and Energy Deposition Distributions*, Vol. 1, Plenum, New York (1975).
18 G. Dearnaley, *Ion Implantation Metallurgy*, AIME, New York (1980).

19 D. K. Brice, *J. Appl. Phys.* **46**, 3385 (1975).
20 H. E. Schiøtt, *Radiat. Eff.* **6**, 107 (1970).
21 J. Lindhard, V. Nielsen and M. Scharff, *K. Dan. Vidensk. Selks., Mat.-Fys. Medd.* **36**, No. 10 (1968).
22 O. B. Firsov, *Sov. Phys.—JETP* **5**, 1192 (1957): **7**, 308 (1958); **9**, 1076 (1959).
23 P. Sigmund, in C. H. S. Dupuy (ed.), *Radiation Damage Processes in Materials*, Noordhoff, Leyden (1975).
24 U. Littmark and J. F. Ziegler, *Phys. Rev. A* **23**, 64 (1981).
25 J. P. Biersack and W. Eckstein, *Appl. Phys. A* **34** 73 (1984); **37**, 95 (1985).
26 A. B. Faik, P. J. Chandler, P. D. Townsend and R. P. Webb, *Radiat Eff.* **98**, 233 (1986).
27 R. Bullough, B. L. Eyre and R. C. Perrin, *J. Nucl. Appl. Technol.* **9**, 346 (1970).
28 R. Bullough and R. S. Nelson, *Phys. Technol.* **5**, 29 (1974).
29 F. W. Clinard and L. W. Hobbs in R. A. Johnson and A. N. Orlov (eds), *Physics of Radiation Effects in Crystals*, Elsevier, Amsterdam, Chapter 7 (1986).
30 M. I. Guseva and Yu. V. Martynenko, in R. A. Johnson and A. N. Orlov (eds), *Physics of Radiation Effects in Crystals*, Elsevier, Amsterdam, Chapter 11 (1986).

4
Photolytic Damage to Materials

4.1 INTRODUCTION—IONIZATION AND SUBTHRESHOLD DAMAGE

Except for very energetic photons (γ-rays) capable of producing Compton or photoelectrons with energies and momentum sufficient to reach the displacement threshold (see Chapter 3), the only direct effect of electromagnetic radiation is the excitation of the electronic system of the material. Excitation may give rise to changes in the state of charge of impurities and/or lattice defects already present in the material but not, in general, to breakdown or damage to the crystallographic lattice. Defects are not formed in metals by such processes but the situation may be different for many organic and inorganic solids (i.e. insulators and semiconductors). Electronic excitation induces an energy transfer from the electrons to the lattice which can lead to defect formation. As an example, it is relatively easy to produce vacancies and interstitials in alkali halides, by irradiating them with electrons, X- or γ-rays and even ultraviolet light with energies close to (or even less than) the band gap value (5–10 eV). There are a variety of mechanisms that can be invoked to explain the production of lattice defects in different materials. Although these mechanisms are often speculative, they illustrate the range of possibilities available to account for excitation-induced damage.

Several general guidelines (e.g. these given by Townsend and Agulló-Lopez[1]) can be given to predict the occurrence of damage.

(1) The electronic excitation energy should be localized at a given lattice site or a few lattice sites.

(2)   The lifetime of the localized excitation should allow for atomic relaxation of the lattice.

(3)   The threshold energy for the production of displaced atoms should be less than that available in the excitation.

(4)   In order to obtain stable damage, the primary damage products (usually vacancies and interstitials) should be separated by several lattice sites so as to avoid recombination and regeneration of the lattice.

One of the possible mechanisms that may illustrate the excitation damage to a simple ionic structure A^+B^- would be the trapping of electrons by the ion A^+ which becomes A^0. The neutral atom, which is not electrostatically bound to the lattice, may then diffuse away, leaving a vacancy at the original lattice site. This mechanism was incorporated in some of the models proposed to account for the photographic process in silver halides. Another simple alternative possibility is the formation, during irradiation of the binary compound, of a multiply ionized ion, e.g. B^{2-}, which would be repelled by its neighbours and ejected to an interstitial position (the so-called Varley[2] mechanism). Electronic excitation routes to damage production must be considered during electron and ion bombardment as electronic excitation is the primary mechanism of energy loss during the implantation of high-energy ions. For light ions such as H^+ or He^+ the electronic term is dominant above about 20 keV.

The possibility of a synergistic effect between collision and excitation damage should be mentioned. Within the region of the collision cascade the subthreshold collisions can generate a perturbed structure which is then susceptible to electronic excitation damage. In other words, both types of damage couple together to determine the final damage structure. Equally, the threshold for dynamic displacement production can be lowered by electronic excitation of the material, if it reduces the strength of the atomic bonding. The interplay between the electronic and nuclear collision damage mechanisms is complex as there are both a contribution to reduced damage thresholds and a contribution to radiation-assisted annealing (as discussed by Chandler et al.[3], for example).

In addition to damage production, there are a number of ionic processes which are markedly influenced by electronic excitation. Examples of these radiation-enhanced defect reactions are[4,5] enhanced self-diffusion in silicon and recombination-assisted dislocation climb in III−V semiconductors.

In this chapter, we focus on damage mechanisms which are well documented and ignore more speculative proposals or those having only historical significance which are mentioned in earlier reviews. In particular, most attention is paid to alkali halides, for which excellent experimental information is available. Processes in other materials are discussed to a lesser extent.

Since intense laser sources are now common tools in research and industrial laboratories, and they are capable of producing damage to materials, a comment on this topic is presented at the end of the chapter. The problem of track formation by energetic ions is also briefly touched upon because this may involve additional processes associated with the high excitation density. In fact, some subtle high-density excitation mechanisms are being invoked for an adequate understanding of these phenomena.

4.2 EXCITATION DAMAGE TO ALKALI HALIDES—MAIN RESULTS

The excitation damage to alkali halides A^+X^- induced by ultraviolet, X, γ and electron irradiation is relatively well understood. The structure of stable damage markedly depends on the irradiation temperature, although in all cases F centres and interstitial halogen centres are produced in the primary damage event.

The simplest steady-state situation corresponds to low-temperature irradiation ($T \leqslant 20-30$ K), where only F and H (and I) centres are produced, together with self-trapped holes (V_K centres), see Tables 1.1 and 1.2 and Chapter 5. At higher temperatures, interstitials become mobile and give rise to a variety of aggregated interstitial centres, including complexes with appropriate impurities. Above room temperature, vacancy centres become mobile and colloids are formed at high dose levels. The main features associated with the damage are as follows.

(1)   It is produced at photon energies down to slightly lower values than the band gap, i.e. down to the excitonic region of the absorption spectrum.

(2)   At low temperatures, an anticorrelation has been found between the efficiency for defect (F-centre) production and that for recombination luminescence emission.

(3)   The efficiency for defect production markedly depends on the particular alkali halide. The dependence can be well illustrated with the so-called Rabin−Klick diagram (Figure 4.1), where the efficiency is plotted against the ratio S/D, S being the anion−anion distance and D the anion (halogen) diameter.

(4)   At high doses and temperatures above the onset for interstitial migration, one of the final stable products of damage is perfect dislocation loops, as revealed by electron microscopy observations[6,7]. In a limited temperature range above room temperature, alkali-metal colloids are also observed[8,9].

All this information has led to a description of the excitation damage to alkali halides as the result of a primary mechanism, causing the formation of

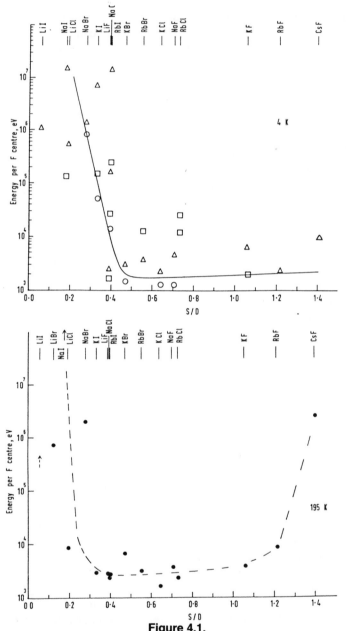

Figure 4.1.
The average energy required to form F centres in alkali halides at (top) 4 K and (bottom) 195 K. Data from X-ray and electron irradiation are included. The dependence on the *SID* ratio noted by Rabin and Klick is interpreted in terms of vacancy–interstitial separation by replacement collision sequences (see Figure 4.5).

vacancy—interstitial pairs, together with secondary thermal and/or irradiation-induced processes. These latter processes, which markedly depend on irradiation temperature, include the migration of the primary defect species and the formation of defect clusters and impurity—defect complexes. Final stages are the occurrence of dislocation loops and (at a high enough temperature) alkali metal colloids.

4.3   PRIMARY DAMAGE—EXCITONIC MECHANISMS

The primary mechanism of damage can be conveniently studied by means of low-temperature (liquid-helium temperature range) irradiation or fast-pulse irradiation experiments at any temperature. With either method, attempts are made to avoid the interfering contribution of thermally induced secondary processes. For the observation of the primary F and H centres the two defects are required to be separated by several lattice sites. Long-term stability implies that low-energy-correlated jumps of the interstitial back to the vacancy are unimportant. Therefore the process of defect separation is an integral part of any model of defect information. The production of damage at energies at or even slightly below the gap, points to an excitonic mechanism as initially proposed by Pooley[10], Hersh[11] and Vitol[12]. The process can be visualized by the early configuration coordinate energy diagram in Figure 4.2. According to the initial suggestions, the irradiation gives rise to the self-trapped exciton (STE) as the intermediate step in the defect production process. Exciton absorption occurs with halogen ions at normal separation but relaxation of an adjacent ion pair to form the trapped exciton centre leads to de-excitation of the STE (electron—hole recombination) via either luminescence emission or through some non-radiative route. This model was stimulated by anticorrelation found between the damage and luminescence efficiencies. It has the advantage of explaining how energy can be provided from an exciton and, through the presence of a long-lived excited state, the momentum required for atomic motion is abstracted from the phonons of the lattice.

   The pulsed irradiation experiments have been particularly relevant in order to establish the exciton model of primary damage. One main result is that F−H pairs, and not F^+−I pairs, are formed directly in the ground state. F^+−I pairs should be generated through a subsequent electron transfer process. Further, it was found[13] by fast-pulsed two-photon excitation of KCl with a mode-locked laser that the time delay in producing the F−H pair is about 10 ps. Similar scales apply to other alkali halides[14]. The time is much shorter than the radiative lifetime of the STE level involved in the luminescence emission, for either the singlet or the triplet. Therefore, it is now clear that the electronic level responsible for the light emission has to

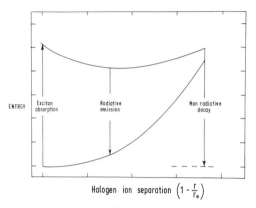

Figure 4.2.
A simple configuration coordinate diagram for exciton absorption and relaxation at a halogen site. Decay is by either luminescence or a non-radiative process which provides kinetic energy for a replacement collision sequence to form a ground-state F centre and an H centre.

be different from that involved in the damage process, at variance with the initial suggestions. In fact, recent calculations suggest that the precursor level for damage lies between those responsible for the σ and π luminescence of the STE.

The remaining problems of the exciton model are the following.

(1) The nature of the precursor damage level.

(2) The de-excitation path leading to the damage.

Some progress has been attained along those directions, as reviewed in recent papers[14]. The energy diagram of the system is complex, when the multicoordinate space corresponding to the various possible dynamic models for relaxation is taken into account. Figure 4.3 typifies[15] the energy level scheme as inferred from some of the proposals advanced for defect production. From these diagrams we basically try to obtain the evolution of the various electronic states during relaxation, the possible crossing of levels and the potential barriers to be overcome on taking the STE from the initial excitation state to the final defect state $F-H$. The complexity of the de-excitation path may be also reflected in the temperature dependence of the primary production efficiency (Figure 4.4). Here, feedback annihilation reactions are also likely to be important and there is no reason to suppose that defects are only formed from a single set of states and that maybe alternative modes of relaxation are possible.

Regardless of the detailed de-excitation path leading to the primary $F-H$ pair, it appears that the separation of the two partners involves a replacement collision sequence along the $\langle 110 \rangle$ axis, in order to generate stable

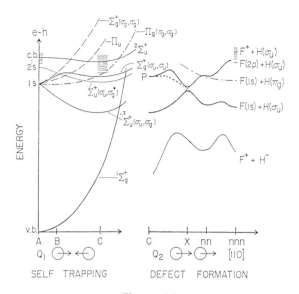

Figure 4.3.
A more detailed analysis of the configuration energy schemes for exciton relaxation: cb, conduction band; vb, valence band. The self-trapping or defect formation can involve alternative states[15].

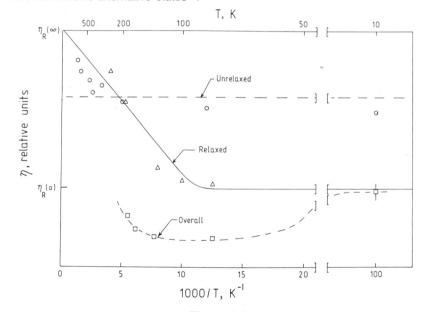

Figure 4.4.
The temperature dependence of primary-defect production efficiency for models of unrelaxed or relaxed states compared with experimental data (after F. Agullo-Lopez and P. D. Townsend, *Solid State Comm.* **33**, 449 (1980)).

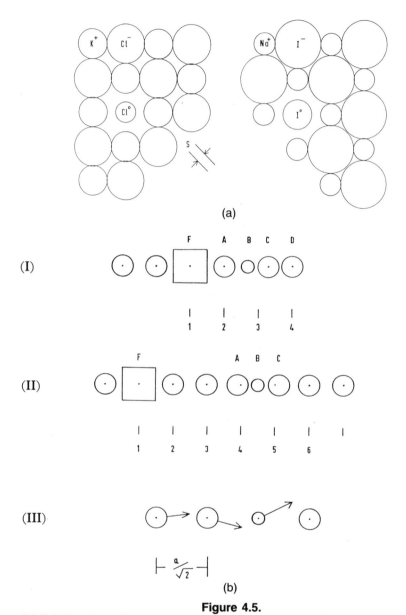

Figure 4.5.
(a) Relative sizes of ions in NaI and KCl. The parameters S and D define spaces along $\langle 110 \rangle$ and the size of a neutral halogen. (b) The separation and stability of F–H pairs formed by replacement collision sequences are shown for (I) close pairs where the limit is $D \approx 2S$, (II) separated pairs where $D \approx 3S$ and (III) an upper limit set by defocusing if $S > D$. See also Figure 4.1.

independent defects. As seen from Figure 4.5, this close-packed direction, which coincides with the orientation of the V_K (as well as the STE) centre is appropriate for efficient energy propagation. The length of the collision sequence depends on the charge state of the moving entity, the energy losses at each replacement and the ability of the surrounding lattice to preserve the directionality of the sequence. A neutral halogen is physically small enough to pass through the surrounding rings of ions and then it may be imagined that the charge redistribution takes place as a separate event. The focusing of off-axis collisions is critically dependent on the dimensions of the lattice and hence the range of the sequence as sensed by the efficiency of defect production or sputtering from the surface. The relationship is described by the Rabin–Klick diagram between F-centre production efficiency and neutral halogen ion size D and the space between halogen ions, S, as described in the next section[16].

4.4 SPUTTERING INDUCED BY IONIZING RADIATION

Ionizing radiation was predicted[17] to induce sputtering of alkali and halogen atoms from alkali halides[17-22]. Essentially, the same results have been obtained by irradiating with low-energy electrons or ultraviolet photons. These processes are, indeed, very relevant to clarify the primary mechanism of damage to alkali halides. The angular distribution of sputtered particles is different for alkali or halogen atoms[19-22]. Alkali atoms are sputtered according to a Lambertian (cosine) distribution which is characteristic of a thermal evaporation process. In contrast, halogen atoms are emitted along some preferential directions, $\langle 110 \rangle$ and $\langle 211 \rangle$, with energies in excess of those for thermal evaporation. This energetic sputtering along the close-packed $\langle 110 \rangle$ direction constitutes direct proof of the occurrence of a replacement collision sequence as a final stage of the exciton decay path. The sputtering along $\langle 211 \rangle$ fits into this same model as again it corresponds to a halogen–halogen bond direction. Transient $\langle 211 \rangle$-directed V_K centres can form and, if they are within one or two steps of the surface, can decay by ejection of a halogen through the surface layer. In bulk colouration mechanisms this $\langle 211 \rangle$ V_K centre is not relevant as the lattice structure would inhibit long replacement collision chains.

The length of the collision chain in the $\langle 110 \rangle$ direction is sensed by the Rabin–Klick diagram (Figure 4.1) where at low temperatures even very close F–H pairs are stable and the easy replacement collision occurs above S/D values of $\frac{1}{2}$. At higher temperatures, well-separated F–H pairs are recorded so that the lower S/D limit moves to $\frac{1}{3}$. At even higher S/D values (greater than 1) the open lattice network causes defocusing of the chains so that the colouration efficiency drops. For sputtering, the chains are always close to the surface and so one is recording virtual H-centre formation (i.e. ejection) hence the S/D parameter reflects only a minor trend in the efficiency of sputtering rather than the dramatic range of colouration efficiency in the Rabin–Klick diagram.

In accordance with these results, the sputtering process may proceed as follows: STE decay generates replacement collisions in the halogen sublattice near the crystal surface and leads to energetic halogen ejection along $\langle 110 \rangle$ or $\langle 211 \rangle$ directions. As a consequence, the surface becomes enriched in the alkali metal, which thermally evaporates with a Lambertian angular distribution. In fact, this evaporation limits the efficiency of the sputtering yield.

Overall the intimate connection between the sputtering process and the colour centre production mechanism is reinforced by the experimental facts that both depend on the halogen size and lattice parameter (the Rabin–Klick diagram) and that the same anticorrelation with the luminescence yield has been found for the sputtering as for the damage efficiency, even at high temperatures[21].

The sputtering process is more complex than suggested from the above simple considerations. In particular, recent experiments have revealed that not only atoms but also halogen molecules are ejected during irradiation and there is some emission along $\langle 100 \rangle$[22].

4.5   SECONDARY PROCESSES—DEFECT MIGRATION

The stable damage observed after or even during irradiation is determined by the production rate of the primary $F-H$ pairs as well as by that of the thermally induced (and radiation-induced) processes involving the migration of the primary defects and the formation of more complex defect species. These are the so-called secondary processes whose relevance markedly depends on irradiation temperature and dose rate. The H centre becomes mobile at low temperatures (about $20-30$ K) and gives rise to di-interstitial centres (H_2 or V_4) or to trapped interstitials at a monovalent (H_A-centre) or divalent (H_Z-centre) cation. These are the centres normally observed at low or intermediate doses. At higher doses the interstitial aggregation process reaches a new stage, where larger clusters are formed which finally develop into perfect dislocation loops. They have been observed and characterized by electron microscopy observations[6] in a variety of electron- and γ-irradiated alkali halides. The loops lie close to $\{100\}$ planes and have the Burgers vector in $\{110\}$ directions ([101] for a (001) loop plane).

The mechanism for the nucleation of the loop from the H-centre clusters is not known. However, the growth of the loops requires both halogen and alkali interstitials. Since damage is restricted to the halogen sublattice, loop growth may occur via a di-interstitial (halogen) molecule digging "its own hole" in the lattice and displacing a cation–anion pair to the edge of the dislocation loop. Catlow *et al.*[23] have investigated the energetics of this process and concluded that it is exothermic by about 2.1–2.6 eV if the displaced ions are attached to an edge dislocation but that it is endothermic by 2.4–2.9 eV if the two remain as interstitial defects. The total area of the loops which grow during irradiation can be used to measure the total number of interstitials produced.

At temperatures at or above room temperature, F centres become suffi-ciently mobile that F_2, F_3, etc., aggregate centres are formed as well as complexes with monovalent or divalent impurity cations (F_A or F_Z centres). The F_2 centres grow approximately with the square of the F-centre concen-tration as expected from a dynamic equilibrium of $F \rightleftharpoons F_2$ reached during irradiation. The equilibrium constant K of

$$[F_2] = K[F]^2$$

depends on the rate of energy deposition and, for proton irradiation, K is proportional to dE/dx for fast ions where electronic stopping predominates[24]. At very high doses the last stage of F-centre aggregation sets in with the formation of alkali-metal colloids[8,9,25]. The formation of colloids by ionizing irradiation or particle bombardment is a rather general phenomenon in ionic crystals, e.g. silver halides, dihalides and azides. The information on colloid

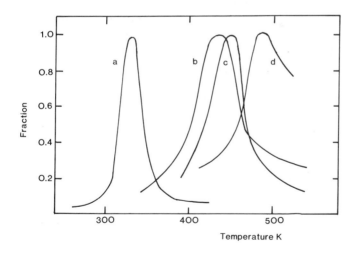

Figure 4.6.
Temperature variations of the efficiency of colloid formation in NaCl (curve a), CaF_2 (curve b), BaF_2 (curve c) and (curve d) SrF_2[8].

growth comes from a variety of techniques including optical and electron microscopy techniques and X-ray scattering. One of the most relevant experimental results is that colloids develop over a relatively narrow irradiation temperature range (Figure 4.6). Essentially, the low-temperature limit of this range is determined by the onset of efficient F-centre mobility, whereas the high-temperature limit corresponds to evaporation from the surface of the colloids. The growth rate of the colloidal band has been shown[9] to be markedly enhanced by prior straining in NaCl, suggesting that dislocations may play a key role in colloid nucleation.

The process of colloid formation in irradiated alkali halides closely resembles that of void growth in reactor-irradiated metals discussed in Chapter 3.

4.6   RECOVERY OF DAMAGE—THERMOLUMINESCENCE

Defects are not a stable feature of the crystal lattice; so migration and annealing change the type and concentration of defects. The decay kinetics of the various centres provide information on the thermally assisted secondary process acting after or even during irradiation. The annihilation processes markedly depend on both irradiation and storage temperature. For very low irradiation temperatures (liquid-helium temperature) where isolated F and H centres are the dominant defects, subsequent thermal annealing involves H-centre migration and recombination with F centres. At higher irradiation temperatures, where H centres are clustered or form complexes with impurities, the recovery process requires interstitial detrapping and recombination with F centres. Several competing channels may be occurring. At irradiation temperatures at or above room temperature, vacancy centres become mobile and recovery mechanisms are less well understood. A yet more complex situation corresponds to heavily irradiated crystals so that dislocation loops and metal colloids have been formed[26].

In addition to the optical absorption data, the annealing stages may be followed by the more sensitive technique of thermoluminescence. Heating of the crystal after irradiation usually produces glow peaks at definite temperatures, which may be correlated with electron−hole recombination occurring during the operation of a given recovery stage. One well-documented case[27] is the glow peak associated with the decay of the V_K centre. From the analysis of the peak shape the activation energy for migration of V_K centres has been determined. In a number of cases[28], thermoluminescence glow peaks have been associated with F and H recombination, not only after low but also high-temperature (room-temperature) irradiation, as illustrated in Figure 4.7. However, luminescence processes which involve impurities could be responsible for the thermoluminescence emission and then correlation with annihilation stages may be obscured. In part, measure-

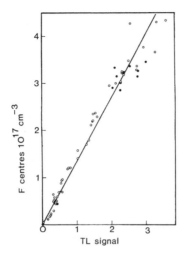

Figure 4.7.
Results of experiments relating the thermoluminescence (TL) emission and the F-centre concentration prior to heating for irradiated (●) samples and for crystals bleached (○) prior to the TL[28].

ments of the thermoluminescence spectra may resolve competing features of the thermoluminescence glow curve and for example the similarity of the spectra taken during each stage of the F-centre annealing supports the model of energy release as interstitials evaporate from trapping sites and diffuse to the F centres.

4.7 KINETIC MODELS—COMPUTER SIMULATIONS

Since a reasonable understanding of the primary and secondary damage mechanisms in alkali halides is now available, a number of kinetic models have been developed to account for the main features of the F growth with increasing irradiation dose. These models involve solution of rate equations[29-32], Monte Carlo simulations[33] or diffusion-controlled reaction approaches[34].

Rate equation models have been mostly used because of their simplicity and have satisfactorily accounted for various relevant features of the F colouring. The source term in the equations is the generation of F−H pairs at a given rate depending on dose rate. For irradiation temperatures at or below room temperature, H centres (not the F centres) are mobile and contribute to the following competing processes.

(1) Recombination with an F centre, leading to recovery of the crystal.
(2) Trapping at some impurity or defect, which is responsible for the nucleation and subsequent growth of interstitial clusters.

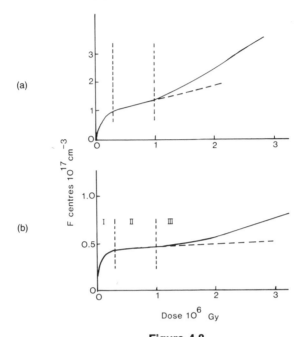

Figure 4.8.
(a) Computer simulations and (b) experimental data for colour centre growth curves in NaCl[32] at dose rates of 600 Gy min^{-1}.

(3) Trapping at some of the already formed clusters. The model, as developed by Aguilar *et al.*[32] satisfactorily simulates the three-stage structure of the F-colouring curve, as illustrated in Figure 4.8. Similar results were predicted by Comins and Carragher[31].
The rapid initial F-centre growth (stage I) is associated with interstitial trapping at impurities. After exhaustion of this process, a flat stage (stage II) is observed where the nuclei for interstitial clustering are being developed. Stage III corresponds to the growth of the interstitial clusters. The model also explains in a semiquantitative way the role of impurity doping and dose rate on the various stages. However, the model is still very simplified and ignores a number of electronic and ionic effects, and particularly the formation of dislocation loops at high-dose levels.

   In order to describe the F colouring at temperatures above room temperature, where anion vacancy migration becomes relevant, a model has been recently proposed by Jain and Lidiard[35]. The main aim of the model is to account for the colloid growth of alkali metal during irradiation. It is based on the approach developed to explain the formation of cavities (voids) in irradiated metals. It applies to the very-high-dose region, after the nucleation stage of both loops and colloids, so that the concentration of these defects is assumed constant. As in the lower-temperature models, it is assumed that primary F and H pairs are generated by irradiation at a constant rate. In addition to mutual F−H recombination, dislocation loops and colloids act as sinks for both F and H centres. Colloids are also sources for F centres. The corresponding rate equations read

$$\frac{dc_F}{dt} = K' - K_1 c_F - K_2 c_F c_H$$

$$\frac{dc_H}{dt} = K - K_3 c_H - K_2 c_F c_H$$

K being the radiation-induced generation rate and $K' = K + K_C$, where K_C stands for the thermal generation of F centres from colloids. The model satisfactorily explains, at a semiquantitative level, several relevant features of the radiation-induced colloid growth in alkali halides and, in particular, the dose dependence of the colloid radius in irradiated NaCl at several temperatures.

4.8 EXCITATION DAMAGE TO SILVER HALIDES—THE PHOTOGRAPHIC PROCESS

The ionization damage to silver halides has several very interesting features because of its relevance to photographic recording and the special situation of these materials which are close to the borderline separating ionic and covalent bonding.

In pure (bulk) silver halides, the existence of stable colour centres similar to those found in alkali halides has not been observed even after prolonged irradiation at low temperatures[36]. In fact, most electron and hole traps are quite shallow and therefore markedly unstable. Holes are self-trapped as Ag^{2+} ions in AgCl, but not in AgBr. At low temperatures (liquid-helium temperature), irradiation induces the formation of shallow electron centres (binding energies of the order of a few tenths of an electronvolt) in both AgCl and AgBr. Two types of centre have been identified: the atomic silver interstitial Ag^0, and the electron bound to some substitutional divalent cations (Cd^{2+}, Pb^{2+}, Ca^{2+}, Fe^{2+}, etc.). Both centres constitute good examples of bound polarons, including coupling to optical phonons and a Coulomb field. At higher temperatures, the above shallow centres become unstable and deeper centres are created during irradiation. The absorption bands of these centres, which depend on the irradiation temperature, appear to correspond to silver clusters containing several atoms. In particular, at 20−30 K they may contain two or three atoms, whereas at or above room temperature the structure of the cluster should be similar to that of metal silver. A very important conclusion to infer from these irradiation experiments is that, at least for AgCl, cation Frenkel pairs are created by the irradiation. Therefore, it is the opposite situation to that found in alkali halides where Frenkel pairs in the anion sublattice are produced. Unfortunately, the details of the creation mechanism are still unknown.

All this information is relevant to an understanding of the physical basis of the photographic mechanism in silver halides. The photographic film consists of silver halide grains (several micrometres in size) embedded in a gelatin medium. The grain—gelatin interphase plays a key role in the defect stabilization process.

Schematically, the photographic process[37] consists of the formation of an embryo or cluster of silver atoms, which acts as the latent image speck. Subsequent chemical processing, the development, which involves a reduction reaction, turns the whole silver halide grain into metallic silver from the initial catalytic speck. The formation of the speck is generally assumed to proceed essentially in accordance with the initial model proposed by Mott and Gurney[38] (Figure 4.9). The incoming photon creates in the crystal grain an electron—hole pair. The recombination of the pair can be avoided if both carriers are appropriately stabilized in the lattice. The hole can reach the grain surface and give rise to a halogen atom which becomes fixed by certain chemical agents in the surrounding gelatin. In contrast, the electron may be trapped at some surface site and turn an Ag^+ ion into an Ag^0 neutral atom. This atom has a low thermal stability, its lifetime being about 1 s in AgBr at room temperature. In order to avoid its annihilation, an additional photoionization event leading to the formation of an Ag_2 dimer is required during the lifetime of the initial Ag^0. Although the lifetime of the dimer reaches several days, it cannot be chemically developed yet. For long-term stability the available information points to a minimum cluster (latent image) of four silver atoms $(Ag^0)_4$ to allow for chemical development of the grain.

The above model explains the so-called reciprocity failures in the Hurter—Driffield curve, which is a plot of the optical density of the developed film against the logarithm of the exposure (integrated flux of incoming photons) (Figure 4.10). At sufficiently low light intensities, the formation of the dimer $(Ag^0)_2$ and consequently of the latent image is not possible. At high intensities, a second photoelectron produced before the initial photoelectron has been trapped at a given site prevents them from being trapped at the same site, so avoiding the formation of $(Ag^0)_2$. This reduces the image formation efficiency.

There are some interesting problems of defect physics still open before full understanding of the photographic process is achieved. First, the nature

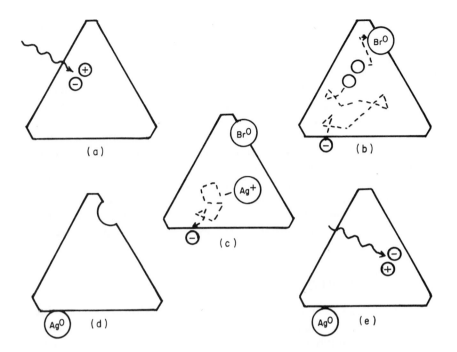

Figure 4.9.
A possible sequence of events in silver halides which leads to the formation of a latent image (the $(Ag^0)_4$ cluster) in the photographic process as a result of exciton production and dissociation.

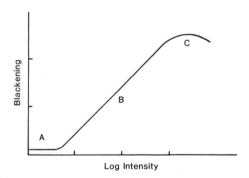

Figure 4.10.
The blackening of a photographic film as a function of light intensity. In region A the rate of arrival of photons is too low to form $(Ag^0)_4$; B is normal response range; C undergoes black−white reversal as the light causes back reactions.

of the surface traps for the photoelectron is not clear nor moreover is the reason why only one or a few of them are operative. The competition of the bulk traps and their relative efficiency is not well understood, although it has to be related to the chemical perfection of the crystal grain. It appears that the charged surface layer of the grain, as well as the chemical sensitization with some chemical elements, such as sulphur, plays an important role in selecting a few active traps on the surface.

The electrons and holes released by the light may originate from the ionization of an exciton, whose separation may be aided by electric fields caused by surface charge. If excitons of pure AgBr were required, the film would only respond to light in the near-ultraviolet region. Sensitization to visible or infrared light is achieved by the addition of dyes. It may be speculated that these offer a low-energy path to exciton generation in the gelatin matrix, and dissociation of the exciton at the AgBr surface then proceeds as outlined above.

4.9   IONIZATION DAMAGE TO OTHER MATERIALS

In addition to alkali and silver halides, ionizing radiation is also capable of inducing photolytic damage (radiolysis) in many other inorganic materials. In particular, photolytic damage has been reported for the alkaline-earth and layered dihalides, polyhalides and azides.

Rather extensive information has been gathered on alkaline-earth fluorides[39] whose defect structure (see Chapter 5) is in some way analogous to that for alkali halides. During irradiation an initial stage is observed, where perturbed F and V_K centres are formed. At higher doses, intrinsic F and H centres are produced, as for alkali halides, although the production efficiency is markedly lower. Unfortunately, the primary mechanism of damage, which might presumably be similar, is not yet well ascertained. STEs are formed in the fluorides but their structure shows some peculiarities[40,41]. In particular, the axis of the STE is centred along the $\langle 111 \rangle$ direction, so that it does not coincide with the $\langle 100 \rangle$ axis of the V_K centres. This should imply a different relaxation mechanism from that operating for alkali halides where the V_K and STE axis are coincident. It appears that the STE relaxes to a close $F-H$ pair, which decays through intrinsic luminescence. It is not clear how well-separated $F-H$ pairs are formed and particularly whether their formation proceeds via the STE as an intermediate stage.

Many layered halides, including lead, bismuth and mercury halides, undergo[42] photolysis or photodecomposition during irradiation with light, X-rays or γ-rays. For lead halides, where more information is available, the final result of the damage is the formation of lead colloids and desorption of halogen gas. The mechanisms responsible for that photodecomposition are not known, although several steps are thought to contribute to the damage. The steps include creation of free carriers by light, trapping of carriers and diffusion of halogen vacancies to allow for the growth of lead colloids. For polycrystalline films, the role of surface defects and environmental conditions appear to be very important, resembling the situation found for silver halides.

In ABF_3 fluoroperovskites such as $RbMgF_3$ or $KMgF_3$, photolytic damage leading to the production of intrinsic colour centres (F, H and V_K) has been observed under X, γ or ultraviolet irradiation[43-45]. The damage mechanism is not definitely known, although presumably it is similar to that operating in the simpler halides. The colour centre production in a non-cubic crystal such as $RbMgF_3$ provides an interesting example for the discussion of exciton relaxation processes and possible focused collisions for energy transfer.

The alkali azides (e.g. NaN_3) also show photolytic damage, although they have received little attention. The quality of the obtainable crystals is still very poor because the materials thermally decompose at a temperature below their melting point. F centres are produced after irradiation with X-rays or ultraviolet light but it is not known whether the formation process is the same as in alkali halides. Anyhow, the operation of a $\langle 110 \rangle$ replacement sequence seems unlikely because the azide group is set skew to this axis; so it would be difficult to transfer momentum or atoms along the chain without large energy losses at each collision. Similar problems are encountered with other azides (e.g. PbN_6) but experimental difficulties are exacerbated as the samples are primary explosives and are triggered by the photolysis.

Other interesting systems showing photolysis are the sodalites[46]. F centres are readily formed by ionizing radiation. Moreover, they can be made photoelectronic, i.e. not only can they be coloured by the absorption of light but also the process can be reversed by a further optical process. Suitable empirical treatments to turn the material photoelectronic include "sensitizing" in a hydrogen or inert gas atmosphere for 30 min. Alternatively the material can be intentionally doped with sulphur, selenium or tellurium but, because of the nature of the doping technique, O^-, O^{2-} and OH^- are also added during the hydrothermal growth process.

Whether the intrinsic photolytic process operating in the halides is also active in other materials such as oxides and semiconductors is a problem of general interest. For oxides, electronic excitation does not generate point defects in pure material; however, surface dissociation and loss of oxygen are common. Recent work on α-quartz and fused silica provided evidence that an oxygen vacancy centre E' and an oxygen interstitial are formed through a photolytic mechanism during an irradiation pulse. The defects created are not stable even at low temperatures. However, these transient

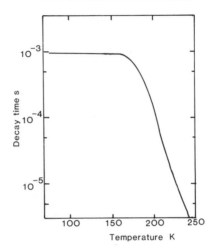

Figure 4.11.
The same decay time dependence with temperature is shown for the 5.4 eV absorption band, the 2.8 eV luminescence and the optical rotation for quartz.

unstable Frenkel defects might be transformed into stable defects when the density of excitation is high enough that they can be again ionized. Such a process may explain the formation of strain centres under intense electron irradiation in SiO_2. The common decay time for the height of the transient absorption band, optical rotation and luminescence intensity associated with the E_1' centre as a function of irradiation temperature are shown in Figure 4.11 for α-quartz. Volume expansion related to the defect formation follows an identical temperature dependence for quartz but no simple correlation exists for silica[47].

4.10 PHOTOCHROMIC PHENOMENA

A number of oxides, as well as other inorganic solids (halides, sodalite, glasses, etc.) undergo reversible changes in colouring by illumination with light of suitable wavelength. This phenomenon, named photochromism, involves only electronic excitation and cannot be considered as lattice or photolytic damage. Photochromic materials possess different absorption spectra depending on illumination conditions and are useful for information storage and display devices.

One of the earlier examples[46,48,49] is $SrTiO_3$ doped with transition-metal ions, which occupy the substitutional position of Ti^{4+}. For example, iron entering in the lattice as Fe^{3+} can be converted into Fe^{4+} by ultraviolet illumination. This introduces a new absorption band at about 800 nm. The process apparently involves electron transfer from Fe^{3+} to an oxygen vacancy introduced in the iron-doped crystal in order to assure charge neutrality. Subsequent optical bleaching regenerates the Fe^{3+} and the crystal recovers the initial absorption. A variety of nickel, cobalt, chromium and iron centres, including oxygen-vacancy-related complexes also present photochromic behaviour. Moreover, interesting photochromic phenomena have been reported for doubly doped crystals, e.g. with iron and molybdenum. These ions enter in the lattice as Fe^{3+} and Mo^{6+}, respectively, so that each Mo^{6+} provides charge compensation for two Fe^{3+}. Initially the crystal is colourless because neither Fe^{3+} nor Mo^{6+} absorbs in the visible region of the spectrum. In the photochromic process, one electron is transferred from Fe^{3+} to Mo^{6+} which become Fe^{4+} and Mo^{5+}. Other oxides such as calcium and barium titanate and titanium oxide show similar effects. Some good examples of photochromic behaviour are found in dihalide crystals (e.g. CaF_2 doped with rare-earth ions (Figure 4.12)) and even alkali halides, where the $F \rightleftharpoons F'$ transition is considered one of the classic photochromic reactions.

In view of the role of silver halides in gelatin in the photographic process, it is expected that they could act in a similar fashion if held in another matrix. Suitable glass matrices typically contain oxides, such as SiO_2 (60%), Na_2O (10%), Al_2O_3 (10%), together with Be_2O_3 (20%) added as a flux. Image formation is induced by traces of silver halides and a sensitizer such as CuO. Silver halide spheres, whose size is in the nanometre range is controlled by the annealing cycle of the glass, separate from the matrix. As in

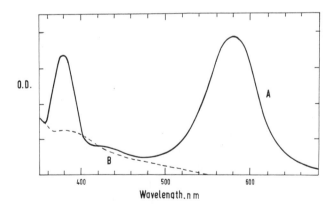

Figure 4.12.
Optical density (OD) of photochromic $CaF_2:CeO_2$: curve A, produced by ultraviolet light; curve B, after bleaching.

the photographic process the operative light-induced reaction appears to be

$$AgCl + h\nu \rightarrow Ag^+ + Cl^- + (e^- + h^+)$$

i.e. with the $(e^- + h^+)^{\star}$ in the form of an exciton. The next stage is $Ag^+ + e^- + Cl^- + h^+ \rightarrow Ag^0 + Cl^0$ but in the glass the Ag^0 clusters can only be temporarily stable as the Cl^0 cannot move into the glass matrix (whereas it was lost in the case of the gelatin). Therefore the process becomes thermally or optically reversible.

The photochromic process is less sensitive than the photographic process as the chemical development of the grains offers a "gain" of about 10^8. However, no chemical development is needed in a photochromic process and so photochromic recording may be preferable in many circumstances.

Photochromism is not limited to inorganic insulators and precisely the same effects are achieved by chemical changes produced in many organic materials. The common organic photochromics have been classified as follows.

(1) *Trans—cis* isomerism caused by rotation about a double bond (azo compounds).
(2) Bond rupture (e.g. spiropyrens).
(3) Transfer of a hydrogen atom to different positions in a molecule (e.g. anils).
(4) Photoionization (e.g. triphenyl methane dyes).
(5) Oxidation—reduction reactions (e.g. thiazine dyes).
(6) Excitation to metastable triplet states.

4.11 EXCITATION AND RECOMBINATION-ASSISTED PROCESSES

In addition to the photolytic mechanisms operating in alkali halides and other materials, there are a number of solid-state reactions or processes which can be induced or enhanced by ionizing radiation. A variety of possibilities can be considered for these assisted reactions that can be classified according to Stoneham[5] as local heating or local excitation models. In the local heating models the electron—hole recombination energy is utilized to increase the vibrational energy in the reaction coordinate, so that the effective energy barrier is reduced (Figure 4.13). They include the recombination-assisted dislocation climb and recombination-assisted annihilation of deep traps in III—V compounds and the dissociation and associated luminescence quenching of hydrogen-related centres in hydrogen-implanted SiC. In the local excitation models, the potential energy surfaces that control the process are different in the ground and excited states of the impurity or defect. This can also alter the activation energy for the reaction

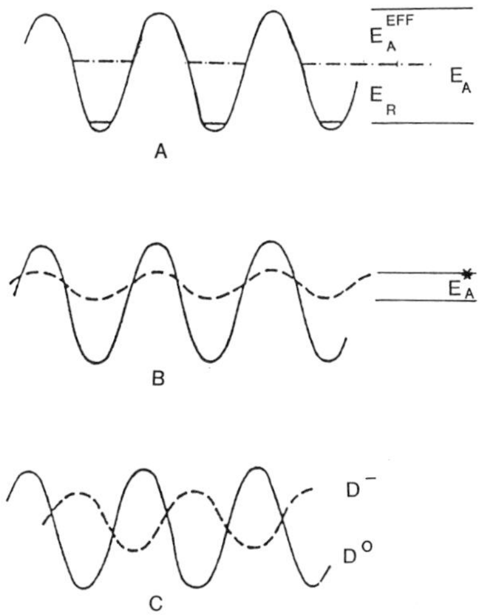

Figure 4.13.
Mechanism of recombination and ionization-enhanced diffusion: (A) local heating changes the activation energy from E_A to E_A^{eff} because of the thermal energy E_R produced by recombination; (B) local excitation model showing a modified energy barrier E_A^* for the excited state; (C) Bourgoin–Corbett model showing different equilibrium configurations for the two charge states D^0 and D^-.

as illustrated in Figure 4.13 or even eliminate it. Typical examples are the enhanced mobility of excited F centres in alkali halides or the Bourgoin– Corbett mechanism in silicon. A few representative processes[4] are now described in somewhat more detail.

(a)   Recombination-assisted dislocation climb.   There is evidence, such as that provided by scanning electron microscopy observations, that the degradation of light-emitting and laser diodes during operation is associated with recombination-assisted dislocation climb. Since dislocation climb re- quires the incorporation of intrinsic defects into the dislocation line, recom- bination-assisted migration of these defects to the dislocation core has to be invoked as the mechanism causing the climb. The energy available in the recombination event can be taken as the gap energy E_g, so that a correlation between the degradation rate of the device and gap energy is expected. This correlation is very nicely illustrated by the data obtained for the family of compounds $In_xGa_{1-x}As$ (Figure 4.14).

(b)   Deep-trap annihilation in GaAs or GaP.   An interesting example of recombination-assisted reactions is provided by the enhancement of the

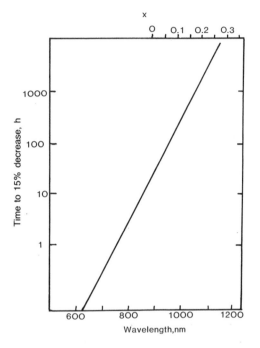

Figure 4.14.
The speed at which light-emitting diodes degrade in output by 15% when oper-
ating at 1000 A cm^{-2} is related to their composition for the set of materials
In$_x$Ga$_{1-x}$As. The peak wavelength during emission is noted.

annihilation rate of deep traps introduced by irradiation in GaAs and GaP.
It has been found that the annihilation rate under forward-bias polarization
of a p—n junction is proportional to the rate of carrier recombination
through the deep centre. This behaviour is illustrated by the annealing rate
of the E$_3$ trap in GaAs. The activation energy for annealing is reduced from
$E = 1.4$ eV to $E' = 0.34$ eV under injection current through the junction.
The difference $E - E' = 1.06$ eV almost exactly matches the energy
released by the recombination of a hole with the electron at the trap.

(c) The Bourgoin—Corbett mechanism.   This local-excitation mechanism[50], which refers to the excitation-assisted migration of a defect centre,
may operate when the lattice configuration of the centre, D, is markedly
altered by the trapping of an additional carrier (electron or hole). If the new
configuration corresponds to a saddle-point configuration for defect jumping
to the next lattice position, this will occur without any energy barrier once
the additional carrier is lost (Figure 4.13). In other words, defect migration

will take place with zero thermal activation energy. This situation may apply to the V^- vacancy centre and the V^{2-} centre in silicon. If this is the case, one can easily understand that the vacancy in silicon diffuses via successive $V^- \to V^{2-} \to V^- \to \ldots$ conversions.

4.12   DAMAGE BY LASER IRRADIATION

There are a large variety of laser sources delivering very high photon fluxes up to about 10^{12} W cm^{-2} in the visible and near-visible spectral regions. These power levels pose severe limitations to the use of appropriate materials in laser systems. A number of "damage" processes could be classified as cases of ionization or excitation damage. They include laser-induced breakdown phenomena in transparent materials as well as surface damage, annealing, melting, amorphization, recrystallization and particle sputtering in absorbing materials. Damage and breakdown processes are particularly relevant in very-high-power laser applications, such as in inertial confinement systems for thermonuclear fusion reactors or some military systems. Atomic rearrangements taking place in surface layers of bulk solids or thin films during laser irradiation are the basis of the laser-annealing methods in the semiconductor technology. All these processes are a form of irreversible damage. In contrast, reversible laser damage refers to the changes in charge distribution (or lattice structure) induced by laser irradiation. These purely electronic effects, usually occurring at much lower power levels, may be analogous to those occurring under X- or γ-irradiation which are responsible for reversible changes in the absorption spectra (photochromic effects) and small corresponding changes $\Delta n/n$ in refractive index. Very large (about 10^{-3}) relative changes in refractive index can be induced by these charge redistribution processes in electro-optic materials via the electro-optic coupling (photorefractive and photovoltaic effects). These effects are finding interesting applications for holographic information storage and retrieval as well as in a variety of coherent optic devices. Relevant damage processes are now described in some detail.

4.13   SURFACE DAMAGE AND ANNEALING

For absorbing materials, the near-surface absorption of the light energy gives rise to a number of effects: enhanced diffusion, melting, defect annealing and recrystallization of damaged regions in ion-implanted semiconductors (laser annealing), disordering and amorphization of crystalline zones and sputtering of ions with changes in material stoichiometry. Much effort has been devoted to laser-induced damage to semiconductors, since the initial observation of reflectivity enhancement of semiconductor surfaces[51,52]. In particular, laser-annealing techniques are now well controlled and have a marked technological relevance in the semiconductor industry for the fabrication of semiconductor devices. Typical examples are the annealing of ion-implanted silicon and the recrystallization of GaAs. The key point in the understanding of these processes is the transfer of the electronic excitation energy to the lattice vibrational modes. Commonly accepted models assume that the excitation energy is converted to heat and eventually leads to local melting of the material. After subsequent cooling, the crystalline or amorphized zones are produced. Different regimes can be considered depending on the irradiation conditions, i.e. wavelength, power and length of the laser pulse. In all cases, the problem of heating and melting is solved by handling the heat conduction equation subjected to appropriate boundary conditions. In this way the time evolution of the temperature profiles in the material and the depth of the melted region can be evaluated. It has to be mentioned here that, even without reaching the melting point, the laser-induced heating may induce important atomic re-arrangements through enhanced diffusion.

Recently, some doubts have arisen about the validity of the thermal model for pulse annealing experiments particularly in the nanosecond and subnanosecond regime[53-56]. It is argued that a highly excited and long-lived dense electron−hole plasma is responsible for lattice instability in covalent semiconductors through softening of the atomic bonds. This softening is the result of the fact that a significant fraction of the bond charge is excited across the gap into antibonding or plane wave states. In a softened lattice, with a correspondingly lower Debye temperature, the melting point is reduced. As an example, for silicon, the reduction has been estimated to be about 40 K for a plasma density $n = 10^{20}$ cm^{-3} or 400 K for $n = 10^{21}$ cm^{-3}. For a critical plasma density ($n = 8 \times 10^{21}$ cm^{-3} for silicon), the material undergoes a second-order transition to a fluid phase where transverse acoustic modes go to zero frequency. The energy is concentrated in the electronic excitation and not in the kinetic (thermal) energy of the lattice. According to this model, recrystallization could occur without the destructive effects of severe thermal gradients when the material passes back through the phase transition to the covalently bonded phase.

A phenomenon closely related to that of surface damage is laser-induced sputtering. Most often a Lambertian and Maxwellian distribution of sputtered particles appears to have been observed, as expected from the thermally induced evaporation of atoms on a heated (melted) surface. However, some experiments, which were carried out at lower light powers so as not to produce melting, suggest that non-thermal processes determine the sputtering yield[57,58]. The main features of this non-thermal sputtering are the following.

(1)   Non-linear dependence of the sputtering yield on laser fluence (integral flux).

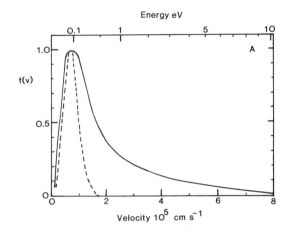

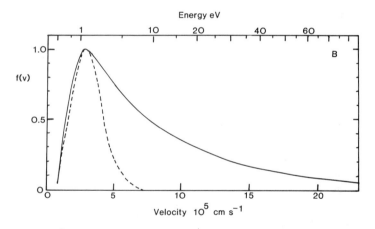

Figure 4.15.
Velocity distribution $f(v)$ for (A) neutral P atoms and (B) P^+ ions laser sputtered from GaP. The dashed curves show the Maxwellian function, $f_M(v)$, which has an effective temperature $T_{eff} = 1.0 \times 10^4$ K fitted to match the most probable velocity. The laser wavelength was 540 nm at a fluence of ~ 0.1 J cm^{-2}.

(2)   Well-defined threshold power, dependent on wavelength.

(3)   Non-Maxwellian distribution for the velocity of the sputtered atoms (Figure 4.15).

Mechanisms have been proposed to account for these effects, although they still lack sound experimental support and are in a rather speculative stage. One such proposal follows the non-radiative recombination model already discussed in connection with damage and sputtering processes in alkali halides. Another model involves the weakening of atomic bonds induced by strong electronic excitation. This mechanism, described above when dealing with laser-annealing processes, facilitates atomic displacement and migration. Specifically, if the electron−hole plasma density is high enough, it may screen the repulsive Coulomb potential between two holes, lowering the energy of a multiply ionized state below that corresponding to the singly ionized state. In fact, this would be a manifestation of the negative U interaction discussed in section 7.3. The multiply ionized atom would be ejected by electrostatic forces.

4.14 LASER-INDUCED BREAKDOWN IN TRANSPARENT MATERIALS

The explanation of damage effects in transparent materials[59] ($\alpha \leq 0.1$ cm^{-1}) is, in principle, more complicated since the linear (one-photon) absorption cannot account for the temperature rise required to reach the melting point. Multiple absorption by impurities or the host material also has to be considered. The efficiency of these processes increases very rapidly with increasing incident power and therefore it easily justifies the occurrence of a threshold power as experimentally found. Unfortunately this mechanism predicts a dependence of the threshold with a high power $n \approx E_g/h\nu$ of the incoming light flux, which is not supported by experiment. Therefore, some additional energy-absorbing mechanism has to be invoked to account for the lattice damage to transparent materials. One such process which enjoys wide acceptance involves the acceleration of the generated carriers by the electric field of the laser light. The accelerated electrons induce new ionizations by impact on other atoms, so that finally a hot microplasma is formed. The microplasma provides an efficient channel for absorbing energy from the electromagnetic field and transferring it to the lattice. This avalanche mechanism is clearly related to that proposed for dielectric breakdown under a direct current (DC) field. Therefore, some correlation between the threshold breakdown fields under DC or high-frequency (light) conditions should be expected, as found for some alkali halides.

In accordance with the above discussion, the avalanche model for laser damage is made up of three stages.

(1) Creation of an initial population of free electrons.
(2) Acceleration and multiplication of the electrons with formation of a microplasma.
(3) Absorption of the electromagnetic (laser) energy by the microplasma and transfer to the lattice.

Stage (1) involves the appropriate linear and non-linear absorption processes necessary to liberate a minimum electron concentration (about 10^{10} cm^{-3})

to trigger stage (2). Certain evidence for this initial threshold concentration is inferred from the morphology of the lattice damage. For laser powers close to the threshold value, the damaged region consists of a row of disconnected microscopic areas in the path of the incoming beam. Each of these areas should be the result of the avalanche generated by one (or a few) of the initially photoionized electrons. At higher powers, the various areas overlap and a continuous damage track is observed.

The acceleration and multiplication processes operating during stage (2) are quite sensitive to the duration τ of the laser pulse. Therefore, the threshold power P_0 would be expected to depend on τ. In particular, experimental data[59] on NaCl are shown in Figure 4.16.

Stage (3) involves the Joule heating of the microplasma by the laser field and the energy transfer to the lattice. Local melting of the material is, then, the final result of the process.

The above avalanche model[60] is in accordance with some experimental data but it requires both further experimental support and a more thorough theoretical formulation. In particular, direct evidence on the formation of avalanches and subsequent microplasmas during laser irradiation is a key point.

4.15 REVERSIBLE LASER DAMAGE—PHOTOREFRACTIVE AND PHOTOVOLTAIC EFFECTS

At significantly lower levels of light intensity than those considered in the above sections, reversible damage can be produced as a consequence of charge redistribution inside the material. This phenomenon, which depends on the existence of impurities and defects, has already been mentioned as one of the effects of the ionizing radiation. The situation presents specific and relevant features for electro-optic (non-centrosymmetric) materials. There, large changes in the refractive index and high electric voltages between the ends of the material (in an open-circuit configuration) can be produced by illumination. These are the so-called photorefractive and photo-voltaic effects, respectively. In some materials, such as $LiNbO_3$, the two effects are intimately related, but in general the photovoltaic effect is just one of the drift mechanisms that may contribute to the photorefractive effect.

The photorefractive effect was first observed by Askhin et al.[61] in 1966 for $LiNbO_3$ as a distortion of a laser beam when it passes through an electro-optic crystal. The effect was correctly attributed to changes induced by the laser light in the extraordinary refractive index n_e whereas n_0 remains essentially unmodified. In contrast, the photovoltaic effect consists of the occurrence of a light-induced current in the direction $+c$ optic axis, when

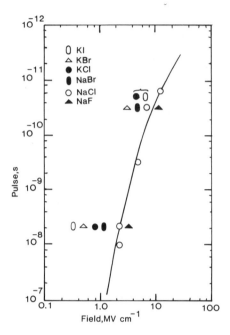

Figure 4.16.
Experimental relationship between the breakdown threshold electric field strength and the laser pulse duration for alkali halides: ⓞ, KI; △, KBr; ●, KCl; ▐, NaBr; ○, NaCl; ▲, NaF; ——, predicted[59] curve which is a fit to experimental NaCl data.

some electro-optic crystals such as $LiNbO_3$ are illuminated in the absence of any applied field. Under open-circuit conditions, a voltage is generated between the ends of the sample.

Most research effort[62] on the photorefractive effect has been devoted to $LiNbO_3$, but there are other promising crystals such as $BaTiO_3$, $KNbO_3$, $LiTaO_3$, $Bi_{12}GeO_{20}$ and $Bi_{12}SiO_{20}$ as well as the ceramic lead zirconate titanate. It is generally agreed that the effect is associated with a light-induced charge transfer between certain impurities or defects acting as donors and other impurities or defects acting as acceptors, although an apparently intrinsic effect has also been reported. In particular, it is generally accepted that two valence states of iron, Fe^{2+} and Fe^{3+}, are the relevant donor and acceptor centres, respectively, in $LiNbO_3$ and probably other materials. The active absorption band attributed to the Fe^{2+} donor peaks at about 2.5 eV and is apparent in the spectra of reduced crystals. The photorefractive effect then involves the following processes: photoionization of Fe^{2+}, directional transport of electrons and trapping by the Fe^{3+} (acceptor). However, some experimental evidence also points to $Fe^{3+} \rightarrow Fe^{2+}$

conversion by hole photoionization to the valence band and hole transport, in oxidized crystals. Anyhow, the redistribution of electrical charge creates an associated electric field distribution, which via the electro-optic coupling causes a refractive index pattern correlated with the light intensity pattern.

For $LiNbO_3$, the directional transport is accomplished without any applied field by the photovoltaic mechanism. However, other transport mechanisms such as diffusion of carriers and drift under an electric field are also active mechanisms and are the only mechanisms operating in crystals not showing a photovoltaic effect.

The processes governing the photoionization, transport and trapping of carriers can be described by a set of non-linear differential equations, which have not been solved in the most general case. Different approximations have led to particular solutions to describe steady-state situations as well as transient response during writing or erasure of a holographic grating.

Although it has been assumed in the above discussion that iron is a relevant photorefractive impurity, the role of other impurities and lattice defects, such as oxygen vacancy centres, cannot be ruled out. In fact, reduction treatments enhancing the photorefractive efficiency very probably introduce F and F^+ centres[63], which may also contribute to the various defect centres for the photorefractive effect.

4.16 DAMAGE ALONG ENERGETIC ION TRACKS

High-energy charged particles leave damage trails (latent tracks) on passing through non-conducting solids such as mineral crystals, glasses and plastics[64]. These trails can be directly seen by transmission electron microscopy under a high magnification or by optical microscopy after suitable etching treatments. The etching treatment involves highly selective dissolution of the material along the damaged track by use of an appropriate chemical reagent.

The damage induced by a high-energy ion in a material provides a very interesting example of ionization damage, since ionization is the dominant energy loss mechanism for energetic charged particles. However, the details of the process are not definitely clear and available models are mostly of a speculative nature. They essentially involve the massive excitation of electrons along the track via the energy loss interactions. The wake of positive ions becomes unstable because of the Coulomb repulsion so that atomic motion and lattice displacements occur along the track (ion explosion), leaving a vacancy-rich core. The long lifetime of the electronic excitation in an insulator (longer than 10^{-12} s) contributes to the feasibility of this mechanism, which requires a certain critical value for the primary ionization rate. Above this threshold an etchable track is formed, because the damaged region has a higher chemical reactivity than the host material does.

In order to obtain a deeper insight into the operative damage mechanisms, recent experimental data[65] on the structure of the tracks in mica, olivine and labradorite are relevant. It has been shown by X-ray scattering that the damage is not continuous along the track but consists of elongated regions of extended defects oriented along the track and separated by regions showing exclusively point defects. The extended damage regions are typically between 10 and 40 Å in size. The average separation between two regions depends on the energy but is roughly $25-50$ Å. These experiments have led to a reasonable understanding of the etching characteristics of latent tracks. However, in order to account for those observations and, in particular, the localization of heavy damage at some separated regions along the ion path, the available damage models have to be revised and possibly modified. One recent proposal[66] involves the Auger decay of K-shell holes as the triggering mechanism for excitation localization along the track. It appears that the cross-section for K-shell ionization of light elements such as carbon, oxygen and fluorine may be consistent with the separation found among extended damaged areas. However, the energy of the Auger electrons (about 500 eV) and their short range (about 10 Å) implies a very high excitation density which can initiate lattice disordering through the bond-weakening mechanism already discussed in connection with non-thermal laser damage. Because of the additional homogeneous excitation developed along the track by the dominant loss mechanism (outer-shell ionization), the local material disorder occurring at the Auger decay site may spread along the track and give rise to the elongated shape of the heavily damaged zone.

REFERENCES

1 P. D. Townsend and F. Agulló-Lopez, *J. Phys. (Paris), Colloq. C6* **41**, 279 (1980).
2 J. H. O. Varley, *J. Nucl. Energy* **1**, 130 (1954).
3 P. J. Chandler, F. Jaque and P. D. Townsend, *Radiat Eff.* **42**, 45 (1979).
4 P. J. Dean and W. J. Choyke, *Adv. Phys.* **26**, 1 (1977).
5 A. M. Stoneham, *Adv. Phys.* **28**, 457 (1979).
6 L. W. Hobbs, A. E. Hughes and D. Pooley, *Proc. R. Soc. London, Ser. A* **332**, 167 (1973).
7 L. W. Hobbs, in M. W. Roberts and J. M. Thomas (eds), *Surface and Defect Properties of Solids*, The Chemical Society, London (1975).
8 L. W. Hobbs, *J. Phys. (Paris), Colloq. C7* **37**, 3 (1976).
9 P. W. Levy, K. J. Swyler and R. W. Klaffky, *J Phys. (Paris), Colloq. C6* **41**, 344 (1980).
10 D. Pooley, *Proc. Phys. Soc., London* **87**, 245 (1966).
11 H. N. Hersh, *Phys. Rev.* **148**, 928 (1966).
12 I. K. Vitol, *Izv. Akad. Nauk, Fiz.* **30**, 564 (1966).

13 R. T. Williams, J. N. Bradford and W. L. Faust, *Phys. Rev. B* **18**, 7038 (1978).
14 N. Itoh, *Adv. Phys.* **31**, 491 (1982).
15 R. T. Williams, *Semicond. Insul.* **3**, 251 (1978).
16 P. D. Townsend, *J. Phys. C: Solid State Phys.* **6**, 961 (1973).
17 P. D. Townsend and D. J. Elliot, *Phys. Lett. A* **28**, 587 (1969).
18 A. Schmid, P. Braunlich and P. K. Rol, *Phys. Rev. Lett.* **35**, 1382 (1975).
   P. D. Townsend, *Phys. Rev. Lett.* **36**, 827 (1976).
19 P. D. Townsend, R. Browning, D. J. Garlant, J. C. Kelly, A. Mahjoubi, A. J. Michael and M. Saidoh, *Radiat. Eff.* **30**, 55 (1976).
20 M. Szymonski, H. Overeijnder and A. E. de Vries, *Surf. Sci.* **90**, 274, 265 (1979).
21 P. D. Townsend, in R. Behrisch (ed.), *Sputtering by Particle Bombardment II*, *Topics in Applied Physics*, Vol. 52, Springer, Berlin (1983).
22 M. Szymonski and A. E. de Vries, in N. H. Tolk, M. M. Traum, J. C. Tully and T. E. Madey (eds), *Desorption Induced by Electronic Transitions*, Springer, Berlin p. 216 (1983).
23 C. R. A. Catlow, K. M. Diller and M. J. Norgett, *J. Phys C: Solid State Phys.* **8**, L34 (1975).
24 L. H. Abu-Hassan and P. D. Townsend, *J. Phys. C: Solid State Phys.* **19**, 99 (1986); *Radiat Eff.* **98**, 313 (1986).
25 A. B. Lidiard, *Comments Solid State Phys.* **8**, 73 (1978).
26 A. E. Hughes, *Comments Solid State Phys.* **4**, 83 (1978).
27 F. J. Lopez, M. Aguilar and F. Agulló-Lopez, *Phys. Rev. B* **23**, 3041 (1981).
28 J. L. Alvarez Rivas, *J. Phys. (Paris), Colloq. C6* **41**, 353 (1980).
29 E. Sonder, *Phys. Status Solidi* **35**, 523 (1969).
30 A. Nouilhat, G. Guillot and E. Mercier, *J. Phys. (Paris), Colloq. C6* **41**, 308 (1980).
31 J. D. Comins and B. O. Carragher, *Phys. Rev. B* **24**, 283 (1981).
32 M. Aguilar, F. Jaque and F. Agulló-Lopez, *Radiat. Eff.* **61**, 215 (1982).
33 F. Hermann and P. Pinard, *J. Phys. Chem. Solids* **32**, 2649 (1971).
34 E. Kotomin, I. Fabrikant and I. Tale, *J. Phys. C: Solid State Phys.* **10**, 2903 (1977).
35 U. Jain and A. B. Lidiard, *Philos. Mag.* **35**, 245 (1977).
36 H. Kanzaki, *Photogr. Sci. Eng.* **24**, 219 (1980).
37 L. Slifkin, The photographic process, in C. H. S. Dupuy (ed.), *Radiation Damage Processes in Materials*, Nordhoff, Leyden (1975); *Sci. Prog.* **60**, 151 (1972).
38 N. F. Mott and R. W. Gurney, *Electronic Processes in Ionic Crystals*, Dover Publications, New York (1964); *Proc. R. Soc. London, Ser.* **164**, 151 (1953).
39 W. Hayes and A. M. Stoneham, in W. Hayes (ed.), *Crystals with the Fluorite Structure*, Oxford University Press, Oxford, Chapter 4 (1975).
40 P. J. Call, W. Hayes and M. N. Kabler, *J. Phys. C: Solid State Phys.* **8**, L60 (1975).
41 R. J. Williams, M. N. Kabler, W. Hayes and J. P. Stott, *Phys. Rev. B* **15**, 725 (1977).
42 M. R. Tubbs, *Phys. Status Solidi (b)* **67**, 11 (1975).
43 J. R. Seretlo, J. Martin and E. Sonder, *Phys. Rev. B* **14**, 5404 (1976).
44 N. Koumvakalis and W. A. Sibley, *Phys. Rev. B* **13**, 4509 (1976).
45 A. Podinsh and W. A. Sibley, *Phys. Rev. B* **18**, 5921 (1978).

46 P. D. Townsend and J. C. Kelly, *Colour Centres and Imperfections in Insulators and Semiconductors*, Chatto and Windus, London (1973).
47 K. Tanimura, T. Tanaka and N. Itoh, *Phys. Rev. Lett.* **51**, 423 (1983).
48 R. L. Berney and D. L. Cowan, *Phys. Rev. B* **23**, 37 (1981).
49 B. W. Faughman, *Phys. Rev.* **134**, 3623 (1971).
50 J. W. Corbett and J. C. Bourgoin, in J. H. Crawford and L. M. Slifkin (eds), *Point Defects in Solids*, Vol. 2, Plenum, New York (1975).
51 M. Bertolotti, in L. D. Laude (ed.), *Cohesive Properties of Semiconductors under Laser Irradiation*, Martinus Nijhoff, The Hague (1983).
52 M. Balkanski, in D. Bauerle (ed.), *Laser Processing and Diagnosis*, Springer, Berlin (1984).
53 J. Bok, *J Phys. (Paris), Colloq. C5* **44**, 3 (1983).
54 J. A. Von Vechten, R. Tsu and F. W. Saris, *Phys. Lett. A* **74**, 422 (1979).
55 E. J. Yoffa, *J. Phys. (Paris) Colloq. C4* **41**, 7 (1980).
56 J. A. Von Vechten, *J. Phys. (Paris), Colloq. C4* **41**, 15 (1980).
57 T. Nakayoma, *Surf. Sci.* **133**, 101 (1983).
58 N. Itoh and T. Nakayama, *Semicond. Insul.* 5, 383 (1983).
59 W. Lee Smith, *Opt. Eng.* **17**, 489 (1978).
60 S. Brawer, *Phys. Rev. B* **20**, 3422 (1979).
61 A. Ashkin, G. D. Boyd, J. M. Dziedzik, R. G. Smith, A. D. Ballman and K. Nassau, *Appl. Phys. Lett.* **9**, 72 (1966).
62 P. Günter, *Phys. Rep.* **93**, 199 (1982).
63 L. Arizmendi, J. M. Cabrera and F. Agulló-Lopez, *J. Phys. C: Solid State Phys.* **17**, 515 (1984).
64 R. L. Fleischer, P. B. Price and R. M. Walker, *Nuclear Tracks in Solids*, University of California Press, Berkeley, CA (1975).
65 E. Dartyge, J. P. Durand, Y. Langevin and M. Maurette, *Phys. Rev. B*, **23**, 5213 (1981).
66 T. A. Tombrello, C. R. Wie, N. Itoh and T. Nakayama, *Phys. Lett. A* **100**, 42 (1984).

5
Point Defects—Halides

5.1 INTRODUCTION

Ionic solids are characterized by an essentially ionic or Coulombic contribution to the cohesive energy. As a general rule, they behave as very good electric insulators with a wide energy gap of $1-12$ eV and include a large variety of compounds such as halides and polyhalides, and binary and ternary oxides. Note, however, that many of them have a certain degree of covalent character. Crystal symmetry is relatively simple, being cubic for many of them and typical structures are face-centred, body-centred, fluorite and perovskite (Figure 5.1). The materials studied most are the alkali and silver halides and, in fact, alkali halides constitute the major examples for this chapter. The study of point defects in simple halides can serve as a reference model to illustrate the situations that can be found in more complex systems. Attention is also paid to some polyhalide materials. Oxides are discussed in Chapter 6.

As for covalent materials, defects in ionic crystals usually present several stable states of charge. Likewise they present important relaxation effects, associated with a strong electron−lattice interaction. Relevant consequences of these effects are the spontaneous Jahn−Teller distortions at defect sites and occurrence of the so-called negative U features, which will be discussed in Chapter 7. Another aspect of that strong coupling is the self-trapping of carriers, i.e. their localization at a given lattice site, giving rise to an electronic intrinsic defect, the "small polaron".

Defects in ionic crystals very often present an effective charge, which

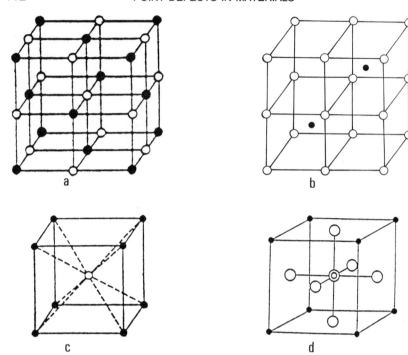

Figure 5.1.
Four examples of halide crystal structures: (a) NaCl; (b) CaF_2; (c) CsCl; (d) $KMgF_3$.

makes them very efficient traps for either electrons or holes. This markedly influences the lifetime of free carriers in the material. In contrast, they play a relevant role in the scattering processes of carriers and therefore in determining the corresponding mobilities.

A typical consequence of the high-energy gap is that electronic transitions between the various levels of a defect centre give rise to observable absorption and luminescence bands that assist in identification of the defect. In such a case, the defect is called a "colour centre". This definition includes impurities, intrinsic defects and impurity–intrinsic complexes.

Finally, the very high electrical resistivity of the ionic materials facilitates the use of magnetic resonance techniques. They offer the most powerful tools for determining the symmetry and structure of defects.

5.2 INTRINSIC DEFECTS IN ALKALI HALIDES

In the simple halides A^+X^- (where A^+ is a monovalent cation and X^- is a halide ion) the basic intrinsic defects are cation and anion vacancies and the corresponding interstitial ions. The thermal production of defects at high temperatures proceeds via either Schottky or Frenkel pairs depending on the particular material. For alkali halides the Schottky pair is energetically favoured and it is the predominant intrinsic defect as inferred from density and lattice parameter measurements. In contrast, for silver halides, the cationic Frenkel pair is the dominant species. We exclusively refer in this section to alkali halides. One of the main advantages of these solids for intrinsic defect studies is that they can be prepared with a very high degree of chemical purity.

Vacancy diffusion (and self-diffusion) has been much studied in alkali halides, both theoretically and experimentally. The activation energy is lower for the cation vacancies, which therefore determine the electrical conductivity of the material. In other words, transport numbers for cation vacancies are very close to unity for alkali halides. Interstitials in alkali halides are generated by irradiation and they are discussed later when dealing with colour centres. Experimental values[1-5] for the formation and migration enthalpies for defects in halides are given in Table 5.1.

5.3 IMPURITIES

5.3.1 Cation impurities

Cation impurities, mainly alkali, alkaline-earth and transition-metal ions, constitute major contaminants in all alkali halides. They lie at a substitutional location for the host cation. A great deal of work on monovalent cations has been concerned with s^2 ions (Tl^+, Ga^+, etc.) and d^{10} ions (Ag^+, Cu^+, etc.) since they present characteristic optical absorption and luminescence bands, appropriate for carrying out detailed spectroscopic studies. Of particular interest is the case of monovalent cations of small radius in relation to the host cation, e.g. Li^+, Cu^+ or even Ag^+. For these ions, the competition among the various contributions to the energy of the defect, electrostatic, repulsive, polarization, etc., is very delicate. Under these conditions, the impurity may shift to an off-centre substitutional position with less energy than that for the exact substitutional (on-centre) position. The effect can be understood in terms of a reduction in the dipolar polarization interaction energy which offsets the small increase in the repulsive energy occurring when the impurity shifts to the off-centre location. One of the best studied examples is $KCl:Li^+$, where Li^+ moves to an off-centre location along a $\langle 111 \rangle$ axis. At each lattice site, there are eight equivalent configurations of Li^+. Tunnelling between these sites leads to a splitting of each vibrational level into four equally spaced levels, if overlapping is neglected and only transitions between adjacent orientations are considered.

POINT DEFECTS IN MATERIALS

Table 5.1.
Enthalpies of formation and migration

Material	$h_f(S)$ (eV)	$h_f(F^-)$ (eV)	$h_m(V^+)$ (eV)	$h_m(V^-)$ (eV)	$h_m(I^-)$ (eV)
NaCl	2.44		0.69	0.77	
KCl	2.54		0.73	0.85	
NaI	2.00		0.58	0.77	
KBr	1.7−2.53		0.67	0.87−0.95	
KI	1.6−2.21		0.67−0.72	1.29−1.50	
RbCl	2.04		0.54	1.45	
LiF	2.34−2.4		0.70−0.73	1.0	
CsCl	1.9		0.6	0.3	
CsBr	1.8		0.36	0.51	
CaF_2		2.2−2.8		0.38−0.47	0.77−0.79
SrF_2		1.7−2.3		0.52−0.58	0.75
BaF_2		1.91		0.47−0.64	0.71−0.74
$SrCl_2$		1.92−2.02		0.27−0.42	0.76−1.04

Formation enthalpies h_f and migration enthalpies h_m for Schottky (S) and Frenkel (F) pairs, as well as various vacancies (V) and interstitials (I), in several alkali and alkaline-earth halides. The signs + and − respectively refer to cation or anion defects.

For the ground level, the energy separation Δ between the A_{2u}, A_{1g}, T_{1u} and T_{2g} levels is $\Delta_6 = 1.15$ cm^{-1} (for ^{6}Li) and $\Delta_7 = 0.82$ cm^{-1} (for ^{7}Li).

An interesting feature associated with off-centre impurities is the occurrence of elastic as well as electric dipole moments which can be experimentally determined. These dipoles can become oriented under an applied electric or elastic field and therefore they contribute to the corresponding linear response coefficient (e.g. dielectric constant) and to the relaxation behaviour. At variance with a gas system, the off-centre impurities provide a unique example of a dilute dipolar system which can be studied down to arbitrary low temperatures. As a consequence, interesting electronic phenomena can be observed. Table 5.2 (taken from Stoneham's[6] book) lists some of the theoretical predictions for off-centre systems together with the reported experimental observations.

It should also be mentioned that small-sized monovalent impurities give rise to local modes in the vibrational spectra of the material. For example Li^+ in KBr gives different local-mode bands for the two isotopes ^{6}Li and ^{7}Li.

The divalent cation impurities present an excess of charge, +e, with respect to the host cation. This induces the generation of a neutralizing cation vacancy per impurity, in order to keep the overall electrical neutrality of the material. Because of the Coloumb attraction the energetically most

Table 5.2.
Off-centre impurity examples

System	Prediction	Observation
NaCl:Li$^+$	Off centre	Probably on centre
NaBr:Li$^+$	Off centre	Probably on centre
KCl:Li$^+$	Off centre	Off centre
KCl:F$^-$	On centre	On centre
NaBr:F$^-$	Off centre $\langle 111 \rangle$	Off centre
KCl:Cu$^+$	Off centre	Off centre
RbCl:Ag$^+$	Off centre	Off centre $\langle 110 \rangle$

favoured location for the vacancy is as first or second neighbour to the impurity. In each case, an impurity–vacancy (I–V) dipole is created with orientation $\langle 110 \rangle$ for nearest-neighbour dipoles and $\langle 100 \rangle$ for next-nearest-neighbour dipoles. Direct evidence for the occurrence of both types of dipole has been obtained from electron paramagnetic resonance (EPR) spectra of transition-metal ions (Mn^{2+}) or rare-earth ions (Eu^{2+}).

The dipole binding energy can be obtained from the slope of the curve of $\log \sigma$ against $1/T$ in the intermediate-temperature region (200–500°C) where the vacancy concentration is determined by the amount of divalent impurities present. Since vacancies can be either free or bound into dipoles, the concentration of the free vacancies contributing to the conductivity includes an exponential term with an activation energy which is equal to the binding energy of the I–V dipole. Table 5.3 contrasts calculated binding energies with experimental estimates. It should be noted that there is a large spread in the reported data. The relative stability of the nearest-neighbour dipole increases with increasing radius of the dopant, whereas a smaller and reverse effect is observed for the next-nearest-neighbour dipole.

The I–V dipoles can reorientate in an electric field and give rise to characteristic Debye relaxation peaks in the imaginary part of the dielectric constant ϵ'' or to well-defined peaks in the thermally stimulated depolarization current or ionic thermoconductivity (ITC) (see Chapter 8). Although the reorientation process is rather complicated and involves several distinct elementary jump events, it can be very approximately described[7] by means of an effective jump rate, with $1/T$ exponential dependence and a unique activation energy. Table 5.4[8] lists the experimental reorientation energy for various systems.

As for monovalent cations, some very detailed spectroscopic studies have been carried out on a number of divalent cations. In particular, optical spectroscopy has been applied to s^2 ions (Pb^{2+} and Sn^{2+}) and optical and

Table 5.3.
Comparison of calculated and experimental binding energies for dipolar
complexes

System	U_1 (eV)	U_2 (eV)	U_3 (eV)	Experimental (eV)
NaCl:Mg^{2+}	0.54	0.55	0.33	0.41±0.07
NaCl:Ca^{2+}	0.59	0.51	0.35	0.44±0.23
NaCl:Sr^{2+}	0.63	0.47	0.36	0.55±0.06
NaCl:Ba^{2+}	0.68	0.45	0.37	0.78±0.03
KCl:Mg^{2+}	0.59	0.64	0.33	
KCl:Ca^{2+}	0.62	0.62	0.35	0.51±0.03
KCl:Sr^{2+}	0.65	0.60	0.37	0.52±0.14
KCl:Ba^{2+}	0.69	0.60	0.39	0.40
KBr:Mg^{2+}	0.55	0.60	0.31	
KBr:Ca^{2+}	0.58	0.59	0.33	0.54±0.07
KBr:Sr^{2+}	0.60	0.57	0.35	0.64
KBr:Ba^{2+}	0.63	0.56	0.36	
NaCl:Mn^{2+}	0.608	0.514	0.371	
NaCl:Cd^{2+}	0.648	0.473	0.380	
NaCl:Pb^{2+}	0.689	0.441	0.387	
KCl:Mn^{2+}	0.630	0.631	0.376	
KCl:Cd^{2+}	0.661	0.602	0.409	
KCl:Pb^{2+}	0.692	0.576	0.415	
KBr:Mn^{2+}	0.586	0.593	0.376	
KBr:Cd^{2+}	0.614	0.567	0.382	
KBr:Pb^{2+}	0.640	0.542	0.387	
RbCl:Ca^{2+}	0.58	0.58		
RbCl:Sr^{2+}	0.62	0.57		
RbCl:Pb^{2+}	0.66	0.57		
RbCl:Ba^{2+}	0.65	0.55		

U_1, U_2 and U_3 correspond to nearest-neighbour, next-nearest-neighbour and third-nearest neighbour energies, respectively.

magnetic (EPR and electron-nuclear double resonance (ENDOR)) spectro-scopy to transition-metal (e.g. Mn^{2+} and Ni^{2+}) and rare-earth ions (e.g. Eu^{2+}). The experimental data have been generally analysed in accordance with the conventional schemes of the crystal and ligand field models[9,10].

5.3.2   Anion impurities

Although anion impurities have been less studied than cationic impurities, abundant information is also available on a number of anion impurities, such as halogen ions X^-, H^- and H^0, OH^-, O_2^-, CN^- and NO_2. One of the most studied is hydrogen. In alkali halide crystals which have been

Table 5.4.
Dipolar reorientation energies

System	Energy (eV)	System	Energy (eV)
$NaCl:Mg^{2+}$	0.64	$KCl:Mg^{2+}$	0.49
$NaCl:Ca^{2+}$	0.69	$KCl:Ca^{2+}$	0.63
$NaCl:Sr^{2+}$	0.72	$KCl:Sr^{2+}$	0.66
$NaCl:Ba^{2+}$	0.75	$KCl:Ba^{2+}$	0.70
$KBr:Sr^{2+}$	0.65	$KBr:Ba^{2+}$	0.68

Note that different experiments differ by up to ± 0.03 eV.

grown with the corresponding alkali hydride, hydrogen enters into the lattice as H^-, substituting for the host halogen anion (U centre). This centre presents a characteristic ultraviolet band associated with a $1s^2 \rightarrow 1sp$ (singlet) transition, i.e. of a similar nature to that responsible for the F band (see later). During illumination at the U band, no fluorescence has been detected, but a conversion from substitutional H^- into interstitial H^- (U_1 centre) presenting a new absorption band is optically induced. The U centre also presents an infrared absorption in the spectral region $10-30$ μm which is associated with a local vibrational mode.

By irradiation (as by illumination), it is possible to turn the U into the U_1 centre and moreover to create the U_2 centre, i.e. atomic hydrogen H^0 in an interstitial position. This centre is particularly interesting because it presents a characteristic paramagnetic resonance spectrum which permits unambiguous identification of the defect. Furthermore, its hyperfine structure, which is associated with the interaction with neighbouring halogen ions, provides very detailed information on the electronic wavefunction of the centre.

Another example which deserves particular consideration is the radical OH^-, since it is a major dopant for all crystals grown in air and moreover it has important effects on the physicochemical behaviour of the material. OH^- occupy substitutional positions at the halogen site and is aligned along the $\langle 100 \rangle$ direction in all studied systems (except for KI where the alignment takes place along a $\langle 110 \rangle$ axis).

The hydroxyl ion presents electronic optical absorption in the ultraviolet spectral region ($5-7$ eV), vibrational absorption in the near infrared ($2.5-3$ μm) and librational absorption in the far infrared (about 300 cm^{-1}), as illustrated in Figure 5.2 for KBr. The ultraviolet band has a maximum at a wavelength λ_m, which shows a Mollwo—Ivey type dependence on the lattice parameter: $\lambda_m = 752a^{0.88}$ (i.e. it scales with the lattice parameter as discuss-

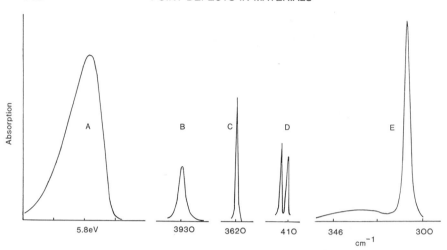

Figure 5.2.
Optical absorption features resulting from OH impurities in KBr. The bands range from an electronic transition in the ultraviolet (band A), to the stretching (bands B and C) and librational (band E) modes in the infrared.

ed in section 5.5.1). Excitation at this band induces a characteristic luminescence as well as photochemical decomposition of the OH^-. The near-infrared band corresponds to the stretching vibration of the OH^- molecule and its frequency decreases by a factor of $2^{1/2}$ on substituting D for H.

The elastic and electric moments associated to the OH^- molecule interact respectively with an applied elastic or electric field so that the initially equivalent $\langle 100 \rangle$ orientations become energetically different.

The strain- or electric-field-induced splitting of the energy levels gives rise to a number of interesting physical effects. First, dielectric and anelastic relaxation effects and in particular the typical Debye-type response for a dipole system can be observed. Other associated phenomena are electro-optical, elasto-optical and electrocaloric effects.

In the electro-optical effect, a dichroism in the optical ultraviolet absorption band is induced by application of an electric field. It can be understood in terms of an oscillator strength which is much stronger for polarization perpendicular to the molecule axis than parallel to it. The same effect can be obtained by the application of a stress field.

A related phenomenon is the change in temperature associated with the onset or removal of an electric field (electrocaloric effect). This paraelectric cooling or heating is the electrical counterpart of the analogous magnetic effect. When the field is applied, dipoles become partially aligned along it. Since the parallel and antiparallel orientations differ in energy by the same amount (except for the sign) with regard to the initial (zero) energy, the alignment induces a decrease in the internal energy of the dipole system. This lost energy is transferred to the crystal, which experiences an increase in entropy and temperature if the experiment is performed under adiabatic conditions. In contrast, when the field is removed, an inverse cooling effect is observed. Absolute (ΔT) values of about $10^{-3} - 10^{-2}$ K have been observed in the liquid-helium temperature range for fields E of about $10^4 - 10^5$ V cm^{-1}.

At zero or low fields, tunnelling transitions between equivalent orientations take place so that degeneracy is removed and an energy level splitting is induced. This splitting (e.g. about 3×10^{-5} eV for KCl:OH) should contribute to the specific heat and can be detected by paraelectric resonance techniques.

Other impurities presenting electric or elastic moments such as NO_2^-, O_2^- or CN^- show similar effects. The CN^- system has been much investigated[11]. The CN^- molecule is oriented along the $\langle 100 \rangle$ directions for rubidium and potassium halides (except for KCl where it lies along the $\langle 111 \rangle$ axis).

The behaviour so far described refers to dilute systems. When impurity concentrations become sufficiently high, interaction effects appear and give rise to interesting cooperative phenomena (discussed later in section 6.11).

5.4 AGGREGATION AND PRECIPITATION OF IMPURITIES— NEW CRYSTALLOGRAPHIC PHASES

I−V dipoles (and therefore the divalent cation impurities) are mobile through the exchange between the impurity and the vacancy, when this is in a nearest neighbour position. As a consequence of their migration, dipoles become associated and give rise to clusters of two (dimers), three (trimers) or more dipoles, which can adopt different configurations inside the host matrix. There is some, although still scarce, experimental evidence on the occurrence of such aggregates by means of EPR and optical spectroscopy, as well as ITC[7]. By EPR, mixed dimers formed from Eu^{2+} dipoles have been identified[12]. In contrast, excitation transfer between Pb^{2+} and Mn^{2+} has been used[13] to ascertain the existence of close pairs of Pb^{2+} and Mn^{2+} dipoles, most probably as dimers. It is worthwhile pointing out that even after standard quenching treatments performed on lightly doped samples (100 ppm or less), the above techniques have revealed the presence of an appreciable concentration of aggregates. Some theoretical calculations on the binding energy of the simplest aggregate have been carried out[14], but unfortunately no comparison is still possible with reliable experimental values.

Above a certain threshold concentration, crystals aged at a given temperature T develop new well-defined crystallographic phases. One typical phase is the dihalide of the divalent metal, which has been identified by X-ray diffraction for a number of systems such as $NaCl:Sr^+$ and $NaCl:Ca^+$. The occurrence of these precipitated phases has also been ascertained by means of electron microscopy and other physical techniques. For example, the presence of $PbCl_2$ precipitates has been inferred[15] from the absorption and luminescence spectra, such as those illustrated in Figure 5.3. The use of optical spectroscopy techniques has been extended to other doped systems $NaCl:M^{2+}$; where Pb^{2+} becomes apparently incorporated into the MCl_2 precipitates formed during appropriate ageing treatments. The main advantage of these optical techniques is their high sensitivity which permits the detection of some precursor phase at the early stages of ageing and allows the kinetics of precipitation on a given sample to be followed in detail.

In other systems, such as $NaCl:Mn$ or $NaCl:Cd$, a new phase, called the Suzuki[16] phase, is formed instead of the dihalide phase. The Suzuki phase has the same cubic face-centred cubic (FCC) structure as the host crystal, a double lattice parameter and a stoichiometry $6NaCl.2MCl_2$ (where M is a divalent cation) as inferred from X-ray diffraction data. It can be visualized as an ordered FCC arrangement of next-nearest-neighbour dipoles, as shown in Figure 5.4. The interest of this phase lies in the fact that it has never been prepared as an isolated material and it contains a high number of intrinsic cation vacancies. The Suzuki-phase precipitates present a characteristic cubic shape, with faces (100), in accordance with a number of electron microscopy observations using decoration or transmission techniques. Raman spectroscopy provides unambiguous identification[17] of the phase, through the occurrence of four Raman peaks corresponding to vibrational modes with symmetry A_{1g}, E_g, T_{2g1} and T_{2g2} (in fact the peak E_g is not generally observed because of its low intensity), as illustrated in Figure 5.5. A survey of the systems in which Suzuki-phase precipitates have been identified is given by de Andrés and Calleja[17].

One of the relevant problems concerning divalent cation precipitation in alkali halides is to determine in what systems and under what conditions a given phase is to be formed. Important steps in this direction have been given, starting from calculations of the relative stability of the nearest- and next-nearest-neighbour dipoles as well as that for the first aggregate (e.g. dimers). Some general trends have already been pointed out.[18]

5.5 COLOUR CENTRES

Charge capture at lattice defects causes optical transitions which absorb light in the previously transparent perfect crystal. Because many of these

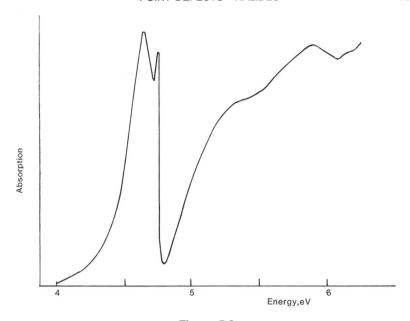

Figure 5.3.
Optical absorption bands induced by $PbCl_2$ precipitates in NaCl. Note the split excitation peak.

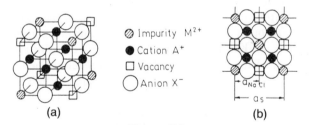

⌀ Impurity M^{2+}
● Cation A^+
□ Vacancy
◯ Anion X^-

(a)   (b)

Figure 5.4.
The Suzuki phase (e.g. MgF_2:6LiF) as seen (a) schematically and (b) projected on the (100) plane. A section of the unit cell is shown.

absorption bands are in the visible region of the spectrum, they are called colour centres.[19] Familiar examples of impurity defects are in gem stones such as ruby or sapphire or in stained glass windows. The term is also used for absorption bands beyond the visible range in the infrared or ultraviolet regions.

5.5.1 Electron centres—the F centre

The simplest and most studied colour centre is the F centre in alkali halides

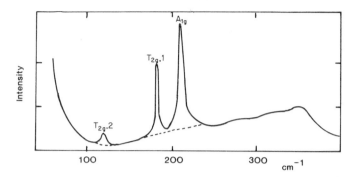

Figure 5.5.
Raman peaks corresponding to T_{2g} and A_{1g} modes for NaCl:Mn.

(from the German word Farbe meaning colour) consisting of one electron trapped at a halogen vacancy. With the present nomenclature the other possible charge states of the vacancy are designated as F^+ (empty vacancy) and F^- (vacancy with two electrons). So the superscript refers to the effective charge of the centre with regard to the host crystal. The properties of the F centre will now be described in some detail as it can serve as a reference case for other centres.

The F centre is paramagnetic and its structure has been studied by EPR or double-resonance techniques (ENDOR). With the latter technique, a very detailed mapping of the electron wavefunction in the ground state has been achieved, through the hyperfine coupling with the neighbouring alkali and halogen ions. In KCl and KBr the spin density distribution up to eight (for KBr or even 13 for KCl) ion shells have been determined, allowing for a very sensitive testing of the theoretically calculated wavefunctions.

Also, the F centre presents characteristic absorption and luminescence bands, whose respective peak energies E_A and E_L are listed in Table 5.5 together with the corresponding width W. Both bands correspond to transitions between the ground a_{1g} (1s) and first excited t_{1u} (2p) states, where the a_{1g} and t_{1g} labels refer to the octahedral symmetry O_h of the centre and 1s and 2p to the corresponding levels in a spherically symmetric potential (hydrogenic model). For all alkali halides the oscillator strength f of the absorption band appears to be close to unity (see Section 9.21).

A simple estimate of the transition energies for F centres in alkali halides yields tolerably accurate results and is now described. In the alkali halides the strongly ionic character of the material allows the lattice to be considered in terms of point ion charges. Replacement of a halogen ion with an electron in the site (i.e. the F centre) suggests that there is a strong localization of the electron density within the vacancy and thus the F centre

Table 5.5.
F-centre parameters in alkali halides

Crystal	E_A (eV)	E_L (eV)	W_A (eV)	W_L (eV)
LiF	5.102		0.596	
LiCl	3.30		0.4	
NaF	3.723	1.665	0.366	0.39
NaCl	2.770	0.975	0.255	0.337
NaBr	2.1			
KF	2.847	1.66	0.228	0.385
KCl	2.313	1.215	0.163	0.261
KBr	2.064	0.916	0.158	0.215
KI	1.875	0.827	0.155	0.185
RbF	2.428	1.328	0.199	0.335
RbCl	2.050	1.090	0.145	0.237
RbBr	1.857	0.87	0.133	0.190
RbI	1.708	0.81	0.121	0.148

is a close analogy of an electron in a box. The familiar Schrödinger equation for a time-independent wavefunction ψ for an electron in a potential well of width a_1 is

$$\nabla^2\psi + \frac{2mE}{\hbar^2}\psi = 0 \tag{5.1}$$

and has boundary conditions of $\psi = 0$ at the edges where x, y and z are 0 or a_1. Wavefunction solutions, normalized for the box volume, are

$$\psi_{lmn} = \left(\frac{8}{a_1^3}\right)^{1/2} \sin\left(\frac{l\pi x}{a_1}\right)\sin\left(\frac{m\pi y}{a_1}\right)\sin\left(\frac{n\pi z}{a_1}\right) \tag{5.2}$$

where l, m and n are non-zero integers. The corresponding energy levels are

$$E_{lmn} = \frac{\pi^2\hbar^2(l^2 + m^2 + n^2)}{2ma_1^2} \tag{5.3}$$

For the trapped electron the transitions are between the energy states (l, m, n); so the lowest possible value is from $(1, 1, 1)$ to one of the $(2, 1, 1)$ set; hence

$$E = \frac{3\pi^2\hbar^2}{2ma_1^2} \tag{5.4}$$

which is the main absorption band energy of the F centre corresponding to an excitation from 1s to 2p.

The strength of the model is that it predicts for a sequence of alkali halides that $E \propto a_1^{-2}$ where a_1 is related to the lattice parameter a. Experimentally it is found for the F centres that

$$E_F = 57a^{-1.77} \qquad (5.5)$$

where E_F is in electronvolts and a in ångströms.

This unifying feature for the series of alkali halides is known as the Mollwo–Ivey law.

The correspondence of a_1 and a for example may be checked by evaluating the "box" dimensions for NaBr which is midway along the range of halides. From the measured E_F value, we find that $a_1 = 6.9$ Å whereas $a = 5.96$ Å. There is thus a good match but, as expected, the simple prediction under-estimates the extent of the wavefunction. ENDOR data suggest that about 90% of the ground-state wavefunction is confined within the vacancy but the remainder extends some ten atomic shells beyond the core. Clearly the excited states are more complex and progression from the present simplistic calculation to an exact system is immensely complex as evidenced by numer-ous publications (e.g. the book by Stoneham[6]).

The simple model also predicts higher-state transitions which are seen as the high-energy features of the F band labelled as K, L_1, L_2 and L_3. These higher states overlap conduction band states and show photoconductivity. The peak spacings can be modified by pressure (i.e. by changing a) but not all bands distort in energy in the same sense and the K and L bands follow a similar pattern with the F and K features moving to higher energies under pressure whereas the L bands reduce in energy.

The simple particle-in-a-box model can be applied to more complex defects. In general, for a series of ionic crystals, evidence of a Mollwo–Ivey-type pattern is helpful in relating similar defects in different members of the series. For example a Mollwo–Ivey plot for F, F_2 and F_3 centres is given in Figure 5.6.

The absorption and luminescence bands of the F centre are rather wide (about 0.1 eV) because of the electron–lattice interaction. The detailed treatment of this interaction requires the use of a many-body Hamiltonian, and consequently rather elaborate theoretical approaches are in order. How-ever, some relevant features of the interaction can be understood in terms of a simple model lent from molecular physics, the so-called configuration coordinate model. The model, which is derived from the Born–Oppen-heimer approximation, allows the energy of the ground and excited states to be written as functions $W(q)$ of a normal coordinate q representing the displacement of the nearby ions (Figure 5.7). These curves, representing the adiabatic potential energy for the ions, can be approximated by a parabola near the minimum q_0 or equilibrium position of the ions. As an example, q may correspond to the totally symmetric or "breathing" mode,

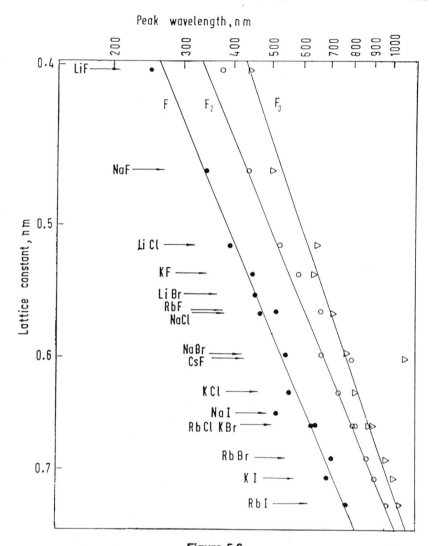

Figure 5.6.
The Mollwo—Ivey law (i.e. $E \propto a^{-x}$) for vacancy defects F, F_2 and F_3 in alkali halides.

which is usually expected to provide the strongest coupling to the electron. The energy levels for the vibrational states of the nuclei in both ground and excited electronic states are drawn as constant-energy lines in the diagram. In fact, they represent the total (electronic and vibrational) energy of the centre for the vibrational and electronic (vibronic) levels that are being

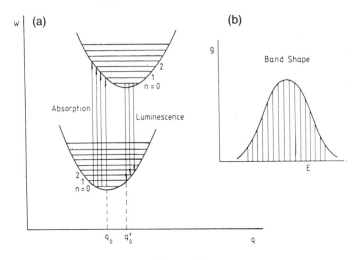

Figure 5.7.
Configurational coordinate diagram showing (a) adiabatic potential energy curves and (b) the resulting shape of the absorption or emission bands as envelopes of the vibronic transitions. In general the emissions are broader and centred at lower energies than the corresponding absorption bands.

considered. The observed absorption or emission bands are the envelope of the various individual transitions between vibronic levels.

According to the classical version of the Franck–Condon principle, the optical transitions between electronic states are vertical, i.e. nuclei do not move during the transition. This may lead, as is apparent in Figure 5.7, to a strong shift (the Stokes shift) between the absorption and emission energies, in accordance with the experimental data summarized in Table 5.5. This situation is typical for the F centre and corresponds to a strong linear coupling between the electron and lattice. If this coupling is small, the minimum of the $W(q)$ curves occurs at roughly the same q_0 value for both ground and excited states, so that a small Stokes shift is observed. This is the situation found for a number of F-aggregated colour centres and gives rise to very striking features in the absorption and luminescence spectra (see section 5.4.4 and 9.2.2).

The linear coupling is represented by the linear term in the adiabatic energy curve for the excited electronic state:

$$W_{ex}(q) = \tfrac{1}{2}M\omega^2 q^2 + E_0 - A\hbar\omega q \qquad (5.6)$$

where the first term $\tfrac{1}{2}M\omega^2 q^2$ corresponds to the adiabatic potential $W_g(q)$ for the ground state (minimum at $q_0 = 0$) and E_0 is the energy of the pure electronic transition. The curvatures, and consequently the vibrational frequencies ω, are the same for both excited and ground states, but the minimum of $W_{ex}(q)$ occurs for a different q value, $q_0' = A\hbar/M\omega$, or equilibrium lattice configuration. If additional quadratic or higher-order coupling terms are included in $W_{ex}(q)$, the curvatures are also different for the two states. The parameter A measures the strength of the linear coupling, but it is often replaced by the so-called Huang–Rhys factor $S = \frac{1}{2}A^2 \hbar/M\omega$, which is the number of vibrational quanta separating the excited-state potential at $q = q_0$ and $q = q_0'$, i.e. at the ground and excited equilibrium configurations.

For the strong-linear-coupling case, the semiclassical configuration coordinate model predicts Gaussian absorption and luminescence bands[20] whose widths are given by

$$W = W_0\left[\coth\left(\frac{\hbar\omega}{2kT}\right)\right]^{1/2} \tag{5.7a}$$

where ω_g and ω_e are the frequencies of the ground and excited states, respectively, for absorption and luminescence bands. W_0 is the bandwidth at 0 K and is given by

$$W_0 = \left(n_e\frac{(\hbar\omega_e)^3}{\hbar\omega_g}\right)^{1/2} \tag{5.7b}$$

for absorption, and

$$W_0 = \left(n_g\frac{(\hbar\omega_g)^3}{\hbar\omega_e}\right)^{1/2} \tag{5.7c}$$

for emission. n_e and n_g are the number of vibrational quanta relative to the potential minima of the excited and ground states, respectively.

At high temperatures, equation (5.7a) reduces to

$$W(T) = W_0\left(\frac{2K}{\hbar\omega}\right)^{1/2} T^{1/2}$$

i.e. the width grows proportionally to the square root of the temperature, as expected from simple classical arguments. For purely linear coupling, $\omega_g = \omega_e$ and then $n_e = n_g = S$ (the Huang–Rhys factor), which is related to the Stokes shift Δ between the absorption and emission peak energies: $\Delta = 2S\hbar\omega$. The agreement between this formula and experimental data is quite good and allows for the determination of the effective frequency ω of the mode coupled to the electron wavefunction.

Although the configuration coordinate diagram adequately represents the transitions within the F centre, this simple picture does not account for the position of energy levels relative to the bands of the perfect solid. The strong lattice relaxation accompanying the electronic excitation of the F centre can be represented in a band scheme as is usually done for donor or acceptors in semiconductor materials. The situation is depicted in Figure 5.8, showing that the relaxed excited state of the F centre lies very close to the bottom of the conduction band (ionization energy of less than 0.1 eV).

In the caesium halides the F band is not a simple Gaussian absorption; for example in CsBr there are two close peaks whereas for CsCl there are three resolved features. The structure is associated with spin—orbit splitting of the excited p state, predicting a double-peaked band, together with a dynamic Jahn—Teller interaction (see section 7.2) with E_g vibrational modes to account for the triplet structure in CsCl. No structure is observed for the F band in the other alkali halides (except RbBr). However, the spin—orbit splitting has been measured by magneto-optical experiments. The experimental values of the splitting are much larger than expected and have a negative sign, so that the $j = \frac{1}{2}$ level lies above the $j = \frac{3}{2}$ level. Both features have been explained by using F-centre wavefunctions properly orthogonalized to the nearest-neighbour ion cores in an extended ion model.

The luminescence emission of the F centre corresponds to the 2p → 1s transition and occurs at much lower energies than those for the corresponding absorption, in accordance with the strong lattice relaxation of the centre. (Emission energies for the various alkali halides are given in Table 5.5.) The lifetime τ has been found to be in the microsecond range at low temperatures, i.e. it is much longer than expected from the oscillator strength of the transition ($f \approx 1$). The data are consistent with a diffuse p excited state, as inferred from theoretical calculations, although mixing effects with the 2s excited level could also play a relevant role. In fact, relaxed 2p and 2s states appear to be very close in energy (about 0.01 eV), so that strong admixtures between them are produced by local electric fields. The partial s character of the excited wavefunction then leads to a longer lifetime.

Two important features of the lifetime data are now mentioned. First, lifetime values depend on temperature as shown in Figure 5.9 for KCl, i.e. are essentially constant at low temperatures and decrease rapidly above a certain temperature. The decrease is easily interpreted in terms of the thermal ionization of the F centre. The lifetime has contributions from thermal excitation (τ_{th}) and radiative decay (τ_p) so that overall

$$\frac{1}{\tau} = \frac{1}{\tau_{th}} + \frac{1}{\tau_p} \qquad (5.8)$$

Because τ_p is temperature independent, the experimental data on the temperature dependence yield the thermal term τ_{th}. Hence the location of the

F-CENTRE LEVELS

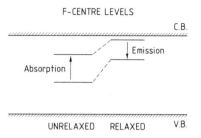

UNRELAXED   RELAXED   V.B.

Figure 5.8.
Schematic diagram showing the unrelaxed and relaxed ground and excited levels for the F band absorption and emission: CB, conduction band; VB, valence band.

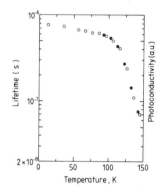

Figure 5.9.
Lifetime dependence of luminescence (○) and photoconductivity (●) for F centres in KCl (after R. K. Swank and F. C. Brown, *Phys. Rev.* **130**, 34 (1963)).

relaxed excited level can be determined with respect to the bottom of the conduction band (see Figure 5.8). Additionally, concentration quenching effects on the radiative lifetime have been observed.

Another interesting effect to be mentioned is the spin memory of the F centre in a magnetic field. It has been found that the spin direction of the electron is maintained through the relaxation processes operating during the deexcitation of the centre.

Finally, we should note that the F and F^+ (empty anion vacancy) centres give rise to absorption bands (respectively named β and α), near the fundamental absorption edge. They are associated with the creation of excitons close to the centres and therefore show a perturbed excitation energy. Similarly, perturbed excitons have also been observed at a number of impurities.

The F centre presents a marked photoconductivity at temperatures T of 80 K or higher, when illuminated in the absorption band. Under these conditions, the electrons excited to the 2p level can be thermally pumped to the conduction band and then contribute to the conductivity of the material. The decay with temperature of the photoconductivity yield follows closely that of the lifetime (see Figure 5.9).

The F centre can trap a second electron and form the F' centre, having a negative effective charge. This diamagnetic centre appears to play a key role in many photochemical processes involving F centres. Unfortunately, the optical study is not easy because the absorption band for the F' centre lies at the F-centre absorption region, although its width is much larger (about 1.5 eV) than that of the F band. The band is associated with transitions from the bound ground state to conduction band states. No other bound state, neither singlet nor triplet, appears to exist. The photoionization threshold is about 1 eV. Near the threshold energy E_0 the absorption coefficient varies as $\mu(E) \propto (E - E_0)^{1/2}$ in accordance with the density of conduction band states. The F' centre becomes thermally unstable below room temperature, so that studies have to be performed at low temperatures. The thermal ionization energy is about 0.5 eV for NaCl.

In this section, we concentrated on a discussion of F centres and, in passing, there was a reference to related defects. In Chapter 1, Tables 1.1 and 1.2 listed a range of alkali halide defects and the commonly accepted models. Some of these complex examples will be discussed in the following sections.

5.5.2   Impurity-related F centres

A variety of F centres bound to monovalent alkali metal (F_A centres) or divalent (F_z centres) cation impurities have been identified and studied. Most work and conclusive results are available for F_A centres, i.e. F centres having an alkali metal impurity as a nearest neighbour and therefore presenting a $\langle 100 \rangle$ symmetry axis. Only alkali ions smaller than the host cation appear to give rise to F_A centres. The structure of the centres has been well established by ENDOR spectroscopy. They can be produced in alkali-metal-doped alkali halides by photothermal treatments involving the migration of anion vacancies (F^+ centres) and their subsequent trapping by the alkali impurity.

The reduction in symmetry from the cubic F to the tetragonal F_A centre causes a splitting of the F-centre band into two components: one polarized along the centre axis (F_{A1}) and the other one perpendicular to it (F_{A2}).

With regard to their luminescence properties, F_A centres are usually classified in two groups designated as $F_A(I)$ and $F_A(II)$. In the $F_A(I)$ centres, such as KCl:Na, the emission band when excited at either of the two absorption components shows a similar Stokes shift to that for the F centres and the lifetime is also quite similar. In contrast, the so-called $F_A(II)$ centres, such as in KCl:Li, present properties radically different from those of the F or $F_A(I)$ centres. The emission band is very markedly Stokes shifted and the lifetime is much shorter than that for the F centre. Another remarkable feature is that the excited states of F_A centres are stable against thermal or electron field ionization, at variance with $F_A(I)$ centres whose thermal ionization energy is close to that for the F centres.

The behaviour of the $F_A(II)$ centres has been attributed to the very strong relaxation of the excited state, which causes a nearest-neighbour Cl^- to shift to an interstitial position between two adjacent anion vacancies (saddle-point configuration) in which the electron wavefunction spreads over the two symmetric wells.

By using the dichroic properties of the absorption of the F_A centre, it is possible to study the reorientation of the centre in the ground as well as the excited states. For the $F_A(I)$ centres, the activation energy for reorientation of the excited state is much smaller than that in the ground state. For the $F_A(II)$ centres, no thermal activation has been found for the excited state, in accordance with the proposal made for the relaxed excited state. The thermal dissociation of the F_A centres can also be optically investigated and the corresponding activation energy measured.

The $F_A(II)$ centres appear as favourable candidates for laser emission and they have already been used in commercial systems. Similar potential applies to F_B centres, i.e. F centres bound to two nearest-neighbour alkali-metal impurities.

In relation to potential tunable laser applications, some effort has been recently devoted to the study of F centres attached to monovalent cations other than alkali metals, e.g. Ag^+, Ga^+, Tl^+ and In^+. For Tl^+-doped alkali halides, which are some of the most frequently investigated systems[21,22], it has been concluded that Tl^+ F-centre complexes with the same structure as F_A centres are formed. However, it appears that for the Tl^+ case the bound electron is about equally shared between the Tl^+ ion and the anion vacancy. The electronic structure of these non-alkali F_A centres appears to be markedly dependent on the electron affinity of the impurity ion. For the sequence $Na^+ \rightarrow Li^+ \rightarrow In^+ \rightarrow Tl^+ \rightarrow Ag^+ \rightarrow Cu^+$, which is ordered according to increasing affinity, an increasing localization of the electron at the impurity is expected for the corresponding F_A centres.

A variety of divalent-cation-related centres named Z or F_Z centres have been found in alkali halides[23]. The simplest Z centre is the Z_1 centre which consists of one F centre trapped at an impurity vacancy dipole as revealed by different techniques, such as EPR and ENDOR, luminescence, electro-optics and magneto-optics. It appears that these centres are associated with impurities having a stable valence state, such as the alkaline-earth cations.

5.5.3   F-centre aggregates

In addition to isolated F centres, the usual colouring procedures, e.g. irradiation or thermochemical treatments, create aggregates of a few F centres, the so-called F_n centres.

The F_2 (or M) centre consists of two F centres in nearest-neighbour positions and it is therefore aligned along $\langle 110 \rangle$ directions. The centre presents two main absorption bands. One of them, the M_1 band, lies on the long-wavelength side of the F band and absorbs light polarized parallel to the centre axis. The other band, made up of two overlapping components (M_2 and M_2'), is hidden under the F band and is excited perpendicular to the F_2 axis. The simplest model for the F_2 centre relates it to a hydrogen molecule in a continuous dielectric. The level scheme is partly illustrated in Figure 5.10, the labelling being in accordance with a C_h symmetry. The M_1 band corresponds to the transition from the singlet ground state $^1\Sigma_g{}^+$ to the singlet $^1\Sigma_u{}^+$ state. The M_2-M_2' band is associated with the transition to the $^1\pi_u{}^+$ level, which is split when allowance is made for the deviation of pure C_h symmetry. In fact, the M_2-M_2' splitting comes entirely from the anisotropy of the host lattice and can be obtained when point ion models are used instead of the continuous model. Other high-energy transitions, similar to the K and L bands of the F centre, have also been observed.

An interesting property of the F_2 centre is the occurrence of a low-lying triplet state (not shown in Figure 5.10), with a long lifetime of a few minutes. EPR and ENDOR measurements have been carried out on this metastable triplet and have confirmed the structure of the centre. Light excitation at the F band region induces the creation of F_2 triplet centres, although the mechanism is unclear.

A single luminescence emission is observed for excitation at the various absorption bands. It is polarized parallel to the centre axis and it is associated with the reverse M_1 transition. It is worthwhile pointing out that, at variance with the F-centre case, the Stokes shift for this transition is rather small (0.35 eV for KCl). Moreover the lifetime is short (6×10^{-8} s for KCl). Consequently, the unrelaxed and relaxed excited states of the M_1 transition are very similar, so that no large expansion of the wavefunction is produced during relaxation. Impurity related F_2 centres, i.e. F_{2A} and F_{2Z}, have been identified.

The F_3 (or R) centre consists of three nearest-neighbour F centres at the vertices of an equilateral triangle. This defect has a $\langle 111 \rangle$ symmetry axis. Two main bands R_1 and R_2 have been assigned to this centre. The R_1 band is associated with transitions from the ground ^2E level to an excited ^2E level, whereby the R_2 band corresponds to the $^2E \rightarrow {}^2A$ transition. Not only $S = \frac{1}{2}$ states have been observed but also states with $S = \frac{3}{2}$, by temporary

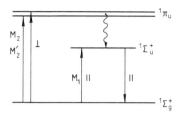

Figure 5.10.
Electronic levels for the F_2 centre in alkali halides. The main absorption and emission transitions are shown.

bleaching experiments similar to those commented on above to produce the triplet F_2 centres.

Ionized aggregated centres, such as F_2^+, F_3^+ or F_2^- and F_3^-, can be produced by suitable irradiation and illumination treatments. Well-defined absorption and luminescence bands have been associated with some of these centres in several alkali-halide crystals.

Further aggregation of F centres leads to the so-called N centres, which are considered as clusters of F centres (F_4 centres). Possible configurations are a parallelogram of vacancies in the (111) plane which may produce N_1 absorption and a tetrahedron of vacancies which may give N_2 absorption.

5.5.4 Vibronic (zero-phonon) lines

It has been already mentioned that the broad absorption bands usually observed for colour centres and impurities are, in fact, the envelopes of a large number of vibronic lines, corresponding to transitions between vibrational sublevels of the ground and excited electronic states. This broad-band behaviour appears for strong electron−lattice coupling, as in F centres, and leads to the typical Gaussian shapes. For some centres, such as rare-earth impurities or some aggregated electron centres, the coupling is weak and optical bands adopt "non-conventional" shapes. In particular, the situation for temperatures T of about 0 K can be more easily analysed, since only transitions from the first $n = 0$ vibrational level of the ground state to the $n = 0$ or a few higher levels of the excited state are involved. The absorption spectrum shows a dominant narrow $0 \rightarrow 0$ line (zero-phonon line) and a few $0 \rightarrow n$ vibronic lines, with energies $E_n = E_0 + n\hbar\omega$, E_0 being the zero-phonon line energy which equals that of the pure electronic transition (higher-order coupling is neglected). In terms of the Huang−Rhys factor S, to measure the coupling strength, the $0 \rightarrow n$ transition probability ($T \approx 0$ K) is given by

$$W_{0n} = \frac{S^n}{n!} \exp(-S) \qquad (5.9)$$

which determines a band shape peaked at $n \approx S$. For $S = 0$ the spectrum simplifies to a single zero-phonon line, whereas for $S \gg 1$ (let us say $S \approx 5$) the system approaches the typical Gaussian shape and loses any vibronic features.

Another interesting property of weakly coupled centres is that at $T \approx 0$ K the emission band is made up of vibronic lines $E_{0n} = E_0 - n\hbar\omega$ and it is the mirror image of the absorption band with regard to the common zero-phonon line as mirror symmetry axis if higher-order coupling is ignored. (An example of phonon line features is shown in Figure 6.2.)

Many of the aggregated electron centres described in section 5.5.3 provide good examples of centres showing vibronic features. Clear examples are the absorption and emission spectra for F_3^- centres in LiF. In contrast, no zero-phonon lines have been observed for F or M centres owing to the rather strong linear electron−lattice coupling.

The width of the zero-phonon lines is very narrow, typically about 1 Å, and it is mainly determined by local strains, i.e. it is inhomogeneously broadened. The small bandwidth makes it possible to measure small shifts or splittings induced by applied fields. These effects have very often been used to determine the symmetry of the centres and the nature of the electronic states. Many appropriate examples are discussed in the literature, including some books and review papers[6,9,24].

The inhomogeneous broadening of the zero-phonon lines allows for the production of hole-burning effects in the line shape, when the crystal is excited with suitable intense and highly monochromatic (laser) light (see Chapter 9). Only centres having a given local environment, whose absorption corresponds to the radiation source, are bleached, whereas the remainder are unaffected (see Figure 9.10). The width of the "hole" is determined by homogeneous broadening mechanisms and it is typically three orders of magnitude smaller than that corresponding to the inhomogeneous line. This method can be used as a tool in very-high-resolution spectroscopic studies and it has been contemplated for practical applications (high-density optical memories).

5.5.5   Colloids

The final stage in the clustering of F centres is the formation of colloidal particles of the host alkali metal. Colloids can be created by additive electrolytic colouring, as well as by irradiation at high doses and temperatures above room temperature. The study of colloids has not been restricted to alkali halides and a variety of other systems such as dihalides, oxides and glasses have also been investigated. The identification of the colloids is generally made through the absorption spectra by using the Mie theory. This theory predicts that, for metallic spherical particles of radius R, the extinction (absorption and scattering) coefficient γ can be written as a multipole expansion:

$$\gamma = \frac{6\pi NV}{\lambda'} \, \text{Im}\left(\sum_{v=1}^{\infty} (-1)^v \frac{a_v - p_v}{2\alpha^3}\right) \tag{5.10}$$

where λ' is the wavelength of light in the host material and $\alpha = 2\pi R/\lambda'$. The parameters a_v and p_v, which describe the contribution to γ of the various electric and magnetic multipoles ($v = 1, 2, \ldots$), are expressed as

$$a_v = (-1)^{v-1}\frac{v+1}{v}\frac{\alpha^{2v+1}}{(1^2)(3^2)\ldots[(W_v-1)^2]}U_v\frac{m'^2-V_v}{m'^2+[(v+1)/v]W_v} \tag{5.11}$$

and

$$p_v = (-1)^v\frac{v+1}{v}\frac{\alpha^{2v+1}}{(1^2)(3^2)\ldots[(2v+1)^2]}U_v\frac{1-V_v}{1+[(v+1)/v]W_v} \tag{5.12}$$

with $m'^2 = m^2/m_0^2$, where $m^2 = \epsilon_1 - i\epsilon_2$, the complex dielectric constant of the metal, and m_0^2 the real (non-absorptive) dielectric constant of the dielectric host. U_v, V_v and W_v are written as an infinite series of powers of α, so that U_v, V_v and $W_v \to 1$ as $\alpha \to 0$. In the limit $\alpha \ll 1$, i.e. $2\pi R \ll \lambda'$, then only the electric dipole term a_1 contributes significantly to γ and so

$$\gamma = -\frac{6\pi NV}{\lambda'} \, \text{Im}\left(\frac{m'^2 - 1}{m'^2 + 2}\right) \tag{5.13}$$

If we now use a free-electron-gas model for the metal, the real ϵ_1 and imaginary ϵ_2 components of the dielectic constant read

$$\epsilon_1 = \epsilon_0 - \frac{\omega_p^2}{\omega^2 + \omega_0^2} \tag{5.14}$$

$$\epsilon_2 = \frac{\omega_p^2 \omega_0}{\omega(\omega^2 + \omega_0^2)} \tag{5.15}$$

ω_p being the plasma frequency and ω_0 the collision frequency of the free electrons. Substitution for $\gamma(\omega)$ gives a band peaked at a wavelength λ_m given by

$$\lambda_m = \frac{2\pi c}{\omega_p}(\epsilon_0 + 2m_0^2)^{1/2} \tag{5.16}$$

and having a half-width

$$\Delta\lambda = \frac{c}{2\sigma}(\epsilon_0 + 2m_0^2) \tag{5.17}$$

where σ is the Drude conductivity of the metal.

The situation described by the above formulae is that prevailing for colloids with small radius ($R \leq 10$ nm) for which λ_m is independent of R. However, for such small-size colloids, quantum effects can be appreciable and appear to give rise to some structure in the absorption band. This structure is associated to the finite spacing among electronic levels (discrete bands).

For larger radii, higher multipole terms contribute to γ and the position of the $\gamma(\omega)$ band depends on R. This behaviour is illustrated in Figure 5.11, where the extinction bands corresponding to different values of R have been plotted for sodium colloids in NaCl as determined from the Mie theory.

In order to apply the Mie theory correctly, we should take into account the contribution of the interband transitions to the dielectric constant of the metal (and not only the intraband transitions). This contribution appears to be important for some systems, such as caesium colloids in CsBr. However, the theory has been developed for spherical particles and experimental

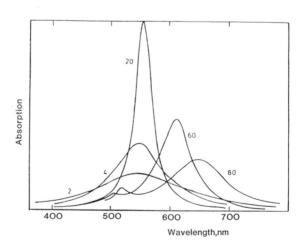

Figure 5.11.
A prediction of the absorption band shapes of colloid centres formed by sodium in NaCl. Note that the peak position moves to longer wavelengths with increasing colloid diameter, which is specified on the curves.

observation shows that colloids very often take other shapes: cylinders, ellipsoids, cubes, etc.

The formation and growth of colloids in alkali halides during appropriate ageing treatments can be monitored by the optical absorption band and therefore provide a convenient example for discussing the theories for nucleation and growth of a solid phase. The process of colloid formation is not well known yet, although electron microscopy suggests that nucleation is heterogeneous on dislocations. However, the possible role of impurities or lattice defects cannot be ruled out. Once the nuclei have been formed, they grow in size by the incorporation of additional F centres. When the concentration of the remaining F centres approaches that corresponding to the thermodynamic equilibrium at the operating temperature, a process known as Ostwald ripening takes place as a final stage in the precipitation process. During this phase the equilibrium concentration of F centres in the vicinity of the colloid surface increases with decreasing F-centre radius and this triggers the annihilation of the small colloids to the benefit of the larger colloids. The theory of this process was initially developed by Lifshitz and Slezov. The results of the theory depend on the active rate-controlling mechanism.

(1)   Condensation of the solute (F centres) on the colloid.

(2)   Bulk diffusion of the F centres.

(3)   Diffusion through dislocations or grain boundaries.

It can be shown that for all three cases the asymptotic behaviour can be described by the time dependence of the average colloid values:

$$R \propto \frac{1}{t^{m+1}} \qquad (5.18)$$

m being a number whose values 1, 2 and 3 are respectively correlated with the three considered cases (1), (2) and (3). For example data for potassium colloids in KCl were fitted by a $R \propto t^{1/3}$ law, although an $R \propto t^{1/4}$ might also be appropriate, particularly at the shorter ageing times[25]. The formation of colloids during heavy irradiation was discussed in Chapter 3 devoted to radiation damage by ionization.

Colloids are of considerable basic interest since they permit study of the physical properties of very small particles and an analysis of their dependence on size. Colloids give rise to photoconductivity, Raman spectra and EPR signals. The EPR spectrum consists of a band with $g \approx 2$ and is associated with the free-electron paramagnetism of the metal. It has been analysed in terms of a theory developed by Dyson which relates the width of the line to the particle size.

The topic has been extensively discussed in the excellent review paper by Hughes and Jain[25].

5.5.6   Hole centres

A variety of hole centres exist which are often referred to as V centres. The schematic structures of the most common centres were illustrated in Chapter 1. The simplest and most studied are the H and V_K centres in alkali halides and we now discuss them in some detail.

The V_K centre consists of a hole shared by two adjacent halogen ions. It has the molecular structure X_2^- and it has been observed not only in alkali halides but also in dihalides, trihalides and possibly in the silver halides. It plays a key role in the mechanisms of photolytic damage to alkali halides as discussed in Chapter 4. The two constituent X ions of the centre have been shifted from their normal positions along the $\langle 110 \rangle$ direction so that the $X-X$ separation is about 30% closer than that corresponding to the perfect lattice. However, this separation is still markedly higher than that for the free X_2^- molecule.

The absorption spectrum consists of two bands, one in the ultraviolet and the other in the infrared spectral region, which can be easily predicted within a simple molecular orbital scheme. In fact, four orbitals σ_u, π_g, π_u and σ_g can be constructed from the p atomic orbitals of the X^- ions. They are arranged as illustrated in Figure 5.12, where the allowed transitions (ignoring spin−orbit coupling) have also been indicated. The ultraviolet absorption band corresponds to the $\sigma_u \rightarrow \sigma_s$ (or $^2\Sigma_u^+ \rightarrow \, ^2\Sigma_g^+$) transition from the ground σ_u state. It presents σ polarization, i.e. the dipolar moment of the transition is perpendicular to the $\langle 110 \rangle$ molecular axis. The infrared band is associated with the $\sigma_u \rightarrow \pi_g$ (or $^2\Sigma_u^+ \rightarrow \, ^2\pi_g$) transition and it is π polarized.

The V_K centre, with a single hole, is paramagnetic and its EPR spectrum presents a very characteristic resolved hyperfine structure with the two constituent halogen ions. This is illustrated in Chapter 10. For KCl the hyperfine structure results from the interaction with the two ^{35}Cl and ^{37}Cl isotopes, each one having nuclear spin $I = \frac{3}{2}$. The width of the hyperfine lines includes the superhyperfine interaction with neighbouring ions of the lattice.

The motion of the V_K centres takes place via discrete jumps (hopping) between adjacent positions. There are four possible jumps, implying rotation of 60°, 90°, 120° and 180° for the molecular axis. The 60°, 90° and 120° jumps induce a reorientation of the centre that can be monitored by the polarization of the absorption bands. By selectively bleaching the centres with polarized light, a partial alignment of the centres can be produced. The kinetics of the subsequent depolarization of the bands provide the information on the dominant jump mechanism. It has been concluded that thermal reorientation occurs via 60° jumps.

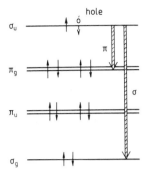

Figure 5.12.
Electronic energy level scheme for the V_K centre in KCl. σ and π emission are shown.

In contrast, the diffusion of the V_K centres can be studied and the activation energy determined by measuring the decrease in concentration during heating after a low-temperature irradiation. Since this process has to do with the recovery of damage in irradiated crystals, it is discussed in somewhat more detail in Chapter 4.

The H centre is centred on an X_2^- ion substituting for a host halogen ion. It has an equilibrium configuration, after lattice relaxation, corresponding to an interstitial atom Cl^0. The orientation of the centre is $\langle 110 \rangle$ for the alkali halides and $\langle 100 \rangle$ for the caesium halides, i.e. along the close-packed directions in all cases. Theoretical calculations indicate that the hole is appreciably shared by the next X^- ions along the centre axis, so that the correct molecular structure is X_4^{3-}. This additional covalent bonding stabilizes the orientation of the centre along $\langle 110 \rangle$.

In accordance with the structural similarity of the V_K and H centres, their absorption spectra have similar bands. As an example, the σ and π absorptions of the H centre in KCl appear at 3.69 eV and 2.86 eV, respectively, whereas those for the V_K centre occur at 3.39 eV and 1.65 eV.

The EPR spectrum for the H centre is similar to that for the V_K centre, except that each hyperfine line is further split owing to the covalent bonding to the other two halogen ions.[26] In addition to the isolated H centre, clusters of H centres can be formed. In particular, optical absorption of the di-interstitial (V_4, H_M or H') centre has been identified. Several different configurations are possible for crystals doped with anion or cation impurities; a large variety of perturbed V_K and H centres have also been seen by the EPR technique. For anion-doped samples, mixed V_K centres such as the $ClBr^-$ or ClI^- are formed in KCl where a Br^- or I^- ion has substituted for one of the Cl^- ions. For cation-doped samples, V_{KA} centres have been detected in which the host cation has been replaced by a monovalent impurity cation. In a sense, it can be considered that the V_{KA} centre has been trapped at the impurity. The situation for crystals doped with divalent cation impurities is more complicated and only recently a detailed structural identification has been achieved for V_{KZ} centres, although their existence had been previously considered.

A similar situation applies to H centres that can be trapped by monovalent or divalent impurities. The best known are the H centres trapped to one (H_A) or two (H_{AA}) monovalent impurities, whose structure is well identified for a number of systems[26].

In addition to atomic halogen interstitials (H centres), there is optical evidence for the occurrence of halogen ion interstitials (I centres), although their detailed structure is not known because of the lack of an associated EPR signal. It has been suggested that it presents a $\langle 100 \rangle$ dumb-bell configuration although a face-centred location has also been invoked. At variance with the H centre the I interstitial is not a primary product of the ionizing irradiation and it is formed as a result of electron capture by an H centre.

5.5.7   The self-trapped exciton

The V_K centre can trap one electron and form a bound V_{K-e} system, which can be considered as an exciton localized at a given lattice site. This self-trapped exciton (STE) constitutes an intermediate stage in the recombination of electrons and holes[27]. When an alkali halide crystal is excited with photons having energies at or higher than the band gap, unrelaxed or free excitons are created. These excitons can move coherently, be ionized or become trapped forming (V_K plus an electron) excitons (STEs).

Within a single-particle scheme, the energy levels for both electrons and holes in an STE are drawn in Figure 5.13. The hole levels are the same as already considered when dealing with the V_K centre. The de-excitation of the STE can take place non-radiatively, giving rise to defect creation (see Chapter 4) or by light emission. The luminescence of the STE is made up of the two transitions marked on the figure, one σ and the other π polarized. The energies of both emissions are markedly lower than those corresponding to the band gap and are sensitive to other defects. The σ transition originates from a singlet state and has a lifetime τ_σ of nanoseconds or more, whereas the π transition originates from the lowest triplet state and its lifetime τ_π is much longer, of the order of microseconds or even milliseconds.

EPR (and ENDOR) experiments have been used to elucidate the electronic structure of the lowest triplet state of the STE. The observed hyperfine structure has conclusively shown that the centre is made up of one

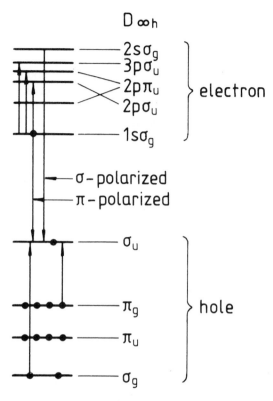

Figure 5.13.
One-electron energy level diagram for the STE.

electron trapped at a Cl_2 molecular ion, with almost the same structure as that of the V_K centre. Optical absorption from the lowest to higher triplets (triplet−triplet absorption) has been detected for the chlorides.

A few studies of the migration of the STE have been carried out. It appears that the activation energy is smaller than that for the V_K centre, in accordance with theoretical estimates which take into account some relaxation of the bonding of the V_K centre by the trapped exciton. This problem is quite relevant to the question of excitation energy migration in alkali-halide crystals and more information is needed.

Another interesting problem is the relaxation from the free-exciton to the STE state. In particular, an energy barrier has been theoretically predicted by Toyozawa (see the paper by Itoh[27]) and it has been invoked to account for the thermal quenching of the free-exciton luminescence, which monitors the free-relaxed-exciton transition. The values for the height of the barrier are in the range 10−30 meV for KI and RbI.

5.6 DEFECTS AND COLOUR CENTRES IN SILVER HALIDES

Silver halides are technologically relevant crystals because of their use for photographic recording. By contrast with alkali halides, it is well established that intrinsic disorder is brought about by cation Frenkel pairs up to the melting point. Mass transport is governed by migration of silver ion vacancies and two types of interstitialcy jump. The associated formation and migration enthalpies for two silver halides are given in Table 5.6.

An interesting feature of these crystals is the marked enhancement of the electrical conductivity σ at high temperatures, with regard to the values predicted by the simple Arrhenius plot. This effect provides a good topic

Table 5.6.
Formation and migration enthalpies for AgCl and AgBr

	Formation or migration enthalpy (eV)	
	AgCl	AgBr
Formation of Frenkel pairs	1.49±0.02	1.26±0.02
Cation vacancy migration	0.31±0.01	0.32±0.01
Collinear interstitialcy	0.02±0.01	0.04±0.01
Non-collinear interstitialcy	0.14±0.01	0.28±0.01

for discussion on temperature effects on thermodynamic parameters. The conductivity anomaly has been attributed to an enhancement of defect formation rates near the melting point T_m. Defect interaction effects, such as Debye-Hückel corrections, do not explain the anomaly.

AgI has the wurtzite structure (β phase) at room temperature but presents a first-order transition to a body-centred cubic (zinc-blende) structure (α phase) at $T_c = 147°C$ (with iodine ions at 0, 0, 0 and $\frac{1}{2}, \frac{1}{2}, \frac{1}{2}$). At T_c, σ increases by about three orders of magnitude from 10^{-3} to $1 \ \Omega^{-1} \ cm^{-1}$. The α phase is a superionic conductor with an activation energy E of about 0.05 eV.

Although the basic structure of the β phase is well established, the detailed configuration of the silver sublattice still poses some questions[30]. The lattice disorder is brought about by active Frenkel pairs as inferred from a variety of experimental data. In fact, in some work it has been suggested[28] that there is a discontinuous sharp increase in the concentration of mobile silver interstitial ions about 50 K below the $\beta \rightarrow \alpha$ transition temperature (i.e. the silver sublattice melting), although this has not been confirmed in later work[29].

Neutron diffraction studies of single crystals have shown that the Ag^+

ions occupy tetrahedral interstices of the iodine sublattice, each Ag^+ having four I^- neighbours. There are 12 such equivalent sites in the defect unit cell and it has been suggested that diffusion of Ag^+ ions occurs by successive jumps from one tetrahedral site to an empty neighbouring site, through the shared triangular face. The activation energy is 0.05 eV, as inferred from conductivity data.

It should be mentioned that addition of RbI to AgI to form the compound $RbAg_4I_5$ induces superionic behaviour almost down to room temperature. This material reaches a conductivity (about 0.27 Ω^{-1} cm^{-1}) which is comparable with that of the sulphuric acid solution used in the lead battery. This property has been used to realize various solid-state electrochemical cells working on the basic arrangement

$$Ag \,|\, RbAg_4I_5 \,|\, I_2$$

As to colour centres in the silver halides, no intrinsic colour centre has been reported except for the self-localized trapped hole in AgCl, i.e. the hole trapped at an Ag^+ ion forming a tetragonally distorted $AgCl_6^{4-}$ centre. It is stable below liquid-nitrogen temperature. It migrates[30] athermally via the small-polaron band below 30 K and by phonon-assisted hopping at higher temperatures with an activation energy of 61 meV.

5.7   DEFECTS AND COLOUR CENTRES IN DIHALIDE AND POLYHALIDE CRYSTALS

The alkaline-earth fluorides as well as some other dihalides with the fluorite structure, such as β-PbF_2 or $SrCl_2$, have been rather extensively studied. Apart from their relative structural simplicity, one of the reasons is the high electrical conductivity ($\sigma \approx 1\ \Omega^{-1}\ cm^{-1}$) that they can reach so that they may serve as one of the model systems for studying fast ionic conduction (superionic conductors).

At low temperatures compared with the melting point, the experimental behaviour has been interpreted in terms of anion Frenkel disorder. It is assumed that the interstitials occupy a centre cubic position surrounded by eight fluorine ions. The migration of the interstitials appears to occur via a non-collinear interstitialcy mechanism. The experimental values[3,4] for the formation enthalpy h_f of the anion Frenkel pair are included in Table 5.1 together with the corresponding enthalpy for the migration of the anion interstitial and anion vacancy. The higher value of the migration enthalpy for the interstitial with regard to the vacancy is counterbalanced by its higher value of the migration entropy, so that vacancy motion dominates at low temperatures whereas interstitials are the mobile species at high temperatures.

At high temperatures, many fluorite dihalides exhibit a broad specific heat anomaly at a temperature T_0 where T_0 ranges from 60 to 87% of the melting temperature. The anomaly is associated with a high degree of disorder in the anion sublattice, which gives rise to a markedly enhanced conductivity σ, reaching values comparable with those of the melting salts (superionic conductor regime). The defect structure corresponding to this situation appears complex and cannot be understood in terms of individual anion Frenkel pairs. Neutron diffraction experiments as well as computer simulations have been carried out, suggesting that the idea of a molten anion sublattice sometimes invoked does not appear appropriate. The mobile anions, which amount to a few per cent, are apparently not located at the cubic-centre positions (as for low temperatures) but are possibly forming clusters with other defects.

As for alkali halides, complexes between impurities and intrinsic defects can be formed. As an example, the association of a trivalent rare-earth impurity M^{3+} (such as Gd^{3+} or La^{3+}) with a fluorine interstitial to form a dielectric dipole has been ascertained by dielectric relaxation measurements. The F^- ion, acting as a charge compensator, can occupy a first-neighbour (nearest-neighbour) centre cubic position leading to a $\langle 110 \rangle$ orientation of the dipole or a second-neighbour (next-nearest-neighbour) position so that the dipole lies along a $\langle 111 \rangle$ axis. The dipole formed by a monovalent impurity such as Na^+ and an associated cation vacancy has also been identified.

An extensive effort has been devoted to understanding the spectroscopy of transition (and rare-earth) impurities in alkaline-earth fluorides, mainly because of its potential relevance for solid-state laser design. Also, colour centres can be produced in these solids by additive colouring and irradiation. Both simple defects and a variety of aggregated centres (including colloids) have been studied. The F, V_K and H centres have been unambiguously identified by EPR techniques. The optical absorption bands corresponding to the F, F_2, V_K and H centres are summarized in Table 5.7 for the alkaline-earth fluorides and $SrCl_2$. The F centre is isotropic for the cubic halides and slightly dichroic for MgF_2, presenting the cassiterite structure (hexagonal). The V_K and H centres can be described as F_2^- molecular ions, which are oriented along $\langle 100 \rangle$, at variance with alkali halides.

Much less information is available[31] on layered dihalides MX_2 or MX_3 (X $\equiv$ Pb, Cd or Hg). The dominant intrinsic defects appear to be Schottky pairs.

Some well-established results have been obtained with regard to colour centres in fluoroperovskites ABF_3, A ($\equiv$Li, Na or K) being a monovalent metal and B ($\equiv$Mg, Mn, Zn or Be) a divalent cation. Several electron and hole centres have been identified in ultraviolet-X- or γ-irradiated crystals. In particular, the F centre has been identified by EPR[32] and even ENDOR[33] in KMgF$_3$. The peak wavelengths for the absorption band of the simplest centres in several trifluoride lattices are included in Table 5.7.

5.8 DEFECTS IN HALIDE GLASSES

A novel topic for defect physics studies is provided by halide glasses[34]. There is a growing interest in preparing materials with a high optical transmission in the medium-infrared region (up to $5-10$ μm), so that the minimum attenuation value (about 0.2 dB km^{-1}) at present attainable with silicon glass fibres could be improved. Heavy-metal halide glasses have now been prepared which appear to be appropriate candidate materials for that

Table 5.7.
Colour centres in dihalides and polyhalides

Material	Peak wavelength (nm)			
	F	F$_2$	V$_K$	H
MgF$_2$	260	370, 400		
CaF$_2$	375	366	320	314, 285
SrF$_2$	449	427	326	325
BaF$_2$	606	550	336	364
SrCl$_2$				
RbMgF$_3$	295 (σ), 325 (π)	230, 285, 387	330	
KMgF$_3$	270	445	340	
KZnF$_3$	230		375	
NaMgF$_3$	290		350	
LiBaF$_3$	280		345	

purpose. Most attention has been paid to the identification of transition and rare-earth impurities at very low concentration levels (parts per billion to parts per million range). Recently some studies on intrinsic defects induced by ionizing radiation have been reported. As an example, for ZrF$_4$-based glasses, some hole centres such as F$_2^-$ molecular ions and F^0 interstitials have been identified by EPR spectroscopy, together with several impurity-related centres. The comparison of the observed defects with those found for halide crystals may be quite valuable in the understanding of the different behaviours of crystal and glassy systems. Moreover, the novelty of octahedral (and not tetrahedral) coordination is provided by these glasses by contrast with the more thoroughly studied oxide-based glasses.

REFERENCES

1 F. Beniere, Diffusion in ionic crystals, in V. M. Tuchkevich and K. Shvarts (eds), *Proceedings of the International Conference on Defects in Insulating Crystals*, *Riga* Springer, Berlin (1982).

2 D. Franklin, in J. Crawford and L. Slifkin (eds), *Point Defect in Solids*, Vol. 1, Plenum Press, New York (1972).

3 A. V. Chadwick, *Radiat. Eff.* **74**, 17 (1983).

4 A. V. Chadwick, F. G. Kirkwood and R. Saghafran, *J. Phys. (Paris), Colloq. C6* **41**, 216 (1980).

5 J. R. Friauf, *J. Phys. (Paris)* **38**, 1077 (1977).

6 A. M. Stoneham, *Theory of defects in solids*, Clarendon, Oxford (1975).

7 F. Agulló-Lopez, J. M. Calleja, F. Cusso, F. Jaque and F. J. Lopez, *Prog. Mater. Sci.* **30**, 187 (1986).

8 C. R. A. Catlow, J. Corish, J. M. Quigley and P. W. M. Jacobs, *J. Phys. Chem. Solids* **41**, 231 (1980).

9 B. di Bartolo, *Optical Interactions in Solids*, Wiley, New York (1968).

10 S. Sugano, Y. Tanabe and H. Kamimura, *Multiplets of Transition Metal Ions in Crystals*, Academic Press, New York (1970).

11 F. Luty, in V. M. Tuchkevich and K. Shvarts (eds), *Proceedings of the International Conference on Defects in Insulating Crystals, Riga*, Springer, Berlin (1982).

12 H. Murrieta, F. J. Lopez, J. Garcia-Solé, M. Aguilar and J. Rubio, *J. Chem. Phys.* **77**, 189 (1982).

13 P. Aceituno, C. Zaldo, F. Cussó and F. Jaque, *J. Phys. Chem. Solids.* **45**, 637 (1984).

14 N. M. Bannon, J. Corish and P. W. M. Jacobs, *Philos. Mag. A* **51**, 797 (1985).

15 C. Zaldo, J. Garcia-Solé, E. Dieguez and F. Agulló-Lopez, *J. Chem. Phys.* **83**, 6197 (1985).

16 K. Suzuki, *J. Phys. Soc. Jpn* **16**, 67 (1961).

17 A. de Andrés and J. M. Calleja, *Solid State Commun.* **48**, 949 (1983).

18 P. Aceitumo, F. Cussó, A. de Andrés and F. Jaque, *Solid State Commun.* **49**, 209 (1984).

19 A. B. Lidiard, in B. Henderson and A. E. Hughes, (eds), *Defects and their structure in Non-metallic Solids*, Plenum, New York (1976).

20 D. Curiè, Luminescence in Crystals, Methuen, London (1960).

21 W. Gellermann, F. Luty and C. R. Pollock, *Opt. Commun.* **39**, 391 (1981).

22 E. Goovaerts, J. Andriessen, S. V. Nistor and D. Schoemaker, *Phys. Rev. B* **24**, 29 (1981).

23 K. M. Strohm and H. J. Paus, *J. Phys. C: Solid State Phys.* **13**, 57 (1980); *J. Phys. (Paris), Colloq. C6* **41**, 119 (1980).

24 W. B. Fowler, *Physics of Color Centers*, Academic Press, New York (1968).

25 A. E. Hughes and S. C. Jain, *Adv. Phys.* **28**, 717 (1979).

26 W. Van Puymbrueck and D. Schoemaker, *Phys. Rev. B* **23**, 1670 (1981).

27 M. Itoh, *Adv. Phys.* **31**, 491 (1982).

28 D. Brinkmann and W. Frendenreich, *Solid State Commun.* **25**, 625 (1978).

29 R. J. Cava and E. A. Rietman, *Phys. Rev.* **30**, 6896 (1984).

30 E. Laredo, W. B. Paul, S. E. Wang, L. G. Rowan and L. Slifkin, *Radiat. Eff.* **72**, 97 (1983).

31 M. R. Tubbs, *Phys. Status Solidi (a)* **49**, 11 (1972); *Phys. Status Solidi (b)* **67**, 11 (1975).

32 T. P. P. Hall and A. Leggett, *Solid State Commun.* **7**, 1657 (1969).
33 R. C. Du Varney, J. R. Niklas and J. M. Spaeth, *Phys. Status Solidi (b)* **97**, 135 (1980).
34 M. Poulain, *J. Non-Cryst. Solids* **56**, 1 (1983).

6
Point Defects—Oxides

6.1 INTRODUCTION

Oxide materials present a mixed bonding character with features that are ionic and others more easily described with covalent bonds. For example, for defects in MgO, ionic bonding is appropriate whereas many defects in quartz appear to be covalently bonded. Therefore the concepts presented for the halides in Chapter 5 are often directly applicable to the defects in oxides. Oxides appear in a variety of crystallographic structures and Figure 6.1 shows examples of the more commonly studied ones. Most early work focused on MgO, SiO_2 (quartz and silica) and Al_2O_3. Whereas later research has moved to the technologically interesting compounds such as $LiNbO_3$, $BaTiO_3$ and $Bi_4Ge_3O_{12}$.

In fact, the diversity of oxide applications ranging from lasers to integrated optics and high-temperature superconductors suggests that the future of defect studies in oxides is a crucial element in the support of these industries.

From the defect physics point of view, oxides present interesting problems associated with the high polarizability of oxygen ions such as the self-trapping of carriers as small polaron states. As mentioned earlier, Jahn–Teller relaxations (see section 7.2) are important for defects, including oxides. Moreover, some oxides show a wide range in compositions, making them very suitable for studying the relationship between non-stoichiometry and point defects. Additionally, other oxides present structural phase changes, and point defects have provided useful probes for studying the

149

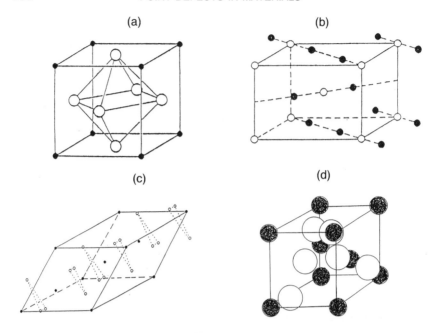

Figure 6.1.
Examples of four oxide crystal structures: (a) WO_3; (b) TiO_2 (rutile); (c) Al_2O_3 (sapphire); (d) SiO_2 (quartz).

microscopic aspects of the transition. All these features are discussed in this chapter.

6.2 ALKALINE-EARTH OXIDES

As for most other oxides, impurity effects complicate the defect studies; for example for MgO the data are few and not always consistent. Usually (but not always) Schottky disorder is assumed for this material and the formation enthalpy of the Schottky pair, as obtained from conductivity data, is rather high, i.e. 3.8−4.8 eV. The experimental enthalpy values for cation vacancy migration are in the range[1] $h_m(Mg) = 1.6-2.13$ eV and those for oxygen vacancy migration $h_m(O) = 1.7-3.8$ eV. Therefore, regardless of the low accuracy of the above values, it seems clear that oxygen is the slowest-diffusing species.

With regard to colour centres the available data for alkaline-earth and some other binary oxides (e.g. ZnO and BeO) with the same face-centred cubic lattice have allowed the identification of a number of centres by paramagnetic resonance and optical spectroscopy[2]. The simplest electron centres are formed by trapping one or two electrons at an oxygen vacancy. In the first case, we obtain the paramagnetic F^+ centre, which is the analogue to the F centre in alkali and alkaline-earth halides, although it possesses an effective charge $+e$ in the oxides. The vacancy with two electrons constitutes the so-called F centre and is neutral and diamagnetic.

F^+ and F centres present characteristic absorption and luminescence bands, whose peak energies are summarized in Table 6.1 for several oxides. The F^+ bands correspond to transitions from the ground A_{1g} to the T_{1u} excited state of the centre. For the F centre the band is associated to the singlet–singlet transition $^1S \rightarrow {}^1P$ ($^1A_{1g} \rightarrow {}^1T_{1u}$). The F^+ centres in oxides present some pecularities with regard to those in halide materials.

(1) Because of the effective charge in the oxide the ground-state wavefunction is more localized in the vacancy.

(2) The surrounding lattice is more strongly distorted and nuclear quadrupole hyperfine effects are more important.

(3) g shifts in the electron paramagnetic resonance (EPR) spectra are larger owing to overlap of the electron wavefunction with the orbitals of neighbouring ions.

(4) Optical bands are markedly asymmetric, at variance with the halides.

One interesting feature to be mentioned is the observation of a clear zero-phonon line in the F^+ absorption and luminescence bands of CaO (Figure 6.2). This would imply a Huang–Rhys factor S of $5-6$ or less, which is not in accordance with the value inferred from the width of the optical bands ($S \approx 15$). This fact, together with the mentioned band asymmetry, has raised some interesting problems on the electron–phonon coupling behaviour in these crystals which are now well understood. The situation has been

Table 6.1.
Optical bands of electron centres in binary oxides

Material	F^+		F	
	E_A (eV)	E_L (eV)	E_A (eV)	E_L (eV)
MgO	4.95	3.13	5.01	2.4
CaO	3.70	3.3	3.10	2.0
SrO	2.99	2.42	2.49	
BaO	2.00			
$\alpha-Al_2O_3$	4.8; 5.4; 6.3	3.75	6.1	~3.0

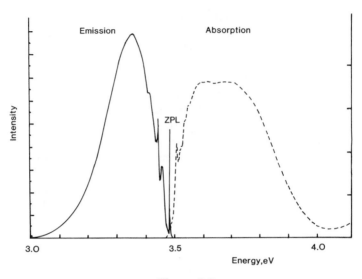

Figure 6.2.
An example of absorption and emission features with resolved phonon features. The curves are approximately symmetric about the energy of the zero-phonon line (ZPL). The example is for the F^+ defect in CaO.

analysed in terms of a dynamic Jahn–Teller coupling to non-cubic modes (E_g and T_{2g}) (see section 7.2).

In addition to the electronic F and F^+ centres, a number of hole centres have been investigated in binary oxides. The so-called V^- centre has been the best characterized by means of EPR spectroscopy. It consists of a hole trapped at a cation vacancy, i.e. an O^- ion in a first-neighbour site to the vacancy. In some oxides, this centre has one optical absorption band in the ultraviolet (2.3 eV for MgO). The cation vacancy with two trapped holes constitutes the V^0 centre. There are also several related centres such as the V_{OH^-}, V_{F^-} and V_a, which respectively correspond to a V^- centre associated with a OH^-, an F^- and an M^{3+} (such as Al^{3+}) ion in a neighbouring host location. Another centre also related to those previously described is the Li^0 or Na^0 centre, consisting of a hole trapped at an O^{2-} ion located as a first neighbour to a substitutional Li^+ or Na^+ impurity. Schematic models for these V centres are illustrated in Figure 6.3, and the absorption bands associated with some of them listed in Table 6.2, for MgO.

6.3 TRANSITION-METAL OXIDES

Transition-metal oxides present very interesting problems related to their electronic structure, defect configurations and non-stoichiometry. The

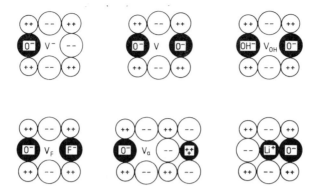

Figure 6.3.
Models for some hole centres in alkaline-earth oxides (after B. Henderson S. E.
Stokowski and T. C. Ensign, *Phys. Rev.* **183**, 826 (1969)).

Table 6.2.
Absorption bands of V-type centres in MgO

Centre	Energy band peak (eV)
V^-	2.3
V^0	2.32
Li^0	1.83
Na^0	1.58
V_{Al}	2.32

various types of physical behaviour are illustrated by the first-row transition-metal rock-salt oxides, where TiO and VO are metallic, whereas MnO,
FeO, CoO and NiO are magnetic semiconductors. The determination of
defect properties in these materials is further complicated by the active
exchange reaction with the surrounding oxygen atmosphere, as well as the
possible coexistence of various charge states for the defects. The question of
non-stoichiometry of these oxides has already been briefly addressed in
Chapter 1. The deviation from stoichiometry, measured by the fractional
parameter x in the general formula $M_{1-x}O$ (where M is a transition metal),
is very important for $Fe_{1-x}O$ $(0.05 < x < 0.16)$, being much smaller for the
other oxides $(x < 10^{-1}$ for MnO, $x < 10^{-2}$ for CoO and $x < 10^{-3}$ for NiO).
One of the key problems in these materials is to relate the non-stoichiometry
to specific defect configurations and much effort is being devoted to this
question in order to understand the full range of physical properties. It is
interesting to point out, in this connection, that a high concentration of

intrinsic defects can exist even for an ideal stoichiometric composition. This is the case for TiO and VO, where about 15% metal and oxygen vacancies are present for $x = 0$, showing that "chemical stoichiometry" is not always synonymous with "perfect" structure.

Calculations by Catlow and Fender[3] and Catlow et al.[4] suggest that the basic defect unit in $M_{1-x}O$ is the 4:1 cluster, already mentioned in Chapter 1, and formed by an M^{3+} interstitial at the centre of a tetrahedon of cation vacancies. The energy gain on clustering is about 2 eV per vacancy. By edge sharing, the 4:1 clusters can associate to give the more complex 6:2, 8:3, etc., clusters, since they are energetically favoured. The clustering of point defects most probably occurs in FeO and MnO at least for $x \geqslant 0.01$.

Non-cubic binary transition-metal oxides, such as WO_3 or MoO_3, present new features with regard to their defect structures. These materials accommodate non-stoichiometry by forming extended planar defects known as crystallographic shear planes. The formation of these planes can be understood in terms of a change in the linkage of the MO_6 octahedra making up the crystal structure[5,6]. In the initial host lattice the octahedra are corner linked but collapse to form blocks of edge-sharing octahedra which are aligned along the crystallographic plane (see Figure 6.4). These planes become oriented along certain preferred directions which depend on the reduction state of the material (i.e. the amount of non-stoichiometry). Moreover, the crystallographic shear planes are found to order over large distances into arrays.

Other oxides, showing high electrical conductivities are those with formula MO_2 (where M is a tetravalent transition or rare-earth cation) and fluorite structure[7,8]. High-conductivity materials are prepared by doping these oxides with divalent M^{2+} as trivalent M^{3+} cations which induce the introduction of compensating oxygen vacancies to keep the overall electrical neutrality. The formation of bound impurity–vacancy pairs has been well established through dielectric and anelastic relaxation studies. The presence of oxygen vacancies in the doped oxides has been ascertained by density and X-ray measurements. The activation enthalpy for oxygen vacancy migration has been determined to be about 1 eV, whereas that for cation vacancy diffusion is about 4–5 eV. At high vacancy (impurity) concentrations, the situation becomes more complex owing to strong interactions among defects, leading to both short- and long-range ordering of defects. Heavy-impurity doping may strongly influence the energetics and stability of the crystal structure. For example, zirconia is doped with CaO or Y_2O_3 in order to stabilize the fluorite structure.

UO_2, also having the fluorite structure, is unusual since it easily incorporates an excess of oxygen in the form of interstitial O^{2-} ions. This is accompanied by the creation of holes which give rise to p-type electrical

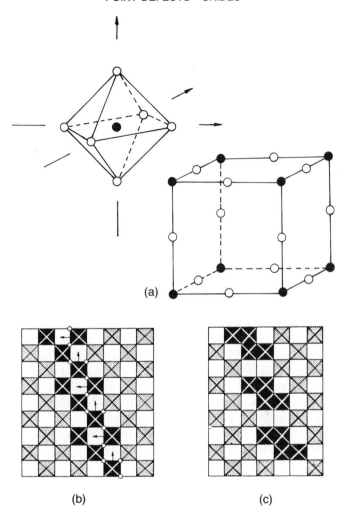

Figure 6.4.
(a) The basic structure of ReO_3 and the ReO_3 octahedra; (b) a block diagram in which the octahedra share corner site oxygen ions. In (b), oxygen vacancies are shown by open circles. These allow edge shearing to fully bonded blocks, shown in (c) as black shaded blocks.

conductivity. Calculations on lattice formation energies have indicated[9] that the defect showing the smallest energy is the anion Frenkel pair (O^{2-} + V_O^{2-}).

For the transition-metal oxides, no colour centres have been reported. In fact, the electronic defect which has been extensively investigated in these oxides is the self-trapped electron (small polaron), although primarily in connection with the electrical transport behaviour[10]. The electron is located within a unit cell owing to the strong lattice distortion it causes, and electron motion takes place (at least at not very low temperatures) via "hopping" from one lattice site to the next equivalent site. Basic evidence for electrical transport by small polarons is the occurrence of a thermally activated low mobility ($\mu < 1$ cm^2 V^{-1} s^{-1}). Unfortunately, electrical studies in these materials are rather difficult and results are influenced by crystal preparation, purity and stoichiometry. Nevertheless, data appear to show conclusively that MnO is a small-polaron material (and possibly also CoO and NiO), the trend being in accordance with theoretical predictions.

Evidence for hole trapping at alkali impurities has also been obtained. The hole appears to localize on one of the cations adjacent to the alkali ion. Hopping activation energies around the impurity have been experimentally determined for CoO and NiO.

6.4 DEFECT STUDIES IN SAPPHIRE

Sapphire ($\alpha - Al_2O_3$) is routinely grown as relatively pure single crystals containing less than 10 ppm of impurities. It has a very wide band gap, is chemically stable and can be coloured by the addition of impurities, by thermal reduction and by energetic electron, neutron or ion beam irradiation. Consequently, colour centre studies are fairly well advanced and are based on 30 years' work. Because the structure is rhombohedral, many of the absorption or emission features are anisotropic; so we can compare, at least to a first approximation, defect models with the observed directional optical features. The studies are further simplified by the fact that there is no colouration by ionizing radiation.

6.4.1 Identification of optical absorption bands

Over the accessible wavelength range of the visible and near ultraviolet the dominant colour centre is peaked near 6 eV. This major feature is not a single absorption band but, as shown in Figure 6.5, is a composite which has never been completely resolved. The figure gives examples for electron irradiation, where the defects are predominantly isolated; high-dose neutron damage where they may aggregate and absorption resulting from ion bombardment where the damage event causes a dense cascade of defects. We are confident that there are strong absorption bands near 4.8, 5.4, 6.1 and 6.3 eV but the component features of the curves do not grow at identical rates with increasing dose and there is evidence of other bands at 5.8−5.9 eV and

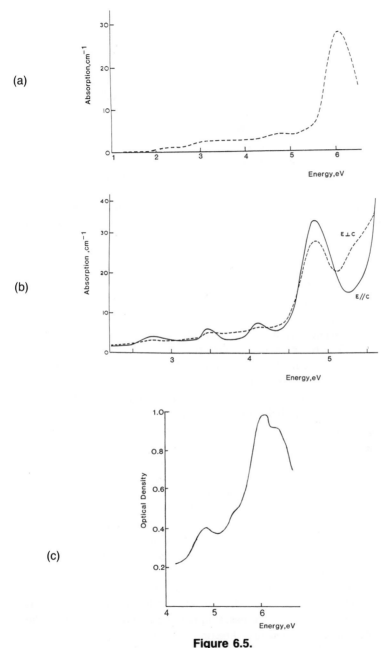

(a)

(b)

(c)

Figure 6.5.
Optical absorption induced in Al_2O_3 by (a) electrons, (b) neutrons and (c) ion beam irradiation. The polarized absorption data in (b) emphasize the 4.8 and 5.4 eV bands.

beyond 6.3 eV. In the heavily damaged material the polarized absorption data show a wide range of optical anisotropy (e.g. $A = \mu_{E\parallel C}/\mu_{E\perp C}$) as well as minor shifts in the peak positions with polarization. Many of these features were recorded[11-13] in the 1950s and 1960s, and progress in identifying the major intrinsic defect colour centres was reviewed by Crawford[14]. Optical absorption associated with impurity ions such as chromium (ruby) or titanium (blue sapphire) or transition metals have been recorded in earlier studies, as for example by Tippins[15]. Table 6.3 lists the current assignments of the absorption bands to specific defect structures. It is apparent that these absorption features are nearly all associated with defects in the oxygen sublattice. The F centre (an oxygen vacancy containing two electrons) and the F^+ centre (an oxygen vacancy containing one electron) are always present. They may be distinguished as the F band appears to be isotropic (see below) whereas the F^+ bands are anisotropic[16] and, as predicted by La et al.[17], the crystal field splitting of the triply degenerate excited state leads to three distinguishable F^+ absorption bands of different anisotropies. The highest of these, at 6.3 eV, is often poorly resolved because of both the overlap with the F band and the instrumental difficulties in this wavelength region. Nevertheless, F and F^+ separation can be sensed by luminescence studies as the F-centre emission is peaked at 3.0 eV whereas all the F^+ bands generate emission at 3.8 eV. The centres also differ in their photoconductivity response with the F band showing photocurrents[18] down to 10 K. Such low-temperature photocurrents suggest that the excited state must lie in or very close to the conduction band. For such an excited state the crystal field splitting of the energy level is less apparent; hence, we can rationalize why the F^+ centre shows three bands but the F centre has only one.

The next member of the set of F-type centres continues with the suggestion that the α centre[19] at 7.0 eV (i.e. O^{2-} vacancy without any trapped electrons) lies closer to the band edge. It is thought to have an f value of 0.2. The direct band gap of Al_2O_3 is about 9.5 eV at room temperature and strong intrinsic absorption sets it beyond 8.5 eV from excitonic absorption. In particular the 9.25 eV absorption is described as an exciton absorption. This reflects the anisotropy of the rhombohedral lattice and hence is stronger for polarization parallel to the c axis than in the perpendicular plane[20].

F-centre aggregates (Figure 6.6) certainly exist in Al_2O_3 and are most readily identified by polarized absorption or luminescence measurements. For example the band at 3.46 eV has a dipole oriented at about 40° to the c axis and could thus correspond to a direction linking two anion vacancies[21,22]. The charge state of the centre is less assured and the alternatives are for two oxygen vacancies which trap one or more electrons. At lower temperatures this feature shows pronounced zero and multiphonon absorption as well as a

Table 6.3.
Optical absorption band defect models for $\alpha-Al_2O_3$

	Absorption (eV)	Emission (eV)
Exciton	9.25	
F	6.03	3.0
F^+	6.3, 5.41, 4.84	3.8
α	7.0	
$F_2(?)$	3.47, 3.37	3.27, 1.49
$F_2(?)$	3.74 or 4.1	
$F_2^+(?)$	2.8, 2.2	1.25
$F_3^+(?)$	3.5	

mirror image luminescence band. The polarization axis of 40° to the c axis is a few degrees away from the expected $(O^{2-}-O^{2-})$ direction but it is possible that the centre is interacting with nearby defects as the structure of the zero-phonon lines is slightly different in the samples irradiated with neutrons or protons.

Irradiation produces a diversity of absorption bands and, because of practical applications of sapphire windows, many studies are made with neutron-irradiated samples. These show a vast range of overlapping bands which can be separated or modified by X-ray irradiation or annealing. Whilst the assignments of the strongest features to F and F^+ models is not in doubt, many of the lower-energy features have been linked to different models. For example, annealing studies[23] lead to models of an F_2 centre as responsible for the 4.1 eV band, an F_2^+ transition for the 2.8 and 2.2 eV bands and F_3^+ for the 3.5 eV band. Figure 6.6 emphasises the complexity of the absorption spectra by comparing the anisotropic absorption bands from just a small spectral range for two neutron-irradiated samples. The irradiation dose was the same in each case but in one case the irradiation temperature was 60°C and in the other 290°C. The 60°C sample was then annealed at 290°C. Quite clearly this leads to a different mixture of the defects although most bands are apparent to some degree in each curve. The annealed samples show better resolution of zero-phonon features.

Even though there are potential disagreements in the models of some features, it is clear that all the models are for oxygen vacancies. Associated with this there must be some oxygen interstitial defects. These can be sensed indirectly by variations in luminescence efficiency[24] between thermo-chemically coloured (i.e. reduced) Al_2O_3, or neutron-irradiated samples. When the F centres are isolated, their luminescence efficiency is high. By contrast the damage track of neutrons or ion beams contains both F centres

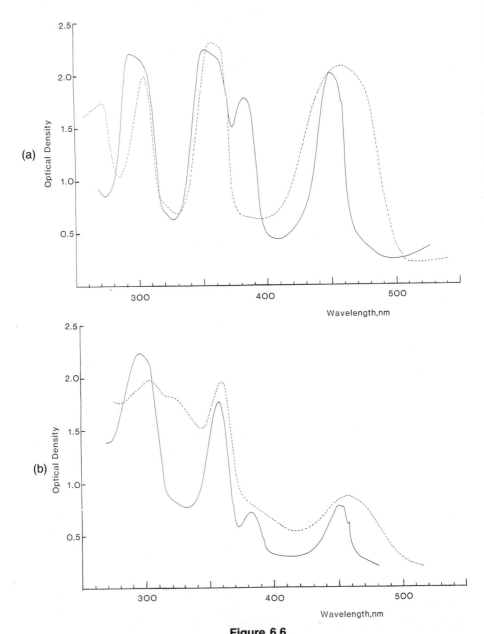

Figure 6.6.
Optical absorption of neutron-irradiated Al_2O_3 for a dose of 3×10^{18} neutrons cm^{-2}. The sample shown by − − − was irradiated at 290°C and that shown by ———— at 60°C followed by a 290°C anneal. Note zero-phonon features and distinct anisotropy of the bands. The optical absorption data of $E \parallel C$ and $E \perp C$ are shown separately in (a) and (b).

and interstitials in close proximity. This produces a concentration quenching of the F-centre luminescence. To test this idea, the crystals can be annealed; the close F-centre–interstitial pairs recombine and the lower F-centre density results in increased luminescence efficiency. Crawford[14] has pointed out that no such concentration-quenching effects are noted for F^+ centres as the wavefunction of the singly charged vacancy site is much more compact.

Point defects on the aluminium sublattice are not resolved optically although some researchers have suggested that the 4.09 or 3.98 eV features relate to interstitial aluminium.

Hole centres have received less attention but a broad band centred at about 3.0 eV is described[25] as a hole trapped at an $O^{2-}-Al^{3+}_{vac}-OH^-_{sub}$ complex; this was termed a V_{OH^-} defect. Nevertheless, hole traps undoubtedly exist and in other types of measurement are discussed. For example, in thermoluminescence data between 100 and 400°C, five such hole traps are believed to participate[26]. Hole-type centres are also firmly established in EPR studies of damaged Al_2O_3 although in many cases such centres are linked to impurity sites.

The data gleaned over the last 30 years suggest that the Al_2O_3 lattice is a useful structure for optical colour centre studies but the difficulty is that, although we can take advantage of the crystal symmetry to interpret polarized absorption or luminescence measurements, there is an excess of features which overlap in energy. For example the "simple" blue luminescence arises from quite different colour centres. Although some of the complexity can be resolved by excitation in different absorption bands, we must resort to lifetime determinations to separate the features.

In summary the major familiar ideas of anion defects (F, F^+, F_2, etc.) are all demonstrated but many other centres still defy positive identification.

6.5   β-ALUMINA

Much attention has been devoted to the "defect" structure of some Al_2O_3-based materials, such as the so-called β-alumina, because of its superionic conducting behaviour[27,28]. β-alumina is an aluminate with the formula $M_{1+x}^+Al_{11-x/3}^{3+}O_{17}^{2-}$ ($0 < x < 0.3$), M^+ ($\equiv Na^+$, Li^+, K^+, Ag^+, etc.) being a monovalent cation. The structure is made up of Al–O spinel blocks, containing four oxygen planes each, separated by bridging layers of M^+ and O^{2-} ions in a hexagonal arrangement (Figure 6.7). The M^+ ions can occupy the five different positions illustrated in Figure 6.7(C). These are termed a Beever–Ross, an anti-Beever–Ross and three midoxygen sites. In the non-stoichiometric material, M^+ ions are disordered among the available positions, becoming very mobile along the bridging planes and

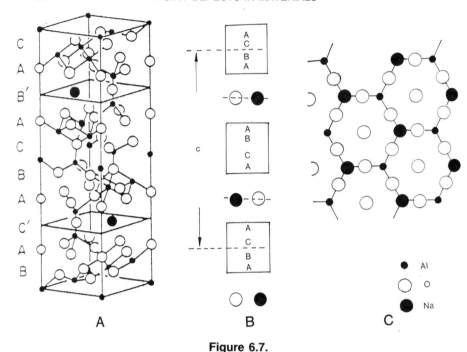

Figure 6.7.
(A) Structure of β alumina (○), oxygen atoms; (●), sodium atoms; (•), aluminium atoms; (B) schematic diagram giving a packing sequence; (C) a cross-section in the conducting plane.

leading to a high ionic conductivity. However, conductivity is restricted to this plane and movement along the c (normal) axis is exceedingly difficult so that the material is highly anisotropic.

6.6   QUARTZ AND SILICA

There have been extensive defect studies in both crystalline and amorphous SiO_2 and the literature has been well reviewed by many researchers[29-32]. Consequently, in this section, only limited comments are made and the major summary of the optical and EPR studies is presented as Tables 6.4 and 6.5 taken from Abu-Hassan[33].

As with sapphire, we are dealing with large-gap stable materials of industrial importance and with a 30 year history of defect studies. The synthetic SiO_2 is of higher purity than natural quartz crystals and is now available with a very low dislocation density. Many metal impurities can be removed by "sweeping" under an electric field as the c axis of the crystal is quite open and provides an easy diffusion path. A common impurity of amorphous

SiO_2 (silica) is OH radicals. The so-called dry or water-free silica types have trade names such as Corning 7943 (oxygen deficient), Suprasil W (about 10 ppm OH) and Spectrosil WF. The "wet" silica material may be Corning 7940, Suprasil 1 (about 1200 ppm OH) and Spectrosil A. These specifications may have varied over the last 20 years; thus, particular caution must be taken in comparing data that are not well defined in terms of material quality.

Since point defect studies involve rather few lattice sites, there are close parallels between quartz and silica as even silica has short-range ordering of the SiO_4 tetrahedra. One major difference is that at room temperature a radiolysis process operates in silica which allows defect formation and an associated compaction of the glass network. At lower temperatures (below 100 K), quartz also shows some radiolytic damage but the vacancy–interstitial pairs are close; so these centres anneal on warming. The presence of extrinsic defects, such as OH radicals, may play a role in this radiolysis.

In addition to the usual notation of vacancy and interstitial, the literature on SiO_2 has acquired some specific terms. These include the following: "a bridging oxygen" which links two silicon ions; "a non-bridging oxygen" which is only bound to one neighbour and hence leaves a dangling bond; a "non-bridging oxygen hole centre" which has a hole on the dangling bond ($\equiv Si$—O^0); a peroxy bond, which is a direct single O–O bond; a peroxy bridge ($\equiv Si$—O—O—$Si\equiv$); a peroxyl radical or superoxide radical ($\equiv Si$—O—O^0 or O_2^-). Some of these defect models are sketched in Figures 6.8, 6.9 and 6.10.

Reference to Tables 6.4 and 6.5 shows that the major oxygen-vacancy-related defect sites are described generically as E′ centres in both quartz and silica. The models of E′, etc., can be substantiated both by EPR, in crystalline samples, and by self-consistency between the behaviour of the various defect interactions. It is also apparent that the role of H^+ or OH^- must be considered even in these intrinsic defect studies. For further details the reader is referred to the reviews and references cited here[29-32].

6.7 TERNARY OXIDES—PEROVSKITES

Ternary oxides belonging to the perovskite or related families (e.g. $LiNbO_3$) are of much interest because of their application to a number of pyroelectric, electro-optic and photorefractive devices. In the latter case, the usefulness of the materials is related to defect properties such as electron or hole photoionization and trapping capabilities giving charge transfer processes, etc., which determine their photochromic and photorefractive response. Therefore, a growing interest is arising with regard to their impurity and defect properties[34,35]. Oxygen vacancies are usually considered as a major

Table 6.4.

Oxygen–vacancy centres (E′ and B) and impurities (A) for crystalline quartz

Label	Spin	Absorption (eV)	Comments	Model
E′$_1$	$\frac{1}{2}$	5.8	Optical bleaching (ultraviolet, 5.6 eV) (less efficient at liquid-nitrogen temperature) Thermal bleaching (100–400°C) Appears only after previous particle irradiation	Oxygen vacancy with an unpaired electron localized in the sp^3 hybrid orbital extending into the vacancy from an adjacent silicon (parallel to the "short" Si—O bond direction)
E′$_2$	$\frac{1}{2}$	5.3	Shows proton hyperfine splitting Produced by γ irradiation	"No definitive model". As E′$_1$ but the hybrid is pointing parallel to the "long" Si—O bond direction Could be thermally populated excited state of E′$_1$ centre
E′$_4$			Shows proton hyperfine splitting	Hydride ion (H$^-$) trapped in an oxygen vacancy with an additional unpaired electron shared unequally between the two silicon atoms neighbouring the vacancy, or neutral oxygen vacancy forming an E′-type defect with a captive hydride H$^-$ ion
E′$_\alpha$			Produced by X-rays of energy 100 keV or less at $T \leqslant 77$ K Present in as-grown silica Anneal at $T \geqslant 200$ K	(1) Radiolytic process: oxygen moves from an undisturbed site into a neighbouring "bonded" position (i.e. no net breakage of the bond) $\equiv$Si—O—Si$\equiv \xrightarrow{h\nu} \equiv$Si—(+O—O)+e$^-$ (2) Momentary rupture of strained Si—O bond $\equiv$Si...O—Si$\equiv \xrightarrow{h\nu} \equiv$Si O—Si$\equiv$
E″$_1$			Optically bleachable	Two neighbouring E′$_1$ centres with various possible geometries

	Para $s = 1$		"No definitive model"
E″$_2$		Thermally unstable (30–100°C)	
E″$_3$	≈5.0	Can be reactivated by ionization if annealed at $T < 120°C$ Not observed in γ-, X- or neutron-irradiated optical spectra	
B$_2$	2.65 ($\tau \approx 10$ ms)	Optically bleachable ($\lambda = 245$ nm) Produced *only* by ion beams (other than H$^+$)	(1) Neutral oxygen vacancy on ($\equiv$Si:Si$\equiv$) i.e. the precursor of E$_\gamma$
	4.3 ($\tau < 10$ ns)	Chemical and radiation annealing Responsible for SiO$_2$–Si interface states in metal–oxide–semiconductor devices Correlated with oxygen deficiency Due to singlet-to-singlet transition	(2) Produced by breaking of Si$-$Si bonds (3) Two-coordinated silicon, i.e. with only two neighbouring oxygen atoms (neutral Si^0)
A$_1$	2.0	Radiation annealing Some workers resolved these into three Gaussian bands:	Occur only at temperatures high enough to support alkali ion diffusion ($T \geqslant 200$ K) Hole trapped in a non-bonding p orbital of an oxygen ion located adjacent to a substitutional aluminium (AlO$_4$)^0
A$_2$	2.7	A$_1$, 1.85 eV (620 nm) A$_2$, 2.55 eV (480 nm) A$_3$, 2.85 eV (355 nm) Optically bleachable by E′$_1$ light (5.8 eV)	Optical absorption due to light-induced transfer of holes from an O^{2-} site to an equivalent site

Table 6.5.

Oxygen vacancy centres (E′ and B) and impurities (A) for amorphous silica

Label	Spin	Absorption (eV)	Comments	Model
E'_β	$\frac{1}{2}$	5.8	Concomitant growth with decay of radiolytic H^0. Similar EPR spectrum to that of the surface E′ centre observed in silica as well as in α-quartz when fractured in high vacuum	(1) Mobile atomic hydrogen (H^0) reacts with pre-existing positively charged three-coordinated silicon (present in the unirradiated as-grown silica). $\equiv Si + H^0 \rightarrow \equiv Si\cdot + H^+$ (2) Due to long-range relaxations of an atomic hydrogen with a neutral oxygen vacancy $\equiv Si\!-\!Si\equiv + H^0 \rightarrow \equiv Si\cdot \equiv Si\!-\!H$ The three-coordinated silicon tunnels out of the plane such that hydride ion projects away from the oxygen vacancy NB H^+ to unpaired spin separation $\geqslant 5$ Å
E'_γ			Caused by knock-on collisions. Sometimes referred to as E'_1	Hole trapped at the site of a neutral oxygen vacancy $\equiv Si\!-\!Si \xrightarrow{h\nu} \equiv Si\cdot Si\equiv + e^-$
B_2		≈ 5.0	Produced in unirradiated impure amorphous silica. Unbleachable optically	Suggested as Al^{3+} substituting Si^{4+} with an adjacent oxygen vacancy
A_1		2.3	Intensity increases with aluminium content. Peak positions vary with type of alkali present. Radiation annealing. Band structure changes with time	Unknown
A_2		2.9	Optically bleachable with E'_γ light (5.8 eV)	

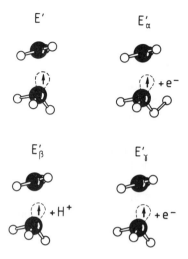

Figure 6.8.
Variants of the E' centre in quartz: ↑, unpaired spins; ●, silicon atoms; ○, oxygen atoms. Associated neighbours or charges are also indicated.

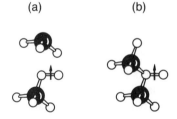

Figure 6.9.
Models of the peroxy radical in quartz; (a) non-bridging; (b) bridging.

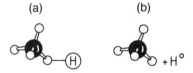

Figure 6.10.
A model of the non-bridging oxygen hole centre in quartz (a) before and (b) after rupture of an OH bond.

defect, their concentration depending critically on impurity doping and reduction state (oxygen partial pressure). Cation vacancies may also be present, as reported for lithium vacancies in $LiNbO_3$, where a marked deviation from exact stoichiometry results from the growth conditions.

Impurities (in particular, transition-metal and rare-earth ions) have been extensively studied in ternary as well as binary oxides. As a general rule, several charge states are possible and have been identified for many impurities; for example, in $SrTiO_3$, iron can exist from Fe^{2+} to Fe^{5+}. However, the problem of lattice location is more complex, since more than one substitutional site is available for the impurity. For perovskite oxides, such as $BaTiO_3$ or $SrTiO_3$, ion-size effects govern lattice location, so that transition-metal ions replace the transition-metal host cation, whereas rare-earth ions substitute for the alkaline-earth cation. In contrast, for the $LiNbO_3$ and $LiTaO_3$ crystals, no such rule has been ascertained and the problem is still unsolved.

For ternary oxides such as $SrTiO_3$, $BaTiO_3$ and $LiNbO_3$, optical absorption bands induced by low-temperature irradiation or thermochemical reduction treatments have been sometimes associated with F or F^+ centres[35]. However, more work is required to ascertain the proposed arrangements unambiguously.

It should also be mentioned that the presence of defects acting as shallow traps may give rise to a high population of carriers in the energy bands of the material and so to a characteristic free-carrier (intraband) absorption (see section 9.2.1). A well-documented case illustrating this behaviour is provided by heavily reduced $SrTiO_3$.

6.8   SELF-TRAPPING OF CARRIERS—SMALL POLARONS

In ionic solids, charge carriers (electrons and holes) are strongly coupled to the optical modes of the lattice and this has an important effect on their mobility. There are a number of theoretical approaches to deal with this coupling, such as the Fröhlich Hamiltonian, variational methods (Feynman) and Green function methods as well as a number of perturbation schemes. The result of coupling between the electron (or hole) and the lattice is that the carrier is accompanied, during its motion, by a lattice distortion which induces, in turn, an additional potential on the particle. In other words, the electron (or hole) is followed by a cloud of phonons, which are being created and annihilated as a consequence of the interaction. Polaron motion may be viewed in two parts, namely a charge plus the accompanying polarization. As a consequence, the total electron energy becomes modified by two terms: one of them constant (self-energy of the polaron) and the other dependent on velocity. The latter can be described as a change in the effective mass of the carrier. We can take, as a basis, the Fröhlich Hamiltonian, describing the electron interaction with the longitudinal optical modes:

$$\mathcal{H} = \frac{p^2}{2m} + \sum_k \hbar\omega a_k^+ a_k + \sum_k (V_k a_k \exp(i\mathbf{k}\cdot\mathbf{r}) + V_k^+ a_k^+ \exp(-i\mathbf{k}\cdot\mathbf{r}) \quad (6.1)$$

with

$$V_{\mathbf{k}} = -\frac{i\hbar\omega}{k}\left(\frac{\hbar}{2m\omega}\right)^{1/4}\left(\frac{4\pi\alpha}{V}\right)^{1/2} \tag{6.2}$$

α being a dimensional constant measuring the intensity of the electron–lattice coupling. $a_{\mathbf{k}}^{+}$ and $a_{\mathbf{k}}$ are the creation and annihilation phonon operators. A perturbative treatment of the corresponding Schrödinger equation valid for $\alpha \ll 1$ leads to a second-order correction ΔE to the energy of the ground state of the polaron $\exp(i\mathbf{k}\cdot\mathbf{r})|0>$, given by

$$\Delta E = -\alpha\hbar\omega - \frac{\alpha}{6}\frac{p^2}{2m} \tag{6.3}$$

$\alpha\hbar\omega$ being the polaron self-energy. The kinetic energy can be now written, taking this into account:

$$E_{\mathbf{k}} = \frac{p^2}{2m_{\mathrm{p}}} \tag{6.4}$$

with the polaron mass m_{p} given by

$$m_{\mathrm{p}} = \frac{m}{1 - \alpha/6} \tag{6.5}$$

For many ionic solids, $\alpha \gtrsim 1$ and the above perturbation treatment is not justified. Then, other theoretical methods should be used, although the basic features of the problem remain valid.

Under certain circumstances, the strength of the coupling may be sufficient to stabilize the electron (or hole) in a given lattice site. In this case, we are concerned with the so-called small polaron[36] since the size of the lattice distortion is comparable with the lattice parameter, at variance with the weak-coupling case or large polaron. In fact, the small polaron has already been described as an electronic defect appearing in a variety of ionic hosts, including halides and oxides. In some cases (e.g. V_K centres) these small polarons present characteristic optical bands and therefore they can be properly classified as colour centres.

Essentially, the localization of a carrier is the result of a delicate balance between the various competing contributions to the total energy. The space localization involves a superposition of all Bloch states of a given band and this, in turn, induces an increase in kinetic energy which can be estimated to be of the order of half the bandwidth. However, other energy contributions, such as lattice polarization and chemical bonding to ligands, favour the localization. A typical illustrative example is provided by the V_K centres in alkali halides. For KCl, the various contributions have been evaluated[37] and are listed in Table 6.6 together with those appropriate for electron self-trapping.

Table 6.6.
Contributions to self trapping energies in KCl

Contribution to the energy	Electron (eV)	Hole (eV)
Localization	−1.9	−0.3
Polarization	+0.3	+0.5
Binding	+0.3	+1.3
Total	−1.3	+1.5

Table 6.7.
Polaron types in oxides

Oxide	Minimum bandwidth (eV)	Possible polaron type
MnO	3.06	Small polaron
FeO	2.18	Small polaron
CoO	1.88	Small polaron
NiO	1.6	Large polaron

It is clear, as experimentally observed, that electrons cannot be localized and that the dominant contribution to hole self-trapping is the bonding energy Cl−Cl in KCl. According to this simple qualitative scheme, small-polaron formation is to be expected for those materials where the electron (or hole) bands are narrow. A favourable case is provided by transition-metal oxides with narrow d bands. As an example, the predicted minimum bandwidth for large polarons in $M_{1-x}O$ oxides are as[38] in Table 6.7. Consequently, the expectations for polaron type are as indicated in the third column, whose trend is in accordance with experiment.

The occurrence of small polarons has a strong bearing on the electrical transport properties of the material. The interpretation of the experiments relies on knowledge of the thermodynamic parameters of the polaron such as the formation energy and activation energy for diffusion. In order to provide some insight for these parameters, we are going to consider a drastically simplified model, which ignores the subtleties utilized by professional theoreticians[39]. If we ignore the question of the bandwidth, the electron (or hole) localization results in a competition between the linear decrease in the electronic energy as a function of the appropriate normal coordinate q describing lattice distortion and the quadratic increase associated with ionic repulsion. Consequently, the total energy $W(q)$ reads

$$W(q) = Aq^2 - Bq \qquad (6.6)$$

The energy zero corresponds to the uncoupled electron (or hole). The minimum energy $W_0 = Bq_0/2 = -B^2/4A$ occurs for $q_0 = B/2A$ and represents the formation energy of the small polaron. It measures the thermal energy required for the annihilation of the polaron, and therefore delocalization of the electron (or hole). In contrast, the decrease in the purely electronic energy due to the coupling is

$$W_{el}(q_0) = \frac{B^2}{2A} = 2|W_0| \qquad (6.7)$$

The two contributions to the total energy can be visualized using Figure 6.11.

The transport behaviour of small polarons is quite complex and several different regimes can be operative depending on temperature. At very low temperatures, self-trapped carriers can tunnel from one lattice site to a neighbouring site, without altering the vibrational state of the crystal. This is the so-called coherent or band-type motion. However, the extreme narrowness of the small-polaron band implies that this type of motion may be eliminated because of disorder which exists in a real crystal. At higher temperatures, a hopping regime sets in, characterized by thermally activated jumps between neighbouring sites. At high enough temperatures, this regime involves multiphonon-assisted jumps and it can be described semiclassically. The energy of the small polaron then varies according to the instantaneous deformation of the lattice as in the simple model used above for Figure 6.11. In order for the carrier to tunnel, its electronic energy at a given site should reach momentarily the same value as that corresponding to the adjacent site (coincidence event). Then the jump rate manifests a thermally activated behaviour with an activation energy

$$E_m = \frac{W_0}{2} \qquad (6.8)$$

W_0 being the formation energy of the polaron already defined. A number of features can complicate the picture, such as the problem of adiabaticity of the process, jump correlation, acoustic mode contribution, etc. A simple, qualitative and clear presentation of this topic has been given by Emin[40], and specific application to experiments on transition-metal oxides by Honig[10]. Not only electrical conductivity but also Hall effect and Seebeck coefficient measurements depend on polaron behaviour and are also used to ascertain the occurrence of small polarons and the operative transport regime.

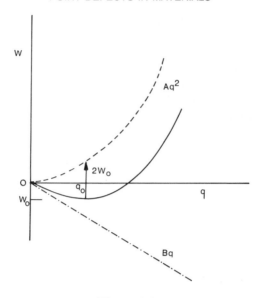

Figure 6.11.
Simple configuration coordinate model for a self-trapped carrier (small polaron).

6.9 POLARON MODELS OF COLOUR CENTRES

There are many examples of colour centres that can be described in terms of a small polaron trapped at a given lattice site or defect. The first theoretical model by Landau for the F centre was a polaron model. Even now, the F centre can be, in a way, considered as a polaron trapped at an anion vacancy. However, in polaron models the emphasis is laid on the electron−phonon coupling and consequently on lattice distortion associated to the trapped charge. A very simple polaron model[41] has been proposed that can be conveniently used to understand the optical properties of some colour centres in insulating crystals, primarily oxides. It may be of utility when more refined or detailed models are not available. The model, put forward by Schirmer, considers one electron (or hole) trapped at a given lattice site, presenting a number of equivalent positions related to each other by the symmetry operation of a point group. A typical example is the E' centre in SiO_2, i.e. an electron associated with one of the two opposite silicon atoms connected to an oxygen vacancy or the V^- centre in MgO, consisting of a hole shared by the six oxygen atoms around a Mg^{2+} vacancy. In principle, all equivalent positions are degenerate. However, when the tunnelling transitions of the electron (or hole) between these positions and the electron−phonon coupling are taken into account, the degeneracy is broken and the situation becomes more complex.

Let us consider, to facilitate the discussion, the simplest case of a linear defect with two equivalent positions, whose wavefunctions are designated as left $|l>$ and right $|r>$ (as in the E' centre). In the basis of these two states, the Hamiltonian of the system can be written as a 2×2 matrix:

$$H = Vq\begin{pmatrix} 1 & 0 \\ 0 & -1 \end{pmatrix} + \mathcal{J}\begin{pmatrix} 0 & 1 \\ 1 & 0 \end{pmatrix} + \left(\frac{\mu\omega_0^2 q^2}{2} + \tfrac{1}{2}p^2\right)\begin{pmatrix} 1 & 0 \\ 0 & 1 \end{pmatrix} \qquad (6.9)$$

where V is the linear electron–lattice coupling coefficient to a normal mode with coordinate q and frequency ω_0. The second term represents the tunnelling Hamiltonian, $\mathcal{J}$ being the exchange coupling coefficient, and the third term represents the kinetic and potential (elastic) energy of the normal mode. The eigenvalues of H can be easily obtained:

$$E_\pm = \pm(V^2 q^2 + \mathcal{J}^2)^{1/2} + \frac{\mu\omega_0^2 q^2}{2} + \frac{1}{2\mu}p^2 \qquad (6.10)$$

where the two first terms represent the adiabatic potential energy surfaces to hold the electronic charge. Therefore, the electronic wavefunctions correspond to the charge trapped at each of the two potential wells characterized by a lattice distortion $\pm q_0$. The minimum potential energy at the wells is E_{J-T} below that corresponding to the undistorted system ($q = 0$) (Figure 6.12). E_{J-T} is the so-called Jahn–Teller stabilization energy. At each well, a full series of vibronic states have to be considered. Then, optical transitions can be induced between the vibronic levels of the two wells (i.e. charge-transfer transitions). At high temperatures, on the assumption of the Franck–Condon approximation and small exchange coupling ($\mathcal{J} \ll E_{J-T}$), the absorption band shape is given by

$$\alpha(\omega) \propto \frac{1}{\omega} \exp\left[-\left(\frac{\hbar\omega - 4E_{J-T}}{\hbar w}\right)^2\right] \qquad (6.11)$$

with a maximum at the photon energy $E = \hbar\omega = 4E_{J-T}(E_{J-T} = V^2/2\mu\omega_0^2)$ and width

$$W = \frac{1}{8E_{J-T}\hbar\omega_0} \qquad (6.12)$$

The parameters of the model, E_{J-T} (or V) and ω_0, can be then obtained from the experimental spectral shape ($E_{J-T} = 1.4$ eV and $\hbar\omega_0 = 0.025$ eV for the E' centre in quartz).

The above model can be extended to more complex situations such as that for the V$^-$ centre in MgO which has octahedral symmetry. The absorption spectrum is now dichroic with two (σ and π) components, separated by an energy $2\mathcal{J}$. For the MgO V$^-$ centre, $E_{J-T} = 1.02$ eV, $\mathcal{J} = 0.19$ eV and $\hbar\omega_0 = 0.06$ eV.

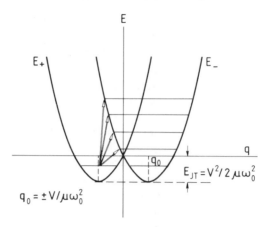

Figure 6.12.
Schematic diagram showing the two potential wells $E_\pm$ and vibronic transitions from the ground level of E_+ to excited levels of E_- ($\mathcal{J} \ll E_{J-T}$).

Similar models could be perhaps used to account for the broad colour centre bands found in other more complex oxides such as perovskites and $LiNbO_3$.

6.10 POINT DEFECTS AS PROBES FOR INVESTIGATING STRUCTURAL PHASE TRANSITIONS

It has been previously mentioned how some optical and paramagnetic probes have been used to monitor the clustering and precipitation processes occurring in cation-doped alkali halides during appropriate ageing treatments. In view of the high sensitivity of the paramagnetic probes to the lattice environment, they can also be used to investigate structural phase transitions. Most work has been performed with S-state ions, Fe^{3+}, Mn^{2+}, Gd^{3+} or Eu^{2+}, because the EPR spectra can be conveniently observed even at high temperatures. As relevant examples of the application of these techniques, we quote the structural phase changes in fluoroperovskites[42,43] and in oxide perovskites[44−46] such as $SrTiO_3$ and $LaAlO_3$. We now consider these two systems in some detail. $SrTiO_3$ experiences a second-order cubic-to-tetragonal transition at about 105 K brought about by a rotation of the TiO_6 octahedra around one of the $\langle 100 \rangle$ axes. $LaAlO_3$ also presents a second-order transition at 800 K although from cubic to trigonal symmetry. It can be described in terms of rotation of AlO_6 octahedra around a $\langle 111 \rangle$ axis. Muller and co-workers[45,46] were able to obtain, from the EPR spectra of the dopant Fe^{3+}, the temperature dependence of the rotation angle ϕ of the octahedra along the tetragonal axis of $SrTiO_3$ and the trigonal axis of $LaAlO_3$. They concluded that the rotation angle ϕ constitutes the order parameter of the transition.

In spite of the success of the method, we must be a little cautious because of the possible local distortions induced by the paramagnetic impurity probe which may disturb the information on the structural details of the "perfect" crystal transition.

Another interesting application of the EPR methods is provided by study[47] of the local position of the Fe^{3+} impurity in the various crystallographic phases of $BaTiO_3$. By using the Newman superposition model to analyse the EPR spin-Hamiltonian parameters, it has been concluded that the Fe^{3+} ion (substitutional for Ti^{4+}) remains at the centre of the oxygen octahedra in all three non-cubic ferroelectric phases (tetragonal, orthorhombic and rhombohedral). In other words, the Fe^{3+} ions do not follow the off-centre shift of the Ti^{4+} ions during the structural transitions of the material to lower symmetry phases. A similar behaviour has been found for Fe^{3+} during the transitions occurring in $KNbO_3$ and $PbTiO_3$. These results have relevance in relation to the role of impurities in structural phase transitions and particularly in illustrating the role of the electronic nature of the metal ion.

6.11 COOPERATIVE EFFECTS AMONG IMPURITIES—DIPOLE AND PROTON GLASSES

As has been already noted, dipole-type defects in dielectric crystals lend themselves very conveniently to the investigation of electrical as well as elastic interactions in a random system and to the elucidation of the possible formation of ordered or disordered arrangements (condensed phase) at low temperatures. The problem is still alive from both a theoretical and an experimental point of view. A number of experiments have been performed for about two decades on OH^--doped alkali halides[48,49] and more recently on CN^--doped alkali halides[50,51] and $KTaO_3$:$Li^{+38,52}$. For the latter system, nuclear magnetic resonance studies have unambigiously shown that Li^+ ions are in substitutial off-centre positions for K^+ and evenly distributed among the six equivalent orientations along the cube axis. The lithium displacement has been estimated to be around 0.86 Å.

For OH^- in KCl, cooperative effects have been detected for concentrations c of 10^{19} cm^{-3} or more and it has been suggested that the electric dipole system may freeze into an electric dipole glass. This has been primarily inferred from the occurrence of a prominent maximum in the dependence of the dielectric susceptibility with temperature, which markedly depends on OH^- concentration. The temperature T_f corresponding to the maximum susceptibility is usually identified as the freezing temperature (T_f $\approx$ 1 K for $\omega \approx 10^2$ Hz, and $T_f \approx$ 32 K for $\omega \approx$ 9.4 GHz). The possibility of an ordinary thermodynamic phase transition has been obtained within a mean-field approach to the problem of a random distribution of tunnelling dipoles. It should occur if the first moment (or width) of the dipole field distribution at $T = 0$ K is greater than the energy separation between tunnelling levels. However, the results showing the strong dependence of T_f on the frequency ω of the electric field and the continuous variation in the dielectric relaxation behaviour with temperature clearly point to a relaxational freezing mechanism and not to a static mechanism. The freezing process appears to be governed by the slowing down of random clusters containing relatively few dipoles. A distribution of relaxation times has to be used to account for the experimental data. As a conclusion, no transition point exists between the paraelectric and the glassy phases so that, depending on the time scale of the measurement, the dipoles may or may not appear frozen.

The same behaviour applies to CN^--doped alkali halides and $KTaO_3$:Li^+. One practical advantage of this latter material is that the "apparent" freezing temperature, as reflected by the maximum in the susceptibility, appears at quite convenient temperatures (about 50 K) for practical electrical frequencies (about kilohertz). The CN^--doped alkali halides, such as $(KCl)_{1-x}$:$(KCN)_x$ constitute a particularly interesting system, since the CN^- concentration can be continuously varied from $x = 0$ (pure KCl) to $x = 1$ for pure KCN. At low x values, the system shows the typical paraelectric and paraelastic behaviour. At intermediate CN^- concentrations, cooperative effects set in and the system behaves as a disordered dipole glass system. For $x \leqslant 1$, the mixed crystal undergoes a well-defined paraelectric-to-ferroelectric transition, where long-range order establishes domains of parallel ordered molecules. This transition remains well defined down to $x = 0.78$ for KCl or KBr.

Another interesting example of random interactions leading to a glassy phase is provided[53] by the hydrogen-bonded mixed compound of rubidium diphosphate and ammonium diphosphate: $Rb_{1-x}(NH_4)_xH_2PO_4$. Pure ammonium diphosphate undergoes a paraelectric-to-ferroelectric transition near 150 K; similarly, rubidium diphosphate also experiences a paraelectric-to-anti-ferroelectric transition near 150 K. However, for x values in the range $0.22 < x < 0.8$, cooling from a high temperature leads to a frozen-in random orientation of protons, resembling magnetic spin glasses. It appears that the NH_4^+ ions form hydrogen H bonds with nearby PO_4 tetrahedra, inhibiting the low-temperature acid–proton ordering.

REFERENCES

1 C. Monty, *Radiat. Eff.* **74**, 29 (1983).
2 B. Henderson and J. E. Wertz, *Defects in the Alkaline Earth Oxides*, Taylor and Francis, London (1977); *Adv. Phys.* **17**, 749 (1968).
3 C. R. A. Catlow and B. E. F. Fender, *J. Phys. C: Solid State Phys.* **8**, 3267 (1975).
4 C. R. A. Catlow, W. C. Mackrodt, M. J. Norgett and A. M. Stoneham, *Philos. Mag.* **35**, 177 (1977).
5 L. W. Hobbs, in V. M. Tuchkevich and K. Shvarts (eds), *Proceedings of the International Conference on Defects in Insulating Crystals, Riga*, Springer, Berlin (1982).
6 A. N. Cormack, C. R. A Catlow and P. W. Tasker, *Radiat. Eff.* **74**, 237 (1983).
7 W. L. Worrel, in S. Geller (ed.), *Solid Electrolytes, Topics in Applied Physics*, Vol. 21, Springer, Berlin (1977).
8 A. J. Nowick, *Comments Solid State Phys.* **9**, 85 (1979).
9 C. R. A. Catlow, in O. T. Sorenson (ed), *Non-Stoichiometric Oxides*, Academic Press, New York (1984).
10 J. M. Honig, in A. Dominguez-Rodriguez, J. Castaign and R. Marquez (eds), *Basic Properties of Binary Oxides*, University of Sevilla Press, Seville (1983).
11 P. W. Levy and G. J. Dienes, *Report on Conference on Defects in Crystalline Solids, Bristol*, Physical Society, London, p. 256 (1955).
12 E. W. J. Mitchell, J. D. Rigden and P. D. Townsend, *Philos. Mag.* **5**, 1013 (1960).
13 P. W. Levy, *Phys. Rev.* **123**, 1226 (1961); *Discuss. Faraday Soc.* **31**, 118 (1961).
14 J. H. Crawford, *Semicond. Insul.* **5**, 599 (1983).
15 H. H. Tippins, *Phys. Rev. B* **1**, 126 (1970).
16 B. D. Evans and M. Stapelbroek, *Phys. Rev. B* **18**, 7089 (1978).
17 S. Y. La, R. H. Bartram and R. T. Cox, *J. Phys. Chem. Solids* **34**, 1079 (1973).
18 B. G. Draeger and G. P. Summers, *Phys. Rev. B* **19**, 1172 (1979).
19 V. N. Abramov, B. G. Ivanov, A. I. Kuznetsov, I. A. Meriloo and M. I. Musatov, *Phys. Status Solidi (a)* **48**, 287 (1978).
20 V. N. Abramov, M. G. Karin, A. I. Kuznetsov and K. K. Sidorin, *Sov. Phys.—Solid State* **21**, 47 (1979).
21 L. S. Welch, A. E. Hughes and G. P. Pells, *J. Phys. (Paris), Colloq. C 6* **41**, 533 (1980).
22 G. P. Pells, *J. Nucl. Mater.* **122−123**, 1338 (1984).
23 K. Atobe, N. Nishimoto and M. Nakagawa, *Phys. Status Solidi (a)* **89**, 155 (1985).
24 M. J. Springis and J. A. Valbis, *Phys. Status Solidi (b)* **123**, 335 (1984); **124**, K165 (1984); **132**, K165 (1985).
25 T. J. Turner and J. H. Crawford, *Phys. Rev. B* **13**, 1735 (1976).
26 M. J. Coteron, A. Ibarra and M. Jimenez de Castro, *Cryst. Lattice Defects Amorphous Mater.*, **17**, 77 (1987).
27 J. H. Kennedy, in S. Geller (ed.), *Solid Electrolytes, Topics in Applied Physics*, Vol. 21, Springer, Berlin (1977).
28 H. Schulz, Crystal structures of fast ion conductors, *Annu. Rev. Mater. Sci.* **12**, 351 (1982).
29 G. N. Greaves, *Philos. Mag. B* **37**, 447 (1978).

30 D. L. Griscom, *J. Non-Cryst. Solids* **40**, 211 (1980); **73**, 51 (1985); *Proceedings of the Thirty-third Frequency Control Symposium*, Electronics Ind. Ass., Washington DC, p. 98 (1979).
31 L. E. Halliburton, *Cryst. Lattice Defects Amorphous Mater.* **12**, 163 (1985).
32 S. McKeever, *Radiat. Prot. Dosimetry* **8**, 81 (1984).
33 L. H. Abu-Hassan, *D. Phil. Thesis*, Sussex (1987).
34 P. Gunter, *Phys. Rep.* **93**, 199 (1982).
35 L. Arizmendi, J. M. Cabrera and F. Agulló-Lopez, in A. Dominguez-Rodriguez, J. Castaign and R. Marquez (eds), *Basic Properties of Binary Oxides*, University of Sevilla Press, Seville (1983).
36 T. Holstein, *Ann. Phys. (NY)* **8**, 343 (1959).
37 T. L. Gilbert, Lecture notes for the NATO summer school, Ghent (1966); quoted in M. N. Kabler, in J. H. Crawford Jr and L. M. Slifkin (eds), *Point Defects in Solids*, Plenum Press, New York (1972).
38 A. M. Stoneham in A. Dominguez-Rodriguez, J. Castaign and R. Marquez (eds), *Basic Properties of Binary Oxides*, University of Seville Press, Seville (1983).
39 A. Zylbersztejn, *Appl. Phys. Lett.* **29**, 778 (1976).
40 D. Emin, *J. Solid State Chem.* **12**, 246 (1975).
41 O. F. Schirmer, *J. Phys. (Paris), Colloq.* C6 **41**, 479 (1980).
42 F. A. Modine, E. Sonder and W. Unruh, *Phys. Rev. B* **10**, 1623 (1974).
43 J. J. Rousseau, A. Leble, J. Y. Buzaré, J. C. Fayet and M. Rousseau, *Ferroelectrics* **12**, 201 (1976).
44 L. Rimai and G. A. de Mars, *Phys. Rev.* **127**, 702 (1962).
45 K. A. Muller, W. Berlinger and F. Waldner, *Phys. Rev. Lett.* **21**, 814 (1968).
46 Th. Von Waldkirch, K. A. Muller and W. Berlinger, *Phys. Rev. B* **5**, 4324 (1972).
47 E. Siegel and K. A. Muller, *Phys. Rev. B* **20**, 3587 (1979).
48 R. C. Potter and A. C. Anderson, *Phys. Rev. B* **24**, 677 (1981).
49 M. Saint-Paul, M. Mesa and R. Nava, *Solid State Commun.* **47**, 183 (1983).
50 A. Loidl, R. Feile and K. Knorr, *Phys. Rev. Lett.* **48**, 1263 (1982).
51 S. Bhattacharya, S. R. Nagel, L. Fleishman and S. Susman, *Phys. Rev. Lett.* **48**, 1267 (1982).
52 V. T. Hochli, *Phys. Rev. Lett.* **48**, 1494 (1982).
53 E. Courtens, *Ferroelectrics* **53**, 227 (1984).

7
Point Defects in Semiconductors and Metals

7.1 POINT DEFECTS IN SEMICONDUCTOR MATERIALS

The need for an understanding of the role of defects in semiconductor devices is obvious and from the academic viewpoint the commercial pressures have generated excellent test systems of very high purity and perfection. It will be seen that in all cases it has been possible to identify several vacancy centres, in various charge states, for these semiconductors together with a limited number of more complex defects involving impurity ions. From the viewpoint of the semiconductor technology the progress is much greater as techniques such as deep-level transient spectroscopy (DLTS) clearly indicate the presence of defects and their position in the band gap, even if an atomistic model for the structure of the defect cannot be offered. This empirical approach is valuable for device production but is not pursued in this chapter.

The electronic structure of semiconductors is often described in terms of covalent bonds and so defect structures are conveniently discussed by reference to broken or dangling bonds, i.e. in a language close to that of molecular physics.

As for ionic materials, atomic relaxation effects play an important role and, in particular, some examples of Jahn–Teller distortions in determining the final structure of the defect are discussed in detail. One main consequence of the atomic relaxations is the difference between the optical and thermal energies required to promote a defect centre to an excited state or to the conduction band. We generally speak of unrelaxed and relaxed electronic states or levels of the defect depending on whether allowance has been made or not for the atomic relaxation associated with it.

Defect centres can often act as either electron donors or electron acceptors, generally characterized by their level in the electronic gap of the material. The description of such a centre in terms of gap levels is misleading, however, since because of the relaxation effects the "donor" and "acceptor" levels should be different from each other.

Many defect centres can adopt several charge states. As a general rule, each additional electron to the centre implies a donor level with a lower ionization energy because of the electronic repulsion energy U. However, the electron lattice coupling causing the atomic relaxation may be sufficient to invert the location of the levels, providing a centre having an apparently negative U energy[1].

After a brief account of the Jahn–Teller effect and U relaxations, in the following sections, we deal with defects in silicon and GaAs as examples of single-element and compound semiconductors. These are definitely the most successfully studied semiconductors. Discussions of earlier work include many general references[2–4] and specific items on germanium[5,6], diamond[7], silicon[8–11] and the III–V[12,13] and II–VI[14] compounds.

7.2   THE JAHN–TELLER EFFECT

The Jahn–Teller effect is now discussed, although examples were mentioned in preceding chapters. One important consequence of the electron–lattice coupling is the spontaneous distortion of a molecular centre to a lower-symmetry configuration, i.e. the Jahn–Teller effect. This occurs when the centre, in a degenerate electronic level, can reduce the total energy, by occupying the lowest split electronic level induced by the distortion. Numerous examples of this effect have been reported for defect centres in both insulating and semiconductor materials. It is responsible for relevant features of the observed optical and electron paramagnetic resonance (EPR) spectra.

The occurrence of a Jahn–Teller distortion can be understood in a simple way by considering the linear and quadratic terms of the electron–lattice Hamiltonian H_{e-L}. For a normal mode coordinate q, the terms are written as $H_{e-L} = Aq + Bq^2$. The energy levels depend on q as illustrated in Figure 7.1, which shows the splitting by the linear term and the increase in energy associated with the quadratic (elastic) term. For the lowest level the combination of linear and quadratic terms give rise to a minimum in the total energy at a value q_0, indicating the amount of spontaneous distortion. The lowering of the energy is designated as the Jahn–Teller stabilization energy E_{J-T}.

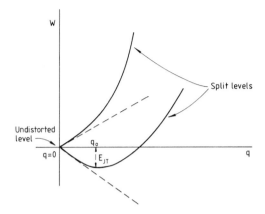

Figure 7.1.
Diagram to illustrate the Jahn−Teller effect: $---$, the splitting from the linear term; ——, the total splitting given by $H_{e-L} = Aq + Bq^2$.

In the more realistic situation the Jahn−Teller effect causes several equivalent minima or distorted positions that keep the symmetry of the centre. The minima are separated by an energy barrier E_B, whose height should be compared with the vibrational quantum energy $\hbar\omega$. If $\hbar\omega \ll E_B$, the distortion is permanent and we are dealing with the "static" Jahn−Teller effect. A reduction of symmetry is effectively observed with suitable experimental techniques (e.g. optical and EPR). If $\hbar\omega \approx E_B$, it is not possible to identify the centre in one of the minima, but the state is a linear combination of those corresponding to each minima. The effect is then "dynamic". Although a lowering of symmetry has not taken place, the Jahn−Teller effect may markedly influence the observed values of physical parameters. The difference between the static and dynamic cases is often subtle and it depends on the experimental method used for the investigation.

7.3   NEGATIVE *U* RELAXATIONS

With a range of possible charge states for the defects it might be assumed that the lowest charge state is favoured. This need not be so. Distortions of the defect structure can favour the capture of a second charge which then appears as though there is an attraction between two electrons (or holes). Such imperfections are termed negative U defects. Several examples are now well documented and the features may be quite general in insulators and semiconductors. Conventionally, we expect that electrons are progressively less tightly bound for the charge sequence D^+, D^0, D^-. If these defects are in a dilute system and interact via the conduction band, then charge exchange is possible which can stabilize the system and lower the total free energy. As an example if

$$D^0 + D^0 \rightarrow D^+ + D^- + energy$$

then the pair of charged defects will appear at the expense of the neutral species. This situation is clearly obtained if the negative U relaxation inverts the D^0 and D^- levels, placing D^0 above D^-. Two silicon defects which fit this pattern are the intrinsic silicon vacancy and the boron interstitial. The boron is incorporated as an equal mixture of B^+ and B^- ions[15], and the vacancies as V^{2+} and V^{2-} (i.e. with two charges per site). This is not confined to silicon and examples are quoted for other semiconductors as well[16].

Such pairing of features may lead to problems in the measurements of the Fermi level and the position of the defect states relative to the energy bands as for example the presence of equal quantities of D^+ and D^- will yield Hall value data for a state which is midway between the two actual levels. Once the problem is suspected, then we can artificially change the equilibrium to reveal the true levels. For boron in silicon this has been done using DLTS measurements whilst the crystal was photoionized to change the B^- population to the $B^0 + e^-$ state. True positions of B^0 and B^+ were then measured[17].

7.4   POINT DEFECTS IN SILICON

Silicon should be the ideal system for defect studies because it is a single-element lattice, and it is prepared in the highest state of purity of any single crystal because of the commercial interest. Positive defect identification is most fruitful by the resonance methods of EPR and electron–nuclear double resonance (ENDOR) and yet again pure silicon is ideal as it has a major isotope ^{28}Si which is 95.3% abundant with a zero nuclear spin. The other isotope, ^{29}Si, at 4.7% abundance with $S = \frac{1}{2}$ is sufficient to give the EPR signals from a well-dispersed dilute system without too many hyperfine interactions.

The resonance techniques have led to an extensive literature on the identification of intrinsic, extrinsic and complex defects in silicon. EPR has been extensively used for defects with unpaired spins and by optical excitation some systems with paired spins have been viewed. Corbett et al.[18] have referenced much of the earlier literature in a review of EPR defects in silicon and discuss intrinsic vacancy defects ranging from monovacancies to pentavacancies and interstitials as isolated features or as dumb-bells. Inclusion of impurity and intrinsic–impurity features leads to a list of some 70 defects separately identified by EPR. Not surprisingly, not all researchers offer the same interpretations of the spectra. The book by Watts[19] gives good coverage of defect structures in silicon.

Progression from EPR to ENDOR to study the distortions of the lattice sites in the neighbourhood of the defect has been very effective and, for example, with shallow donors in silicon it has been possible to resolve the extent of the wavefunctions over some 30 neighbouring atomic shells[17,20]. This analytical precision gives a unique description of the defect model but as yet it is unmatched by theoretical modelling as the scale of a calculation involving some thousands of atoms is too great. The ENDOR detail also suggests that the more localized, but specific, methods such as perturbed angular correlation (PAC) or extended X-ray absorption fine-structure spectroscopy (EXAFS) still leave much to be desired. Nevertheless, this "ideal" silicon system fully reveals the difficulties of defect discussions in insulators and semiconductors. It is clearly inappropriate to discuss "point" defects as though they exist in isolation in perfect crystal lattices. There are no crystals which have the requisite perfection; moreover most crystals contain so many grain boundaries, dislocations and extrinsic impurities that a truly isolated point defect is very rare.

7.4.1 Impurities

The purity of silicon has been cited as an example of the perfection which can be achieved with a sufficiently determined effort (see Chapter 1). Yet even in this case, silicon is grown by two processes. In Czochralski growth a seed is in contact with molten silicon in a furnace; in the alternative floating-zone method, the furnace walls are at lower temperatures as the liquid phase is maintained by surface tension between two solid silicon regions. For semiconductor work, both methods are useful and produce electrically active impurity concentrations of shallow donors at levels as low as 0.1 ppb. This specification ignores non-electrically active sites. In particular, the Czochralski growth can give oxygen contamination of 10 ppm and even the floating-zone method gives 0.1 ppm from oxygen in the vacuum system. "Perfection" is thus a relative term depending on the application of the material and for defect studies there are no perfect test systems.

7.4.2 Oxygen in silicon

Oxygen is such a common impurity in silicon that it is worthy of special mention. Interstitial oxygen acts as a trapping site for vacancies and then moves onto the lattice site. The defect is termed the A centre. In the neutral charge state with just two electrons there is no EPR signal. Optical pumping produces an EPR signal from the unpaired spin state. Since ^{16}O has a zero spin nuclear moment, details of hyperfine interactions are only obtained by doping with the isotope ^{17}O $(I = \frac{5}{2})$[21].

Capture of an additional electron at the site produces an EPR signal from the ground state. Somewhat surprisingly the use of ^{17}O suggests that the oxygen is not directly coupled to the EPR signal. This is understandable from Figure 7.2 in which one sees that the oxygen bridges on a pair of the silicon neighbours and the oxygen atom is pulled off centre towards the silicon atoms. The extra electron does not fit into this part of the defect but instead is localized on the other pair of silicon atoms. The electron level resembles a normal lattice bond and falls at 0.17 eV below the conduction band. Because the electron does not strongly interact with the oxygen nucleus the connection between the oxygen atom and the EPR centre is not proven at this stage. However, there is an infrared vibration at 828 cm^{-1} (12 μm) observed in parallel with the EPR signal. This frequency is characteristic of an Si$-$O$-$Si vibration and is confirmed by the use of ^{18}O which causes a mass shift in frequency to 791 cm^{-1} (9 μm) as expected. Conclusive EPR$-$ infrared correlation is found by applying a uniaxial stress to the silicon at high temperatures. The bent Si$-$O$-$Si bonds distort but recover once the stress is released. The recovery kinetics of both the EPR and the infrared signals follow identical Arrhenius plots with an energy[22] of 0.38 eV.

7.4.3   Intrinsic defects

In order to produce simple intrinsic defects, i.e. isolated vacancies, interstitials or divacancies in dilute concentrations, precisely the same approach has been used with silicon as is discussed for metals such as copper, namely irradiation with relativistic electrons to displace lattice atoms in crystals held at temperatures where other defects are immobile. As with the metals the same general pattern is seen; the interstitials are mobile at very low temperatures, and vacancies and divacancies are liberated closer to room temperature. The annealing stages have substages corresponding to association of intrinsic and impurity features but these are more complex in the covalent silicon lattice than in metals as the silicon defects exist in a range of charge stages. Particular substages appear dominant in n- or p-type material. For example, after γ irradiation at 4 K, direct recovery between vacancy$-$interstitial pairs proceeds at 70 K in n-type silicon but at 150$-$200 K in p-type silicon. The advantage of optical measurements is useful in sensing the more complex or longer-range diffusion effects; for example the photoconductivity spectrum of pure silicon gives a peak near 3.8 μm. The width of this feature depends on the energy of the damaging electrons. At lower energies where the collisions only nudge the interstitials from the vacancy,

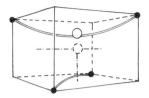

Figure 7.2.
Substitution of oxygen into a silicon site leads to relaxation and new bonds linking Si−O−Si and Si−Si as shown: ●, silicon atoms; ○, oxygen atoms; ◌, silicon vacancy.

the peak is narrow whereas, for energetic electrons where the vacancy− interstitial pair is well separated, the photoconductivity peak broadens.

(a)   Vacancy centres.   It might be expected that the single silicon vacancy would be a simple example to model; however, the vacancy is surrounded by four equivalent neighbouring atoms. Charge capture into this environment can then induce a relaxation which lowers the symmetry of the defect as expected for a Jahn−Teller distortion. For silicon, V^{2+}, V^+, V^0, V^- and V^{2-} charge states are all believed to exist with V^+ and V^- recorded by EPR. Jahn−Teller-type distortions are stabilized by the presence of im- purities and, with the addition of germanium, V_{Ge}^+ and V_{Ge}^- centres appear.

In Figure 7.3 a sketch of the model of the simple vacancy site for silicon in which the four dangling silicon bonds pointing to the vacancy give rise to a ground a_1 electron orbital and a triplet excited orbital t_2 (Figure 7.4). Five different charge states can occur from V^{2+} to V^{2-}; their electronic struc- tures are obtained by distributing the available electrons between the a_1 and t_2 orbitals. For the neutral vacancy V^0, four electrons (one from each surrounding silicon) are available and adopt the $a_1^2 t_2^2$ configuration. Since the a_1 and t_2 electrons have paired spins, V^0 cannot be studied by EPR spectroscopy. In contrast, V^+ ($a_1^2 t_2$) and V^- ($a_1^2 t_2^3$) vacancies have a net spin and are EPR active (see Chapter 10). In fact, the situation is more complicated because of the Jahn−Teller distortion undergone by the t_2 triplet level. Depending on the vibrational modes to which the electrons couple, tetragonal and trigonal distortions are possible, both leading to the splitting of t_2 into a ground singlet and a doublet relaxed level. For the V^+ centre, the distortion is tetragonal (D_{2d} symmetry) with equivalent orienta- tions along the $\langle 100 \rangle$ axis. The Jahn−Teller stabilization energy E_{J-T}, i.e. the lowering in energy with regard to the unrelaxed structure, is 0.19 eV (see Figure 7.1). Reorientation of the centre can take place with an activa- tion energy of 0.013 eV. The V^0 vacancy also presents tetragonal symmetry,

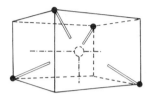

Figure 7.3.
Model for the vacancy in silicon, showing the four dangling bonds pointing to the vacant site.

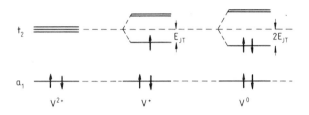

Figure 7.4.
Electron energy level diagram for various charge states of the vacancy in silicon (after J. Bourgoin and M. Lannoo, *Point Defects in Semiconductors II*, Springer-Verlag (1983)).

but the Jahn−Teller stabilization energy is about double that for V, i.e. $2E_{J-T}$ (Figure 7.4). The negative vacancy V^-, with an EPR-active configuration, shows a mixed trigonal−tetragonal distortion. Reorientations about either the trigonal or the tetragonal axes are separable as they show distinct activation energies of 0.008 eV and 0.072 eV, respectively. Finally, the V^{2-} vacancy ($a_1^2t_2^4$ configuration) is not observable by EPR in the ground state but can be optically excited and viewed whilst in the high state. In the ground state the defect is mobile with an activation energy of 0.18 eV.

The vacancy centre in silicon provides an interesting example of the negative U centres mentioned in section 7.3. The unrelaxed donor levels for the V^+ and V^0 vacancies are at 0.32 eV and 0.57 eV, respectively, above the top of the valence band, whereas the V^{2+} centre has a resonant level lying within the valence band. The Jahn−Teller relaxed donor levels

$$E^+ \text{ (relaxed)} = E^+ \text{ (unrelaxed)} - E_{J-T}$$

$$E^0 \text{ (relaxed)} = E^0 \text{ (unrelaxed)} - (4E_{J-T} - E_{J-T})$$

$$= E^0 \text{ (unrelaxed)} - 3E_{J-T}$$

lie at 0.13 eV (V^+) and about 0.0 eV (V^0) above the top of the valence band

(Figure 7.5), therefore being inverted with regard to the unrelaxed locations. A consequence of this level arrangement is that either V^{2+} or V^0 should be dominant vacancy defects in silicon, depending on the location of the Fermi level. In fact, V^+ centres tend to disappear via the energetically favoured reaction

$$V^+ + V^+ \rightarrow V^0 + V^{2+}$$

As an illustration of a second intrinsic silicon defect, Figure 7.6 gives the Watkins–Corbett[23] model of the divacancy. Two silicon atoms are removed from adjacent sites and the neighbouring atoms relax and form new bond linkages, shown here as bent bonds. Electrons resonate between the two vacancies. For the covalently bonded material, several charge states can be accommodated and, if there is a single electron, the divacancy has an effective positive charge $(VV)^+$. Analysis of the hyperfine interactions gives some 60% of the electron density directed along the axis of the line between the two vacancies whilst a further 20% of the electronic charge is spread uniformly over the four nearest neighbours.

An EPR signal is also derived from the charge state of $(VV)^-$ where there are two electrons along the divacancy axis and a third electron is forced into an antibonding orbital.

(b) Interstitial defects. Much less structural information is available for self-interstitial than for vacancy defects in silicon. A major reason is that no EPR spectra have been so far associated with any charge state of the self-interstitial. Moreover, the interstitials produced during irradiation appear to diffuse rapidly through the lattice and presumably form complex centres with impurities and other defects. So, although a variety of interstitial defects have been invoked to account for the annealing features of the radiation damage, the situation is still very unsatisfactory. It should be

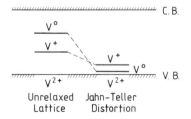

Figure 7.5.
Location of the unrelaxed and Jahn–Teller relaxed donor levels for the V^0 and V^+ states of the vacancy in silicon.

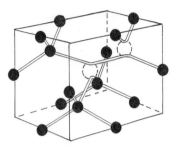

Figure 7.6.
A model of the divacancy in silicon. Note the new long-range bond linkage.

mentioned that some detailed calculations suggest a split configuration along $\langle 100 \rangle$ for the isolated self-interstitials, similar to the case of metals.

7.4.4 Association of defects

Intrinsic–impurity defect interactions are well documented in silicon where the rapid interstitial motion at 4 K is attributed to radiation-enhanced diffusion. It is not clear whether this is via an excited state, an exchange between excited interstitials and lattice atoms, or a charge exchange which allows the moving entity to switch between +, 0 and − states. When substitutional impurities are present, the interstitials move onto the lattice site and displace the impurities by reactions such as

$$Si_{int} + Al_{sub} \rightarrow Al_{int} + Si_{sub}$$

7.4.5 Models for defect-assisted diffusion in silicon

Diffusion processes are of major importance in silicon technology as they are implicit for many methods of impurity doping. Even if impurity ions are introduced by implantation, diffusion may occur in the subsequent annealing stages as the radiation damage is removed. Intuitively, we expect the mechanism of diffusion to depend on the type of lattice site that the impurity will occupy in equilibrium. The familiar dopants of boron, aluminium, gallium and phosphorus are all substitutional and several small ions, e.g. lithium, occupy interstitial sites. Neutral impurity species of group IV elements (carbon, germanium and tin) will presumably behave in a similar fashion to the host silicon.

At first sight, it might be imagined that the only intrinsic point defects which modify impurity diffusion rates are silicon vacancies. However, the silicon self-interstitial is an equally fundamental defect but only relatively recently has the suggestion[24] been accepted that it can play a role in high-temperature diffusion. The reason for disregarding silicon interstitials was that, when they are formed as isolated point defects or as vacancy–interstitial pairs, they are mobile in both n- and p-type silicon below 150 K. However, this ignores the possibility that self-interstitials can be formed by other routes and, in association with impurity ions, may remain as stable impurity–interstitial complexes to a high temperature.

The numerous discussions of defect-assisted diffusion most obviously include impurity–vacancy pairs as this provides (Figure 7.7) a low-energy route for movement of the impurity into the vacancy site. Subsequent movement of the vacancy is followed by the impurity. If there is a vacancy or impurity concentration gradient, the net effect is an enhanced diffusion rate along the gradient. Interstitial-assisted diffusion processes can be summarized[13] either by the Frank–Turnbull[25] mechanism or by the "kick-out" mechanism[26]. A variant of the Frank–Turnbull process for charged interstitials is termed the Longini[27] mechanism. Figure 7.8 gives a sketch of how the two mechanisms may proceed. In the Frank–Turnbull route there is a balance between the interstitial impurity A_i, the vacancy V and substitutional impurities A_s with

$$A_i + V \rightleftharpoons A_s$$

The kick-out model assumes that the interchange of impurities between interstitial and substitutional sites is accompanied by the formation of interstitial silicon I. The impurity and interstitial do not dissociate but allow a diffusion progression by repeated interchange:

$$A_i \rightleftharpoons A_s + I$$

The two processes predict, as for example discussed by Gosele[13], remarkably different changes in the diffusion coefficient for substitutional impurities which migrate by an interstitialcy mechanism. In the Frank–Turnbull

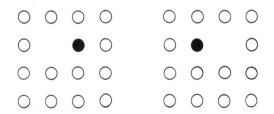

Figure 7.7.
Vacancy-assisted diffusion. The labelled atom (●) moves to the left as the vacancy jumps to the right in this sketch.

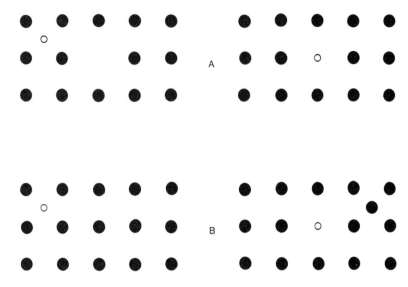

Figure 7.8.
(a) The Frank–Turnbull and (b) the kick-out mechanisms of interstitial diffusion:
○, initial interstitial atom; ●, atoms originally on lattice sites.

model the ingress of a surface impurity proceeds with a concentration profile against depth that is a classical error function. Thus the form is the same as normal random jumps but the effective diffusion coefficient is higher for the A–V complex.

By contrast the kick-out diffusion model leads to a strongly concentration-dependent diffusion coefficient. One consequence is that the effective self-interstitial diffusion coefficient increases with decreasing concentration. The resulting profile is flatter than the error function.

7.4.6   Examples of defect-modified diffusion

Two impurities commonly found in silicon devices as the result of fabrication treatments are gold and oxygen. These arise from the surface contacts or SiO_2 passivating layers. Gold produces energy levels in midgap and so can be viewed either favourably or not depending on the specific device. For example it is useful if a short lifetime is required for the carriers but gold is a problem if high minority carrier concentrations are needed.

Gold diffuses into silicon fairly readily and typically reaches depths of 10 μm in 1 h at about 900°C. The measurements of the concentration profile yield[28] an effective diffusion coefficient and the form of the profile allows the alternative enhancement mechanisms to be distinguished between. At 900°C the profile supports the kick-out model but at lower temperatures of 750°C there is an error function profile predicted by the Frank–Turnbull model. Consequently it must be concluded that *both* mechanisms are operative but dominate in different temperature ranges. This supports the idea that, in the diffusion of impurities at high temperatures, the role of self-interstitials must be considered.

Oxygen impurity effects are equally complex and do not have a universal effect on all impurities. In some cases there is an enhancement of the diffusion rate whereas for other elements the rate decreases, even in a single sample[13]. In general the presence of surface oxides enhances the subsurface diffusion of boron, indium, aluminium, gallium or phosphorus but hinders antimony migration. It may be argued that the antimony moves by a coupled $Sb-Si_{vac}$ motion whereas all the others are assisted by self-interstitials. If self-interstitials are generated during oxide growth, they will annihilate vacancies and suppress the antimony movement but favour that of the other ions.

7.5 AMORPHOUS SILICON

Amorphous silicon[29] is gaining widespread technological relevance, such as in photovoltaic cells for solar energy applications. These applications rely on the possibility of impurity doping so that junctions with different conductivity characters could be manufactured. In early amorphous silicon prepared by evaporation or sputtering, the electronic behaviour could not be substantially modified by doping, therefore preventing its consideration as a useful semiconductor material. After 1975, amorphous silicon prepared by the "glow discharge" method to decompose silane (SiH_4) has permitted an efficient control of the Fermi level and conductivity behaviour by appropriate doping. The reason for that lies in the much higher density of states in the material prepared by evaporation or sputtering. This implies an effective pinning of the Fermi level and inhibits the influence of impurity doping.

7.6 DEFECTS IN GaAs

GaAs is a semiconductor of major importance as it is a direct-gap material and hence can be used efficiently to generate light in light-emitting or laser devices. Variations in p–n junction structures with doping to form hetero-junction or quantum well devices are now routinely used for a wide range of near-infrared optical components. There is also an effort to use GaAs in purely electrical structures which could act as higher-speed replacements to silicon devices. Because of these industrial applications, much effort has been devoted to the production of pure GaAs and defect studies of the material have evolved in parallel. The defect work has generated an extensive literature[12,30,31] but in the main it has only reached the stage that different defects can be identified but specific atomistic models are not possible. This is in marked contrast with silicon and the reason is that GaAs is not a suitable candidate for EPR studies. All three isotopes (^{75}As at 100% abundance, ^{69}Ga at 60% and ^{71}Ga at 40%) have a nuclear spin I of $\frac{3}{2}$. Thus, defect signals are masked or unassignable. Recently, detailed ENDOR work and optical excitation are helping to overcome the problems for some defects[32]. In this section therefore, we concentrate on a brief catalogue of defect levels sensed by electrical measurements and impurity features detected by infrared absorption. From electrical data and variations in stoichiometry by control of the arsenic gas overpressure during growth, it is generally agreed that the major intrinsic defects result from the arsenic sublattice. Vacancies and interstitials in the gallium sublattice are thought to recombine rapidly and so do not contribute to device properties.

7.6.1 Electrical measurements with GaAs

Electrical measurements used include simple resistivity and DLTS. DLTS is sensitive and has the advantage of resolving different defects as signals which appear at characteristic temperatures. For simplicity, megaelectron-volt electron beams have often been used to generate point defects and, in general, some eight or nine electron traps are revealed in samples irradiated at about 4 K. Table 7.1 lists these. A further three levels (P_1, P_2 and P_3)[33] are detected in samples annealed to 500 K (the so-called stage III) and these have characteristic DLTS peaks near 200, 280 and 350 K. Table 7.2 lists the hole traps. It is of course not possible to sense all these traps in a single specimen and the electron features are only detected in n-type material and the hole traps in p-type material. The E traps are insensitive to the type of material and the position of the Fermi level. However, the relative strengths of the signals from the traps may be influenced indirectly by the Fermi level as this can alter the mobility of interstitial arsenic. For example lowering of the Fermi level can favour interstitial trapping at impurity sites such as B or C or formation of antisite defects[34]. The H traps differ for crystals grown by liquid-phase or vapour-phase epitaxy and as a result of changes in the Fermi level. Many of the same traps are seen after ion beam or neutron irradiation[35,36].

The electron traps are thought to demonstrate at least four types of $As_{vac}-As_{int}$ pairs and the signals are insensitive to the presence of impurities, whereas the hole traps are all influenced by impurities and so are more complex centres. Traps E_1 and E_2 arise from the same defect centre. E_7 and

Table 7.1.
Electron traps detected after electron irradiation of GaAs

Label	$T_{observed}$ (K)	Depth (eV)	E_{anneal} (eV)
E_1	20	0.045	1.5–1.6
E_2	60	0.14	1.5–1.6
E_3	160	0.30	1.5–1.6
E_4	310	0.76	1.5–1.6
E_5	360	0.96	1.5–1.6
E_6			
E_7	40		0.7
E_8	80		
E_9	110		

Table 7.2.
Hole traps detected after electron irradiation of GaAs

Label	$T_{observed}$ (K)	Depth (eV)	E_{anneal} (eV)
H_0	50	0.06	
H_1	150	0.29	≈1.5
$H_{2(A)}$	190	0.41	
$H_{2(B)}$	340	0.71	

E_9 concentrations correlate with resistivity changes in the annealing stages I and II (at 230 and 280 K). The displacement thresholds determined during electron irradiation vary from about 9 to 20 eV. These values are sensitive to crystal orientation and for example irradiation along the $\langle 111 \rangle$ direction produces different results for the As–Ga and the Ga–As directions. This is consistent with the displacement of arsenic atoms. Production of traps E_7 and E_9 requires large threshold energies and so these are tentatively ascribed to As_{vac}–Ga_{vac} pairs. Additionally the activation energy for annealing of traps E_1–E_5 is similar at about 1.5 eV; hence it may be inferred that they are all removed by a single diffusing species. The prime candidate for this is an As_{int} which is thermally liberated.

Thus, to summarize, the defects in the arsenic sublattice generate a recognizable set of defect states but we cannot justify any detailed models for the defects from the electrical measurements.

7.6.2 Optical detection of impurity sites in GaAs

Whilst there has been slow progress in identifying intrinsic defects in GaAs, there have been excellent results from optical absorption measurements for many of the standard impurities. Dopants which enter the crystal during growth are boron, carbon and nitrogen. Another major impurity is silicon which is ion implanted to generate conducting channels. The technique which has offered the successful identification is that of localized vibrational modes (LVMs). The principles are well documented[37-39] and we are seeking the resonances determined by the ion masses and the coupling between the ions. For many impurity ions there are a multiplicity of lines caused by several isotopes. These features can be enhanced by doping with selected isotopes. Limitations in recording these LVMs are set by the restrahl band of the GaAs host and the infrared free-carrier absorption. Although we can sometimes minimize the free-carrier signals by suitable control of the Fermi level, the more common approach is to use fast (about 2 MeV) electron irradiation damage to introduce deep-lying defect levels to remove the free carriers. This is of course not a perfect approach as it inevitably causes the production of arsenic and gallium vacancies and interstitials which interact with the impurity ions and form even more complex defects. For GaAs, such defects frequently include antisite occupancy. In recent years, improvements with Fourier transform infrared absorption techniques have provided very high resolution (less than 0.05 cm^{-1}) from rapidly acquired spectra. At least 50 impurity-related defect complexes have now been identified in GaAs[37-39].

As examples of the sites and the clarity of the resolution, it should be noted that boron can occupy either the gallium or the arsenic site and the LVM resonances for $^{10}B_{Ga}$, $^{11}B_{Ga}$, $^{10}B_{As}$ and $^{11}B_{As}$ occur at 540 cm^{-1}, 517 cm^{-1}, 628 cm^{-1} and 601 cm^{-1}, respectively. Even for the heavier impurities where isotopic mass differences are less obvious, the LVM is adequate. In an example of double doping with both silicon and germanium, it was noted that the two impurities could associate onto a pair of lattice sites to form a $(Si_{Ga}-Ge_{As})$ complex. High resolution of the longitudinal vibrational modes allows separation of the four (^{76}Ge, ^{74}Ge, ^{72}Ge and ^{70}Ge) isotopes. The tendency of impurities to pair together or to associate with intrinsic defects is quite common with many examples such as $B_{Ga}-Se_{As}$ or $B_{Ga}-Te_{As}$.

In samples which have been electron irradiated, the interaction of impurities with intrinsic defects is readily apparent and may be one route for the production of intrinsic antisite defects. For example a possible exchange of an As_{int} with B_{Ga} could generate As_{Ga} and B_{int}.

The LVM data do not readily lead to models of the defects sensed by DLTS or conductivity and none of the methods yet used has led to an understanding of the structure of the so-called EL2 trap. This is a midgap level which is frequently a problem in GaAs devices. It is definitely associated with growth conditions and has variously been ascribed to impurities, antisite point defects and dislocations.

7.6.3 Magnetic resonance detection of GaAs defects

EPR, ENDOR and optically detected variants of these techniques have been applied to GaAs defect spectroscopy. They confirm that there are a wide range of defect sites involving antisite occupancy and impurity–intrinsic complexes. However, at present, there are not many instances where the resonance methods can be directly linked to defects sensed by DLTS or optical absorption, but the detailed structures derived from the resonance signals will undoubtedly clarify the problems in future developments. For example (see sections 7.6.2 and 10.5) the $EL2^+$ defect gives a signal from an arsenic antisite–arsenic interstitial pair.

7.7 POINT DEFECTS IN METALS

Industrial metallurgy does not have the stringent requirements associated with semiconductors or the insulators discussed previously; hence there is a limited availability of pure metals. From a defects viewpoint this leads to less definitive measurements and more caution must be exercised in separating intrinsic and intrinsic–impurity features. There are also fewer techniques which are generally applicable to offer a view of individual defects and much of the earlier work was concerned with electrical measurements which are sensitive to a summation of all the defects in the sample. For example electrical resistivity changes are caused by electron scattering at every type of defect including vacancies, interstitials, impurities and larger-scale features. Consequently the approach has been to use annealing techniques to remove less stable defects selectively and to follow the corresponding recovery of damaged material towards the resistivity of the original crystal. More recently a variety of hyperfine techniques have been applied to metals (see Chapter 10) which have given some more specific information on individual defect types.

As will become evident, the simple vacancy- or interstitial-type defect exists in all types of material, including metals, and in some techniques, e.g. field ion microscopy or transmission electron microscopy can be directly "viewed".

In order to proceed with the study of point defects in metals, the two main production methods, quenching and irradiation, will be successively discussed. Subsequently the problems of alloying and associated defects are considered.

7.8 POINT DEFECTS PRODUCED BY QUENCHING

As a general rule, metals are particularly suitable for quenching studies, because of their high thermal conductivity and good plasticity. It is assumed that, in a rapidly quenched (cooled) sample, vacancies which were in thermal equilibrium at high temperatures are frozen in. Much of the information on vacancy defect properties, as well as on migration and clustering behaviour, has been obtained from quenched samples.

7.8.1 Vacancy properties

Numerous studies have been carried out in order to investigate the thermal creation of vacancies and to measure their thermodynamic parameters such as the formation enthalpy h_f and the migration enthalpy h_m. Evidence for the generation of vacancies in quenched crystals has been inferred from a variety of techniques including density, lattice parameter, stored energy and transport measurements, and even from direct observation by field ion microscopy.

Most often, electrical resistivity has been the physical property measured after fast quenching from a high temperature. From the relationship $\Delta \rho = \Delta \rho_\infty \exp(-h_f/kT)$ between the increase in resistivity and the quenching temperature, h_f can be determined. More recently, positron annihilation experiments are providing[40] reliable data in good agreement with those obtained from resistivity measurements or other techniques (e.g. nuclear magnetic resonance (NMR)). Table 7.3 gives the values of E_f for various common metals determined by positron annihilation techniques.

It is now well documented that vacancy migration is the dominant mechanism accounting for self-diffusion in metals[41−44] (occasionally with some contribution of the divacancies at high temperatures near the melting point). The overall activation energy Q for self-diffusion is $h_f + h_m$, where h_m is the activation energy for vacancy migration. Values of h_m are included in Table 7.4. A simple guideline gives $Q = 34T_m$ cal mol^{-1} (or $30kT_m$ eV), T_m being the melting temperature in kelvins. Errors are 10% or less although greater differences exist for a number of body-centred cubic (BCC) metals[44].

7.8.2 Vacancy clustering and annealing

The interaction between isolated vacancies gives rise to the formation of divacancies and larger clusters. In particular, the equilibrium concentration of divacancies may become relevant at high temperatures near the melting point. Divacancies appear to be quite mobile in some metals and have been invoked to account for deviations from the Arrhenius behaviour in diffusion studies. Unfortunately unequivocal evidence for divacancies and reliable values for the binding energy are not available.

The annealing of the excess vacancies introduced by quenching has been often investigated in isothermal experiments at intermediate temperatures.

Table 7.3.
Formation enthalpies of
vacancies in metals determined
by positron annihilation

Metal	Enthalpy h_f (eV)
Aluminium	0.66±0.02
Silver	1.16±0.02
Gold	0.97±0.01
Copper	1.31±0.05
Nickel	1.7±0.1
Vanadium	2.1±0.2
Niobium	2.6±0.3
Molybdenum	3.0±0.2
Tantalum	2.8±0.6
Tungsten	4.0±0.3

Table 7.4.
Activation enthalpy h_m for vacancy diffusion

Metal	h_m (eV)	
	From irradiation experiments	From quenching experiments
Silver (FCC)	0.66	0.64
Aluminium (FCC)	0.59	0.65
Gold (FCC)	0.71	0.83−0.89
Copper (FCC)	0.70	0.72
Nickel (FCC)	1.04	1.27−0.90
Platinum (FCC)	1.43	1.42
Iron (BCC)	0.55	
Niobium (BCC)	1.3	1.35
Tantalum (BCC)	0.7	0.7
Tungsten (BCC)	1.7	1.78
Cadmium (HCP)	0.3	0.38
Zinc (HCP)	0.4	0.44

Values taken from Johnson and Orlov[43].

Not only do isolated vacancies anneal by clustering but also they can disappear after being trapped at dislocations and contribute to dislocation climb.

As a final stage of vacancy clustering, dislocation loops can be formed by collapse of planar vacancy aggregates. These loops are generally of a sessile type, which grow by incorporating new vacancies. The surface area of the loop is roughly proportional to the concentration of involved vacancies.

In some metals, e.g. gold, vacancy collapse gives rise to a tetrahedral arrangement of dislocation segments along $\langle 110 \rangle$ edges, the faces being the close-packed $\{111\}$ planes. These tetrahedra, observed by transmission electron microscopy, are very stable.

7.9 DEFECTS INDUCED BY PARTICLE IRRADIATION

Bombardment with energetic electrons offers a controllable method of generating and studying simple defects such as interstitials and vacancies. Therefore, our discussion concentrates heavily on this method. To displace an atom from a stationary lattice site under adiabatic conditions requires some 25 eV for most metals and this is imparted by electrons of about 1 MeV in a typical case (see section 3.2). The formation of Frenkel pairs during high-energy particle irradiation has been primarily inferred from simultaneous measurements of lattice parameter a and volume V changes induced by irradiation. As discussed in Chapter 8, the observation of the equality $\Delta a/a = \frac{1}{3} \Delta V/V$ is a proof of Frenkel pair creation. Unfortunately there is no clear experimental evidence on the structure of the interstitial defect. However, abundant theoretical work has systematically provided support that the dumb-bell configuration is the most stable. Moreover, some experimental data appear to be in accordance with this model. The vacancy and interstitial partners can be isolated from, clustered to or associated with impurities or other defects, so that a variety of defects are expected in irradiated metals. Irradiation and electrical resistivity measurements at 4 K have the advantage that at this temperature the defects do not move and the thermal scattering contributions to the resistivity are minimized. As seen from Figure 7.9, the influence of all defect scattering centres is included in the resistivity shift $\Delta\rho$, which is the sum of the terms from the various defects (the Matthiessen law), i.e.

$$\Delta\rho = \frac{mv_{\mathrm{F}}}{e^2} \sum_i c_i \sigma_i \qquad (7.1)$$

where c_i and σ_i are the concentration and scattering cross-section of the ith defect type, m the electron mass and v_{F} the velocity at the Fermi energy (see Chapter 8).

7.9.1 Annealing studies using resistivity data

In order to separate the different contributions to $\Delta\rho$, annealing techniques have been used to remove less stable defects selectively and to follow the corresponding recovery of damaged material towards the resistivity of the

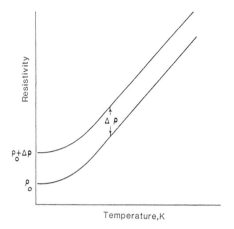

Figure 7.9.
The temperature dependence of resistivity for pure and defective metal.

original crystal. In principle, by assuming σ_i to be constant in equation (7.1), we can estimate the relative concentration of defects from the fractional area under the annealing curve. Unfortunately, this is far from exact as it is obvious that, if annealing occurs through interstitial−vacancy recombination, both types of scattering centre are simultaneously removed.

The second problem is that, if defects associate into clusters or dislocation loops, then the total number of scattering centres changes during annealing. A further note of caution is that the number of annealing stages does not reveal all the defects and a single type of defect may anneal by several routes and each route will appear as an annealing stage. For example, a defect formed from several atoms may move by either linear or rotational processes and each type of motion will be characterized by a diffusion energy.

In spite of the above comments, the annealing technique has been moderately successful and five major annealing regimes have been identified in nearly all metals, independent of their crystal structure. Indeed we can scale the annealing curve by reference to the melting temperature of the metal and demonstrate the universal character of the annealing stages. Figure 7.10 shows the stability curve for the resistivity enhancement caused by electron irradiation of copper at 4 K. The stages of annealing are labelled I−V. These features have been reviewed in detail in many places[45−47].

Reference to Figure 7.10 shows that stage I is made up of several substages, commonly labelled I_a-I_e, and the example for copper is typical of that for most metals. The relative size of the substages can be altered by

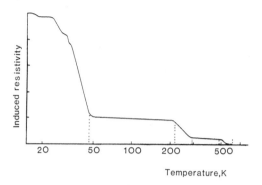

Figure 7.10.
Annealing stages for copper. The additional resistivity was induced by electron irradiation at 4 K.

the presence of impurities, by the electron energy during the damage and by the presence of dislocations. Details for copper are given in Table 7.5.

When defects are formed by the minimum energy required to displace lattice atoms, the defects must be interstitials and vacancies, probably paired at nearby sites. With first-order reaction kinetics and an estimated number of jumps close to unity, it is suspected that the annealing is a direct jump of the displaced interstitials back to the associated vacancy (i.e. it is the interstitial which moves). The variations between stages I_a, I_b and I_c may refer to alternative interstitial structures or to directions of separation in the lattice. This is consistent with the small changes in their relative concentrations with impurity or electron energy. On the assumption that the interstitials move, then stages I_d and I_e correspond to separated interstitial— vacancy pairs. In these cases the recovery may be by tens or hundreds of jumps. Greater movement through the lattice allows more sensitivity to impurities or dislocations and obviously more distant pairs are formed if the electron energy is raised to offer more spare energy to the struck atoms to diffuse away from the original lattice site. Finally, it should be noted that stage I_e shows second-order kinetics. This suggests that the long-range diffusion is a random process and the probability of recombination is a function of the concentrations of both vacancies and interstitials. However, since both were initially formed in equal concentrations, then the apparent order is two. Small deviations from the second order develop if there is significant trapping at impurities or if the interstitials aggregate into larger units.

The discussion so far is based on experimental data but the model may be supported by reference to theoretical calculations of vacancy and interstitial structures in metals and hence the formation and migration energies may be

Table 7.5.
Stage I annealing of copper

	I_a	I_b	I_c	I_d	I_e
Temperature (K)	16	28	32	39	53
Activation energy (eV)	0.05	0.085	0.095	0.12	0.12
Reaction order	1	1	1	1	>2
Number of jumps	1	1	1	10	10^4
Impurity effect	—	Small reduction	—	Large reduction	Large reduction
Dose effect	—	—	—	—	Moves to lower T
Increasing e⁻ energy	Increase		Reduce		Increase

estimated. Computation requires an accurate knowledge of interatomic potentials and these are a major problem. Alternatives have been reviewed in many places[48-50]. The two main methods involve an "empirical potential" or a "pseudopotential" calculation. Despite the computational difficulties there is good agreement that migration energies for interstitials are about 0.1 eV whereas, for vacancies, larger energies of about 1.0 eV are required. This firmly supports stage I as interstitial motion. Note, however, that it does not preclude the fact that interstitial movement may be important in other annealing stages.

Johnson[49] considered six models of interstitial sites in the face-centred cubic (FCC) lattice at a more detailed level (Figure 7.11). The calculated migration energies were typically 0.1 eV and differences were sensitive to the choice of interatomic potential. It is also important to note that the presence of a nearby second defect, such as a vacancy, impurity or interstitial, can distort the lattice and stabilize otherwise less favoured sites.

For the hexagonal close-packed (HCP) metals there appears to be more identifiable substages and nine have been reported for both zinc and cobalt. The same features are detected in both the recovery of the resistivity and the more specific measurements of magnetic relaxation. Although stage I occurs below 50 K for the majority of metals, a notable exception is found for the BCC structure; for example for iron this stage I is delayed[51] to 120–140 K. A possible reason for this is that the interstitial is in the form of a dumb-bell along ⟨110⟩ which is presumed to be the most stable interstitial configuration. Indeed if dumb-bells exist in BCC metals (along ⟨100⟩), then their motion will be inhibited at low temperatures as the mechanism of diffusion towards vacancies is a complex mixture of a lateral

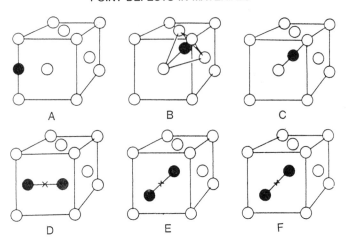

Figure 7.11.
Self-interstitial configurations in an FCC lattice: (A) octahedral; (B) tetrahedral; (C) crowdion; (D) (100) split (dumb-bell); (E) (111) split; (F) (110) split. Interstitial is shown as (●).

movement and rotation towards the next interstitial site, both of which require about 0.25 eV. The model is sketched in Figure 7.12.

If we proceed to the higher annealing stages of the electrical resistivity, the stage II for copper, at 50–100 K, is also related to interstitial defects but in this case it is assumed the interstitials are not free but associated with impurities and/or interstitial loops. Addition of impurities enhances specific substages in this temperature range and computer calculations predict that small interstitial aggregates could be mobile. Precise calculations are not possible as the variables of lattice distortions caused by strain fields are difficult to quantify.

Historically there have been arguments over the origin of stage III and stage IV. The alternative models were movement of vacancies, divacancies or a particularly stable form of the interstitial. Experimentally, we can maintain the simplicity of point defect generation with fast electrons but vary the parameters of crystal purity, dislocation density or the initial vacancy concentration. The latter is achieved by quenching in vacancies from a high temperature. There is consistency between the stage III activation energies derived from the various treatments. For example for copper the migration energy for quenched defects is 0.72 eV which is close to the energy of 0.70 eV determined after electron irradiation. The thermal formation energy of vacancies is found to be 1.31 eV and thus there is consistency with the experimental value for self-diffusions of 2.07 eV, which should be the sum of the formation and migration energies.

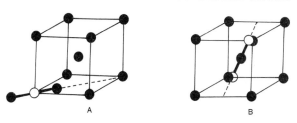

Figure 7.12.
Sucessive stages in the migration of a split $\langle 110 \rangle$ interstitial in a BCC lattice.

Although there is general agreement that stage III is the result of vacancy motion for most FCC, BCC and HCP metals, there are a few examples where the possibility of divacancy motion cannot be ignored. Computer estimates of vacancy and divacancy motion only differ by 10%.

Resistivity data produce somewhat ill-defined stages IV and V in some metals and these are associated with dislocation loops or aggregates. Such multiple-defect structures are not favoured by electron irradiation and consequently the higher annealing stages have greater significance for neutron- or ion-beam-irradiated samples. However, these do not produce identical annealing curves. This is to be expected as ion implantation generates a damage cascade along the ion track with a very high defect density. In general, the core of the track is rich in vacancies whereas the outer region contains the interstitials.

7.9.2 Annealing studies using spectroscopic data

So far, mostly resistivity data have been considered but the more specific hyperfine and resonance techniques (see Chapter 10) may lead to a direct model for the defects. Their use is not yet widespread and there are often problems of sensitivity and limited applicability. In particular, most methods rely on the radioactive decay of a few suitable isotopes. For example, most PAC data are obtained with samples doped with indium. The method senses the electric field gradient at the indium nucleus and so information is provided on the indium site in the lattice. The occurrence of indium interstitials and indium vacancy complexes has been ascertained. In contrast, isolated intrinsic defects cannot be studied with these methods.

PAC data have permitted the interpretation of stage I of the annealing curve of copper as arising from indium interstitial motion and stage III as indium vacancy motion. Stage III models of vacancy migration have been strongly supported, and for gold the same PAC signals appear in both quenched and electron-irradiated samples.

Additional proof of the vacancy character of stage III comes from NMR work of substituted impurities (nickel, palladium or platinum) in copper. NMR spectra are taken in parallel with the resistivity measurements and the monovacancy NMR signal disappears in step with the resistivity recovery.

The use of several concurrent techniques may help to resolve different defect structures, although sometimes the comparison is not straightforward. For example, data obtained by EXAFS and ion beam channelling experiments have been contrasted in studies of interstitial sites in chromium doped with silver. Irradiation with 2.8 MeV electrons[52] gave EXAFS signals which were interpreted as an Al−Ag dumb-bell on a single lattice site. The orientation of the pair was probably along $\langle 111 \rangle$ and certainly not $\langle 100 \rangle$. On annealing, the defect signal had faded by 120 K (i.e. stage I) and a second signal which was apparent disappeared by 185 K (near stage III). Ion beam channelling of similar alloys[53] with silver dopant levels of 0.02 or 0.1 at.% gave anneal stages near 200 K for both impurity concentrations and the structure was assessed to be a mixed-atom $\langle 100 \rangle$-oriented dumb-bell. For the lightly doped sample the 200 K anneal stage was less sharply defined and a second channelling signal appeared as the first declined. Higher anneal stages were noted at 280 K.

At first sight there is disagreement between the two types of measurement, even though similar alloys were used. However, for the EXAFS work the electron irradiation only generates the simplest of point defects whereas the channelling work used beams of helium or krypton to introduce damage. In this case the struck atoms have considerably more energy after displacement and are able to migrate through the lattice. Further, a significant part of the collision cascade obtained with ions may favour more complex defect formation. The spectroscopic defect probe techniques are localized and only sense simple defects. The distortion of sites by more extensive, or adjacent, defects is not taken into account. It must thus be assumed that there is no real divergence of the interpretation from the various methods of analysis.

7.10   POINT DEFECTS IN METALLIC ALLOYS

Impure metals and alloys present a more complex pattern of neighbouring lattice sites and therefore the variety of possible point defects increases. In addition to the isolated impurities, we may have vacancies or interstitials from each constituent metal species, complexes with impurities and all permutations from a random arrangement of the alloy atoms. Consequently, there have been fewer definitive studies of such systems in spite of their applied interest. However, the presence of suitable impurities in a given metal may aid the investigation of defect structures by allowing the use of hyperfine techniques such as Mössbauer spectroscopy.

Impurity atoms may adopt different lattice locations depending on size, chemical nature and lattice type. Light atoms such as hydrogen, carbon or nitrogen occupy interstitial positions and have high diffusivities. These impurities have a marked technological importance, particularly in iron, and a vast literature exists on them. In the FCC metals, the octahedral sites are favoured whereas, for the BCC metals, tetrahedral sites are normally occupied. In some special cases, e.g. V (BCC), both octahedral and tetrahedral sites have been observed.

The energetic equivalency between octahedral (or tetrahedral) sites may be broken by applied stress, leading to a reordering of the impurity atoms. This is the case of the so-called Snoek reorientation of carbon atoms in the stress field of a moving dislocation, which gives rise to a well-known internal friction peak of iron. This effect has also relevance as a possible hardening mechanism of the material.

At high concentrations the solute atoms may not be randomly distributed but instead form an ordered array. A new type of defect may then exist as an antisite occupancy. In metal alloys this type of defect may be energetically quite favourable as the lattice bonding does not inhibit it. Clearly the problem requires a knowledge of local strain energies and electron energy density terms to predict impurity location.

It should also be noted that in alloys or impurity-doped materials (including non-metals) there will be random fluctuations in the distribution of the constituent atoms. If, for example, the impurity ions are larger than those of the host, there may be considerable strain energy from the distortion of the lattice. To minimize this strain energy, it can then be energetically favourable to form impurity pairs or clusters, which may lead to precipitation of the impurity and secondary alloy phases. In contrast, non-uniformity in the material is a common feature of crystalline solids containing dislocation lines since they act as sinks for impurities and point defects. With so many conflicting features, studies of point defects in metallic alloys are not ideal academic exercises but several reviews of the literature exist[54,55].

In applied metallurgy, we must often take a pragmatic view and utilize the consequences of specific impurities or certain heat treatments even if we lack the detailed understanding of the atomic processes. Since it is clear that defects control the properties of chemical reactivity, strength and corrosion resistance, considerable progress can be made by empirical development of new alloys. Addition of defects by irradiation can significantly degrade the material and this is particularly obvious for steels in nuclear reactors. However, in the more recent area of surface treatments by ion implantation, the implantation dose, implant temperature and the impurity ion are under control and the modification can be in the favourable sense. For example, surface hardness can be increased, and friction, wear and the chemical reactivity of corrosion process can be reduced. The improvements in surface properties are typically by factors of $5-10$ but in some cases some spectacular changes have been achieved.

The conditions under which two (or more) atomic species can be combined to produce a stable alloy are not simple and are discussed later.

7.11   RADIATION-ENHANCED DIFFUSION

At room temperature, impurity atoms and clusters in metals have normally moved to sinks so that relatively few defects are mobile. However, the addition of vacancies or interstitials can significantly raise the diffusion constants. This occurs by a bonding between the various mobile defects with the more stable entities. The association of a vacancy with a large impurity offers a relaxation of the lattice which reduces the strain energy, and the pair is then mobile as a unit. The assisted diffusion coefficients may increase by 100 or more. This is demonstrated in Figure 7.13 which shows the temperature dependence of the diffusion coefficients of nickel and beryllium in copper to be similar to the self-diffusion coefficient of copper[56]. Irradiation with a beam of copper ions allows the impurities to move. The radiation-enhanced self-diffusion no longer follows a simple Arrhenius plot. The continuous curve is based on a calculation of radiation-enhanced diffusion[57,58]. A very similar function is seen for the experimental data for nickel and beryllium impurities. The data are supportive of the theoretical treatment.

During ion implantation there is the possibility that the implant ions are mobile because of the defects associated with the implantation process and thus the observed implant distribution may not match simple theories. To separate movement which occurs during impurity injection from radiation-enhanced diffusion, a double implantation procedure may be used. In the following example, nickel was doped to 6 nm depth by implantation of manganese at 30 keV. The stability of this impurity was then assessed by secondary-ion mass spectrometry after a further implantation[59] with 75 keV Ni^+. The flux was sufficiently small that collision cascades could be considered as independent events; thus all motion derived from defects liberated from the cascades. The peak of the manganese distribution moves with nickel ion dose until it peaks at the same depth as the defect concentration caused by the nickel beam. Thus the manganese ions diffuse up the vacancy concentration gradient. Indeed the form of the final impurity distribution is

$$c_{Mn}(z) = kc_{vac}^{P}(z)$$

That is, it matches a higher-power form ($p \propto 2.8$) of the vacancy distribution.

Not all impurities move in the same sense and the use of ion implantation can equally demonstrate inward or outward diffusion, phase separation or impurity mixing. To continue with examples of implants in nickel, Figure

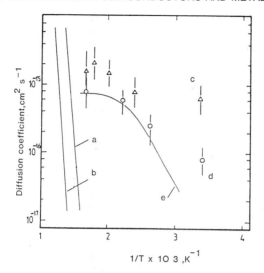

Figure 7.13.
A comparison of thermal diffusion coefficients in copper for beryllium (line a) and nickel (line b) with data for radiation-enhanced diffusion during implantation of beryllium ($\triangle$ data, c) and nickel ($\bigcirc$ data, d). Curve e is the calculated self-interstitial value during irradiation.

7.14 shows results after a 75 keV nickel bombardment of a single crystal of nickel which, prior to irradiation, was uniformly doped with 0.16 at.% Mn and 0.06 at.% Si. However, the figure shows that after implantation the impurity movement has occurred to depths at least ten times the projected range of the implant (18 nm). As in the previous example, the manganese has moved up the defect concentration and peaked some 18 nm below the surface. By contrast the silicon impurities have moved away from the region of maximum damage. The form of the redistribution is sensitive to the implant temperature and for an ambient case the secondary maximum near 140 nm was absent. For the silicon motion a model of silicon interstitialcy motion was proposed at the higher temperatures whereas Si$-$Ni interstitial complexes were favoured to describe the room-temperature data[60,61].

7.12 ALLOY FORMATION

Ion implantation of metals to produce modified surface properties frequently involves doses in excess of 10^{17} ions cm^{-2}, which, with ion ranges of the order of micrometres, leads to average doping levels of 1 at.%. At the peak of the implant distribution it may be much higher. An understanding of the possible alloys which may result, or how the impurities will be incorporated

POINT DEFECTS IN MATERIALS

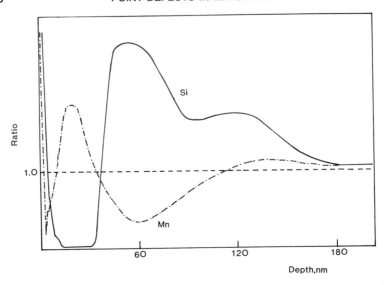

Figure 7.14.
Measured depth profiles of silicon and manganese solutes in a nickel crystal after irradiation with Ni^+ ions at 500°C.

into the metal, is not always obvious. This is particularly true for atomic mixtures which have not been considered in conventional metallurgy or under similar non-thermodynamic conditions. Phase separation and impurity motion have already been mentioned. Guidelines for alloy formation were first summarized by Hume-Rothery[62]. Firstly, for metal A to dissolve into metal B, there should be only a small disparity in the atomic size (less than 15%). Slightly oversize atoms tend to be stable whereas larger ions cause too much lattice strain to be incorporated as single atoms and clustering or new phases are needed to reduce this effect. Conversely, undersize atoms are mobile and so do not form uniform alloys. Secondly, it should be noted that metals A and B tend to be from adjacent regions of the periodic table if they are miscible. This occurs because the Fermi levels of the constituents should be similar, probably within ±0.2 eV. The quantitative guidelines were advanced by Hume-Rothery to predict the structure of alloys in terms of the electron-to-atom ratio. Alloy formation is therefore favoured for elements with similar electron affinities whereas for examples with a large difference in the levels a compound A^+B^- of a more ionic character develops. For the metallic alloys the electron density values determine the phase which develops so that, at electron-to-atom ratios of 1.36 and 1.48, FCC and BCC phases, respectively, are favoured. Phase transitions occur in an initial FCC system such as the Cu−Zn alloy at values of 1.50, 1.62 and

1.75 to the β (BCC), γ and ϵ (HCP) phases, respectively. In terms of a band model these critical electron-to-atom ratios correspond to when the Fermi sphere makes contact with the Brillouin zone boundary.

In this view of the occupancy of electron states, electrons above the Fermi level move to "corner" states of the Brillouin zone. If the energy of the system then rises excessively, the lattice can minimize the free energy by restructuring to accommodate these corner states within a new zone pattern.

These guidelines have been refined and extended by Miedema and co-workers[63-65]. Miedema and co-workers commence with a model in which the alloy is built up of Wigner–Seitz cells with the structure of the metals in isolation. Electron transfer minimizes the total energy of the system, and the cell size must be included to achieve close packing. This leads to maximum contact at the Fermi surfaces of the metals. Hence, in estimating the formation enthalpy ΔH, major terms arise from the difference $\Delta\phi$ in chemical potential and difference in electron density at the cell surface. The latter term is given by $(\Delta n_{W-S}^{1/3})^2$ where n_{W-S} is the electron density of Wigner–Seitz cells. Overall the formation enthalpy for an alloy $A_c B_{1-c}$ is determined by

$$\Delta H = f(c)[-Pe(\Delta\phi)^2 + Q(\Delta n_{W-S}^{1/3})^2]$$

where $f(c)$ is a function of the concentration of metal A. P and Q are constants. For polyvalent non-transition metals it is necessary to introduce a third term $-R$ which takes account of the p-type character of electron orbitals. The two major terms may be computed from measurements of chemical potential and compressibility.

A plot of $\Delta\phi$ against $\Delta n_{W-S}^{1/3}$ distinguishes stable alloys (ΔH negative) from unstable or metastable systems. Figure 7.15 indicates four main regions of alloy formation: the stable area with a negative ΔH; a forbidden area $\Delta H \gg 0$ where solubility is limited to less than 10%; an area near the origin where ΔH is nearly zero so that compounds may form even at higher than 10% doping levels but many of these systems are only stable at higher temperatures; the region with $\Delta H \geq 0$ which precludes alloy formation. The predictions of the model have been tested for at least 500 binary alloy pairs and more recently extended to more complex systems of metallic mixtures using ion beam doping. This has further proved the usefulness of the model.

Overall the ion-implanted examples show the most complexity with alternatives that the surface layers develop point defects, rearrange by radiation-enhanced diffusion, separate phases or develop non-uniform concentration profiles. To find general guidelines for this range of problems is difficult but it is suggested that amorphous binary alloys develop if the components have different crystal structures and are both present in high concentrations (e.g.

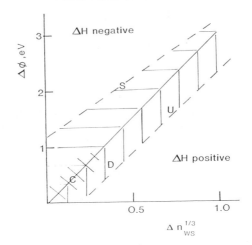

Figure 7.15.
The regions of alloy formation. Region S is the region of stability where the enthalpy term ΔH is negative. Region U is the region of immiscible system where ΔH is positive and, in this region, the solubility of metal A in B is less than 10%. In region C, near the origin, $\Delta H \approx 0$ and some alloys exist with more than 10% solubility. Region D with $\Delta H \geq 0$ precludes alloy formation at low temperatures, even for metals of the same crystal structure.

greater than 30%). Alloys in a metallic crystalline structure are favoured for simple lattice arrangements which follow the rules of Miedema and co-workers. During bombardment the simple structures stay crystalline and more complex lattices move to an amorphous state[66].

REFERENCES

1 P. W. Anderson, *Phys. Rev. Lett* **34**, 953 (1975).
2 F. A. Kroger, *The Chemistry of Imperfect Crystals*, North-Holland, Amsterdam (1973–1974).
3 D. Shaw (ed.), *Atomic Diffusion in Semiconductors*, Plenum, New York (1973).
4 S. Mahajan and J. W. Corbett (eds), *Defects in Semiconductors II*, *Materials Research Society Symposium*, Vol. 14, Elsevier, New York (1983).
5 E. E. Haller, W. L. Hansen and F. S. Goulding, *Adv. Phys.* **30**, 93 (1981).
6 E. E. Haller, *Festkörperprobleme* **XXVI**, 203 (1986).
7 J. E. Field (ed.), *The Properties of Diamond*, Academic Press, London (1979).
8 R. B. Fair, in F. F. Y. Wang (ed.), *Impurity Doping Processes in Silicon*, North-Holland, Amsterdam, p. 315 (1981).
9 J. C. C. Tsai, in S. M. Sze (ed.), *VLSI Technology*, McGraw-Hill, New York, p. 169 (1983).
10 W. Frank, U. Gosele, H. Mehrer and A. Seeger, in G. E. Murch and A. S. Nowick (eds), *Diffusion in Crystalline Solids*, Academic Press, New York, p. 64 (1984).

11 W. Langheinrich, in O. Madelung, M. Schulz and H. Weiss (eds), *Landolt–Bornstein*, Group III, Vol. 17C, Springer, Berlin, p. 118 (1984).

12 D. Pons and J. C. Bourgoin, *J. Phys. C: Solid State Phys.* **18**, 3839 (1985).

13 U. Gosele, *Festkorperprobleme* **XXVI**, 89 (1986).

14 H. Hartmann, R. March and B. Selle, *Curr. Top. Mater. Sci.* **9**, 1 (1982).

15 J. R. Troxell and G. D. Watkins, *Phys. Rev.* B **22**, 921 (1980).

16 I. A. Drabkin and B. Yu. Moizhes, *Fiz. Tekh. Poluprovodn.* **15**, 625 (1981).

17 G. D. Watkins, in J. Narayan and T. Y. Tan (eds), *Defects in Semiconductors II*, Materials Research Society Symposium, Vol. 2, Elsevier, New York, p. 21 (1981).

18 J. W. Corbett, R. L. Kleinhenz and N. D. Wilsey, in J. Narayan and T. Y. Tan (eds), *Defects in Semiconductors II*, Materials Research Society Symposium, Vol. 2, Elsevier, New York, p. 1 (1981).

19 R. K. Watts, *Point Defects in Crystals*, Wiley, New York (1977).

20 J. L. Ivey and R. L. Mieher, *Phys. Rev.* B **11**, 849 (1975).

21 K. L. Brower, *Phys. Rev.* B, **4**, 1968 (1971).

22 D. R. Bosomworth, W. Hayes, A. R. L. Spray and G. D. Watkins, *Proc. R. Soc. London*, Ser. A **317**, 133 (1970).

23 G. D. Watkins and J. W. Corbett, *Phys. Rev.* **138**, A543, A555 (1965).

24 A. Seeger and K. P. Chik, *Phys. Status Solidi* **29**, 455 (1968).

25 F. C. Frank and D. Turnbull, *Phys. Rev.* **104**, 617 (1956).

26 U. Gosele, W. Frank and A. Seeger, *Appl. Phys.* **23**, 361 (1980).

27 R. L. Longini, *Solid-State Electron.* **5**, 126 (1962).

28 N. A. Stolwijk, B. Schuster, J. Holzl, H. Mehrer and W. Frank, *Physica B* **116**, 335 (1983).

29 W. E. Spear, Doped amorphous semiconductors, *Adv. Phys.* **26**, 811 (1977).

30 J. C. Bourgoin and M. Lannoo, *Point Defects in Semiconductors: Experimental Aspects*, Springer, Berlin (1983).

31 A. Mircea and D. Bois. *Defects and Radiation Effects in Semiconductors, Inst. Phys. Conf. Ser.* **46**, 82 (1979).

32 J. M. Spaeth, *Proceedings of the Fourth Conference on Semiconducting III–V Materials, Hakone, 1986*, Ohmsha, Japan, p. 299 (1986).

33 D. Pons, *Physica B* **116**, 388 (1983).

34 R. B. Beall, R. C. Newman, J. E. Whitehouse and J. Woodhead, *J. Phys. C: Solid State Phys.* **17**, 2653 (1984).

35 G. Martin and S. Makram-Ebeid, *Acta Electrons.* **25**, 133 (1983).

36 A. A. Rezazadeh and D. W. Palmer, *J. Phys. C: Solid State Phys.* **18**, 43 (1985).

37 R. C. Newman, *Proceedings of the Materials Research Society, Proc Europe, 1986*, Editions de Physique, Les Ulis p. 99 (1986).

38 R. C. Newman, *Festkörperprobleme* **XXV**, 605 (1985).

39 A. S. Barker and A. J. Sievers, *Rev. Mod. Phys.* **47**, Suppl. 2 (1975).

40 R. W. Siegel, *Annu. Rev. Mater. Sci.* **10**, 393 (1980).

41 N. L. Peterson, *J. Nucl. Mater.* **69–70**, 3 (1978).

42 A. D. Le Claire, *Diffusion in Body-Centred Cubic Metals*, American Society for Metals, Metals Park OH (1985).

43 R. A. Johnson and A. N. Orlov (eds), *Physics of Radiation Effects in Crystals*, North-Holland, Amsterdam (1986).

44 N. L. Peterson, *Comments Solid State Phys.* **8**, 93 (1978).

45 M. W. Thompson, *Defects and Radiation Damage in Metals*, Cambridge University Press, Cambridge (1969).

46 M. T. Robinson and F. W. Young (eds), *Fundamental Aspects of Radiation Damage in Metals*, US ERDA Conference Publication No. 751006, Energy Research and Development Administration (1975).

47 P. Erhart, K. H. Robrock and H. R. Schober, Basic defects in metals, in R. A. Johnson and A. N. Orlov (eds), *Physics of Radiation Effects in Crystals*, North-Holland, Amsterdam, Chapter 1 (1986).

48 I. M. Torrens, *Interatomic Potentials*, Academic Press, New York (1972).

49 R. A. Johnson, *J. Phys. F: Met. Phys.* **3**, 295 (1973).

50 A. M. Stoneham and R. Taylor, *Handbook of Interatomic Potentials II*, *Metals*, *Report* No AERE-R 10205, Atomic Energy Research Establishment, Harwell (1981).

51 H. Schultz, in J. I. Takamura, M. Doyama and M. Kiritani (eds), *Point Defects and Defect Interactions in Metals*, North-Holland, Amsterdam, p. 183 (1982).

52 W. Weber and H. Peisl, in J. I. Takamura, M. Doyama and M. Kiritani (eds), *Point Defects and Defect Interactions in Metals*, North-Holland, Amsterdam, p. 368 (1982).

53 L. M. Howe and M. L. Swanson, in J. I. Takamura, M. Doyama and M. Kiritani (eds), *Point Defects and Defect Interactions in Metals*, North-Holland, Amsterdam, p. 53 (1982).

54 V. F. Zelensky and E. A. Reznichenko, in R. A. Johnson and A. N. Orlov (eds), *Physics of Radiation Effects in Crystals*, North-Holland, Amsterdam, Chapter 8 (1986).

55 F. E. Luborsky (ed.), *Amorphous Metallic Alloys*, Butterworths, London (1983).

56 V. Naundorf, M. P. Macht, H. J. Gudladt and H. Wollenberger, in J. I. Takamura, M. Doyama and M. Kiritani (eds), *Point Defects and Defect Interactions in Metals*, North-Holland, Amsterdam, p. 935 (1982).

57 R. Sizman, *J. Nucl. Mater.* **69–70**, 386 (1978).

58 C. Abromeit and R. Poerschke, *J. Nucl. Mater.* **82**, 298 (1979).

59 A. D. Marwick and J. E. Hobbs, *Report* No. AERE-R 11527, Atomic Energy Research Establishment (1984); *Proceedings of the Fourth International Conference on Ion Beam Modification of Materials, Cornell University, 1984 Nucl. Instrum. Methods* (1984).

60 R. C. Piller and A. D. Marwick, *J. Nucl. Mater.* **71**, 309 (1978).

61 D. R. Harries and A. D. Marwick, *Philos. Trans. R. Soc. London, Ser A* **295**, 197 (1980).

62 W. Hume-Rothery, *Electrons, Atoms, Metals and Alloys*, Dover Publications, New York (1963).

63 A. R. Miedema, F. R. de Boer and R. Boom, *Calphad* **1**, 341 (1977).

64 A. R. Miedema, F. R. de Boer and R. Boom, *J. Less-Common Met.* **41**, 283 (1975); **45**, 237 (1976); **46**, 67, 271 (1976).

65 A. R. Miedema, F. R. de Boer and R. Boom, *Philips Tech. Rev.* **33**, 149, 196 (1976); **36**, 297 (1976).

66 H. Wiedersich, in R. A. Johnson and A. N. Orlov (eds), *Physics of Radiation Effects in Crystals*, North-Holland, Amsterdam, Chapter 4 (1986).

8
Experimental Techniques I—General

8.1 INTRODUCTION

Every physical property of a material provides, in principle, a method for studying and characterizing structural defects, through their effect on the measured values of that property. However, the practical usefulness of a method essentially depends on its sensitivity and resolution. The sensitivity is the magnitude of the change in the measured value induced by the occurrence of a unit concentration of defects. The resolution measures the ability to distinguish between two types of defect.

The ideal technique would be capable of yielding a detailed picture or image of the defect, i.e. that would permit us to "see" it. In most cases, however, the information is rather indirect and the experimental data require some elaboration. In principle, the techniques may be classified into two categories: integral or spectroscopic techniques. In integral techniques, the various types of defect contribute additively to the measured magnitude, so that the effect of each type of defect cannot be, in principle, separated. The success of such a technique lies in the careful design of experiments wherein a given type of defect is predominant and in the availability of a satisfactory theoretical model relating the concentration of defects to the induced change in the value of the measured property. In spectroscopic techniques, each type of defect induces a separable effect on a given parameter of the technique. However, the difference between integral and spectroscopic methods is often quite subtle and it depends on experimental conditions or the sophistication of the techniques.

Since a large variety of techniques are available, a brief or sketchy account will be given for the more conventional ones. In this chapter, we concentrate on integral measurements and, in the next two chapters, we emphasize the spectroscopic techniques. The more familiar methods of optical absorption or electron paramagnetic resonance (EPR) have yielded detailed and specific defect models. Several newer or more novel methods are mentioned even though they have not yet been widely demonstrated.

8.2   IMAGING AND DIFFRACTION METHODS

These methods involve diffraction of electromagnetic or particle waves through the defective material. The observed diffraction pattern is, in some way, related to the size, morphology and distribution of the defects. These methods, when having adequate resolution, permit the "direct" observation of the individual defects. However, a detailed theoretical analysis of the imaging processes is required in order to produce a meaningful correlation between the image and the observed defect structure.

One method, which takes advantage of the interference between the electron waves diffracted by the atoms surrounding another X-ray-absorbing atom, is the so-called extended X-ray absorption fine-structure spectroscopy (EXAFS). Although the instrumentation set-up is rather different, it may be considered as a high-resolution microscopy technique, providing a very detailed geometrical picture of the material at a local level.

8.2.1   High-resolution microscopy

The observation of point defects in materials is today possible with the help of high-resolution microscopy techniques. Consequently, direct evidence is now available on the occurrence and basic structure of a few simple defects, such as vacancies and interstitials.

The optical microscope is capable of a resolution of about $0.2-0.3$ μm and consequently is only appropriate to study large defects, such as micro-cavities, bubbles or large precipitates of impurities. The resolution is set by the diffraction of light through the instrument aperture; thus, for greater resolution, instruments operating at shorter wavelengths, i.e. with electrons, X-rays or ions, must be used.

8.2.2   Electron microscopy

In the transmission electron microscope[1-4], the resolution is essentially limited by diffraction and spherical aberration of the objective lens. For small apertures of the electron beam, diffraction is dominant whereas, for larger apertures, the spherical aberration becomes the dominant resolution-limiting effect. It can be shown that under some approximations the least resolvable distance can be expressed as

$$\Delta X = \mathcal{H} C_s^{1/4} \lambda^{3/4} \qquad (8.1)$$

where $\mathcal{H}$ is a constant, C_s is the spherical aberration constant and λ the electron wavelength. As an example for a microscope voltage V of 10^5 V, which is a typical value, $\lambda = 0.037$ Å and the resolution ΔX attainable according to equation (8.1) is about $2-3$ Å. This resolution has been obtained in modern electron microscopes having good mechanical and electrical stability for appropriate operational conditions and sample thickness, together with adequate image interpretation models. In this connection, we should be aware[2] of the complex and often confused meaning of the word resolution. We are here talking about "interpretable resolution" in the sense that an adequate theory is available to support the correctness of the image interpretation. In contrast, we may consider the so-called instrumental resolution, which may be higher and refers to the fineness of detail that can be consistently recorded in an image. This latter concept is representative of the mechanical and electrical stability of the instrument as well as of some lens aberrations.

To improve resolution further and to reach the $1-2$ Å range, we should move to microscopes working at higher voltages (up to 1 MeV or even more). For an E of 1 MeV the wavelength is 0.0036 Å, giving an optimal resolution of 0.6 Å. In addition to the enhanced resolution, there are additional advantages such as the possibility of using thicker samples, and a reduced chromatic aberration. However, megaelectronvolt electrons will displace lattice atoms and in all materials the ionization of the lattice will cause damage in many insulators. Hence it should be realized that electron microscopy is not a passive measurement technique. To enhance resolution and to inhibit some of the photolytic damage processes[5], low temperatures are often used. An example of the resolution obtainable with modern transmission electron microscopy (TEM) techniques is shown in Figure 8.1.

There are several possible configurations appropriate for high-resolution microscope and the scanning transmission electron microscope, (STEM), in either the bright field or the dark field modes. In the bright field image the defects appear of low intensity on a bright background because the beam is diffracted away from the usual direction. In the scanning transmission electron microscope the configuration is in a way opposite to that in the transmission electron microscope, the objective lens being placed before the

Figure 8.1.
A TEM lattice image of defects near the edge of a sample of GaAs.

specimen. The incident electron beam focused on the specimen and scanned across its surface. Then, the transmitted or altered beam is recorded with a detector. For more details about these techniques the literature[1,2,6] should be consulted.

A different type of electron microscope working at a low voltage (about 20 kV), which is very much used, is the scanning microscope (SEM). It has a great depth of field, although a resolution markedly lower ($\Delta X \approx 200$ Å) than that obtained with the conventional (transmission) microscopes. Therefore, it has a limited application to point defects although it is a very useful tool for the observation of large-size defects. In this microscope, the electron beam is focused and scanned across the sample, in order to explore a given zone of the material. Variations in the operating potential of the scanning electron microscope allow the sample to be imaged via cathodo-luminescence, surface potential, secondary-electron emission or X-ray production and hence a picture of impurity distribution, and of course faults in the structure[7], can be obtained.

8.2.3 Field ion Microscopy

An instrument which has provided an atomic resolution (about 1 Å) and has allowed, for the first time, the observation of point defects is the field ion microscope[8]. Unfortunately, its usefulness is severely restricted to a few materials under special conditions, as is described below. The first microscope was built by E. W. Müller in 1951 and he has continued to be a major contributor to the technique.

The field ion microscope arose from the field electron microscope, in which a fine metal tip acts as a cathode inside an evacuated chamber. The anode is a screen opposite to the tip and is biased at a high voltage. The very high electric fields (about $30-50$ MV cm^{-1}) reached at the tip ionize the metal by a tunnel effect, and the ejected electrons are projected onto the screen, yielding an image of the metal tip. Basically, the field ion microscope is a similar instrument with a cooled tip having a typical radius of about 1000 Å, inside a chamber containing gas (helium, hydrogen, etc.) at a low pressure. Much higher fields are now reached at the immediate proximity of the tip. The very high field causes field-induced ionization of the gas molecules and the resulting positive ions are projected onto the screen. The gas pressure is about 10^{-3} Torr, so that the ion mean free path is sufficiently long for them to reach the screen without deviation. However, as ionization occurs at regions of maximum electric field, the ions produce an image of the tip potential on the fluorescent screen.

A major problem for the applicability of the technique lies in the preparation of a tip with an adequately small radius and excellent structural perfection. Chemical polishing and electrode polishing provide a starting point, but the perfection of the tip is improved by field-induced surface evaporation of atoms lying at surface roughness. The final result is a smooth and crystallographically perfect tip. The field-induced evaporation technique can be used to peel off successive atomic layers of the material offering the possibility of bulk studies. However, the technique is limited to materials which do not evaporate under the fields needed to ionize the gas used for imaging.

With this technique it has been possible to observe directly a variety of defects such as grain boundaries, stacking faults, dislocations and several types of point defects: impurities, vacancies, interstitials and small clusters.

8.2.4 Extended X-ray absorption fine-structure spectroscopy

One of the most interesting and promising tools for investigating the atomic structure of point defects, particularly impurities, is EXAFS. It consists of measuring the absorption spectrum of a material for X-rays with photon energies in the vicinity of an absorption edge (mostly the K edge) of any of the elements making up the material. The energy spectra near the edge show a typical modulated structure, the so-called fine structure, which contains the structural information and extends up to several hundred electronvolts above the absorption edge[9,10] (Figure 8.2). The modulation in the absorption is caused by coherent backscattering of the X-ray-induced photoelectrons by neighbouring atoms. The final state of the excited electron after X-ray absorption is a superposition of an outgoing wave from the absorbing atom with the waves backscattered from neighbouring atoms. As a consequence, the absorption cross-section includes a smoothly varying term (as for free atoms) and an oscillatory part arising from the interference effects. This component is markedly dependent on the interatomic distance and momentum of the primary photoelectron, and it constitutes the extended X-ray absorption fine-structure spectrum. It is described by a function of the k vector of the primary photoelectron: $k = (2mE)^{1/2}/h$, where $E = h\nu - E_k$, $h\nu$ being the energy of the incident X-ray and E_k the absorption (K) edge. The interest of this function lies in the fact that its Fourier transform is a radial structure function $\phi(r)$, exhibiting a number of peaks related to the space distribution of the atoms which neighbour the absorbing atom (Figure 8.3). Therefore the location of these atoms can be very precisely determined (about 0.01 Å) through the Fourier transform of the EXAFS function. Since the range of the coherently backscattered electrons is only a few ångströms, the structural information around the absorbing atoms is very local and only includes a few coordination shells. Moreover, the technique can be applied to amorphous as well as to crystalline materials. In summary, the information that can be obtained from

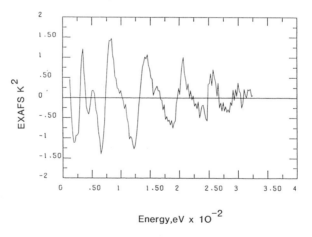

Figure 8.2.
A spectrum obtained by EXAFS for rubidium in RbBiF$_4$ taken at 80 K.

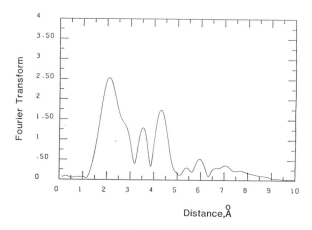

Figure 8.3.
The Fourier transform of the EXAFS data shown in Figure 8.2.

EXAFS experiments includes coordination number, bond lengths and chemical nature of the coordinating atoms.

The standard experimental procedure consists of exposing the sample to an X-ray beam and measuring the absorption spectrum. Generally, the X-rays from a synchrotron are used because of the very strong intensity available in a wide-range continuous spectrum, in comparison with X-ray tubes.

As an example, EXAFS has been applied[11] to study the local structure around the silver atom in the α (superionic) and β phases of AgI. It has been observed that silver atoms in the superionic phase (α) are slightly off centre and shifted about 0.1 Å towards the faces of the iodine tetrahedra.

For defect studies where the absorbing atoms are in a small concentration, an X-ray fluorescence technique is appropriate[12]. Instead of absorption, the X-ray emission intensity as a function of X-ray incident energy is detected. In this way, the dominant background absorption of the host atoms is avoided.

Another interesting possibility is to obtain the extended X-ray absorption fine-structure spectra through the Auger electrons arising from the de-excitation of the absorbing atom. Since the range of these electrons is only a few ångströms, the method opens the way to surface studies of either perfect or adsorbed layers.

8.2.5 Tunnelling scanning microscopy

Before ending this summary of microscopy techniques, we should mention the novel scanning tunnelling microscope, developed[13,14] at the IBM laboratory, Zurich, which has found very successful applications in the study of surface topography. The principle of the microscope involves the tunnelling electronic current between a sharp metal tip and a conducting solid surface lying at a few ångströms distance from each other. A small voltage (of the order of millivolts) is applied between the tip and the surface to induce the tunnelling current. In order to keep the current fixed during the scanning motion of the tip, its position must be adjusted to keep a constant distance to the surface and so the vertical displacements which are made provide an image of the surface topography. A very ingenious, although simple, piezo-electric system was used to position the tip at a close enough distance from the surface. The vertical resolution of the microscope is extremely high (about 0.1 Å) and therefore more than appropriate to detect and measure monatomic steps, and point defects such as adsorbed atoms. The lateral resolution is worse than 2–3 Å but still sufficient to see the corrugation associated with the atomic structure of the surface (either the clean original surface or that of any adsorbed layer). An illustrative example is the observation[15] of single oxygen atoms adsorbed on the (110) face of nickel. Although the initial experiments were performed in an ultrahigh vacuum, the techniques can also be applied under ordinary pressure conditions.

8.3 DIFFRACTION METHODS

X-ray and neutron diffraction techniques are the standard tools for investigating the structure of crystalline (and even amorphous) materials and determining their lattice parameters. The presence of defects disturbs the diffraction pattern of the perfect crystal, which can be used to gain information on the structure and concentration of these defects. A variety of methods, such as X-ray diffuse scattering, small-angle X-ray diffraction, X-ray topography and anomalous transmission, have been developed, mostly for extended defects, including precipitates, dislocations and boundaries. For precipitated phases, standard X-ray diffraction methods can be used to identify their crystallographic structure, if the total concentration is high enough (>1%). All these methods have benefited from the availability of synchrotron sources (and high-flux reactors) because of their high intensity and monochromaticity in comparison with conventional X-ray tubes. Transient and kinetic studies can then be performed *in situ*.

The small-angle X-ray diffraction technique is described next, since it has been shown to be particularly useful for studying aggregates of point defects and precipitates. A rather similar technique of light scattering can also be used.

8.3.1 Small-angle scattering

The existence of heterogeneities in material can be detected through the X-ray (or thermal neutron) diffraction spectrum at small scattering angles[16,17]. To this end, we use a well-known effect in optics, i.e. the central diffraction cone caused by an aperture or obstacle of size D has an angle 2θ given by $2\theta \approx \lambda/D$. Consequently, in accordance with this rough formula, the size of the material heterogeneities could be detected by measuring the angular spread of the diffraction beam. The method is only applicable to a certain range of D values. If D is too large, the scattering angle is too small to permit detection whereas, for sufficiently small D, the diffraction angle is large, but at the cost of a low differential intensity. In practice, the observation is restricted to scattering angles $2\theta \leqslant 2°$, which correspond to defect sizes larger than about 10 Å (for $\lambda \approx 1$ Å).

The basic theory of the method is contained in the book by Guinier and Fournet[16]. They showed that for randomly oriented particles (even non-spherical in shape) the distribution of scattered intensity can be well approximated by a Gaussian function:

$$I_\theta = I_0 \exp(-K_\lambda R^2 \theta^2) \qquad (8.2)$$

where I_θ is the intensity at the scattering angle 2θ or Bragg angle θ (I_0 is the intensity for $\theta = 0$), R is the scattering mass radius or mean-square distance from the centre of "gravity" of the scattering points and $K_\lambda = 16\pi^2/3\lambda^2$ involving the wavelength λ of the incoming radiation (X-rays or thermal neutrons). The validity of the Guinier law is usually tested by plotting log I_θ against θ^2. The slope of the plot is given by KR^2, which allows the determination of R and therefore the obstacle size. All other conditions being equal, I_0 is proportional to the particle volume.

An illustrative example of the usefulness of X-ray and neutron small-angle diffraction is provided by the application of these techniques to the formation of colloids by irradiation in ionic crystals[18]. Identification of markedly non-spherical shapes, such as platelets, has been reported[19] for some systems. The detection by small-angle scattering of the so-called Guinier–Preston zones[17] in Al–Cu and Al–Ag alloys is a routine method in metallurgy.

8.4 SPECIFIC VOLUME AND LATTICE PARAMETER MEASUREMENTS

A simple method for investigating point defects involves the measurement of the change in density or specific volume induced by them. In some particular cases, the technique may allow for the determination of the formation volume of the defects or their concentration. In order to analyse these experiments, the continuum (elastic) model for defects is appropriate[20,21]. For a monatomic material containing a relative concentration c of non-interacting defects, the associated change in volume is given by

$$\frac{\Delta V}{V} = c\frac{v_f}{v_0} = c\frac{v_R}{v_0} + 1 = c\{\chi_R + 1\} \qquad (8.3)$$

for vacancies

$$\frac{\Delta V}{V} = c(\chi_R - 1) \qquad (8.4)$$

for interstitials and

$$\frac{\Delta V}{V} = c\chi_R \qquad (8.5)$$

for substitutional impurities; these are particular expressions of the Vegard law. In all cases, the change in volume refers to the perfect crystal containing the same number of atoms as the defective one. v_f is the formation volume of the defect, v_0 being the atomic volume, and v_R the contribution associated with the atomic relaxation of the material. The ratio $\chi_R = v_R/v_0$ is the relaxation coefficient. For vacancies and interstitials, the change in density is equal to that in volume: $\Delta\rho/\rho = \Delta V/V$. However, for substitutional impurities the appropriate relationship is

$$\frac{\Delta\rho}{\rho} = c\left(\frac{\Delta M}{M} - \frac{v_R}{v_0}\right) = c\left(\frac{\Delta M}{M} - \chi_R\right) \qquad (8.6)$$

where ΔM is the difference between the mass of the impurity and the mass M of the host atom. Equations (8.3)–(8.6) apply to defect concentrations that are low enough that solid-solution (non-interacting) behaviour can be assumed.

For defects of known concentration, the measurement of $\Delta\rho/\rho$ permits the determination of the relaxation coefficient χ_R. As an example, there is a linear change in density induced by divalent impurities in alkali halides as a function of concentration. From a fitting to equation (8.6), χ_R can be determined.

Sensitive methods for measuring changes in density have been developed. Measurements of the equilibrium height reached by the sample inside a long tube filled with a suitable liquid subjected to a temperature gradient (flotation method) are often used. Sensitivity limits are in the range 10^{-5}–10^{-7}.

If the distribution of defects is uniform (homogeneous), the shape of the material is not modified, and the change Δl in any linear dimension l is given by $\Delta l/l = \frac{1}{3}\Delta V/V$. Therefore, we can use length instead of density measurements, allowing for higher sensitivity. Sensitive transducers of length changes could lead to measurements of 1 in 10^8.

It is often very useful to combine the above methods to determine changes in length or density, with the measurement of the change induced in the lattice parameter of the crystal. Under the above-mentioned assumption of a homogeneous defect distribution, the relative change $\Delta a/a$ in lattice parameter is related to the relaxation induced by each defect, according to

$$\frac{\Delta a}{a} = \frac{1}{3}c\chi_R \tag{8.7}$$

By comparing this formula with that corresponding to the change in density, we obtain the following. For vacancies,

$$\frac{1}{3}\frac{\Delta\rho}{\rho} - \frac{\Delta a}{a} = \frac{1}{3}c \tag{8.8a}$$

For interstitials,

$$\frac{1}{3}\frac{\Delta\rho}{\rho} - \frac{\Delta a}{a} = -\frac{1}{3}c \tag{8.8b}$$

For substitutional impurities,

$$\frac{1}{3}\frac{\Delta\rho}{\rho} - \frac{\Delta a}{a} = 0 \tag{8.8c}$$

These formulae facilitate the identification of the nature of the defects involved by distinguishing between various relevant possibilities.

Sometimes (e.g. with irradiation), defects are created in pairs. For Frenkel pairs, the sum of equations (8.8a) and (8.8b) leads to

$$\frac{1}{3}\frac{\Delta\rho}{\rho} - \frac{\Delta a}{a} = 0 \tag{8.9}$$

which provides a practical criterion for deciding whether the crystalline disorder is brought about by Frenkel pairs. Figure 8.4 shows the relative changes in length and lattice parameter in deuteron-irradiated copper at low temperatures[20]. They are equal within experimental error.

8.5 MECHANICAL METHODS

The elastic, anelastic and plastic behaviour of materials are sensitive to the presence of point defects. The influence on the elastic parameters is rather

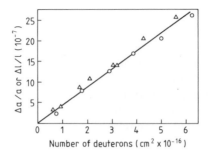

Figure 8.4.
Comparison of the relative change in $\Delta a/a$, ($\bigcirc$), and sample dimensions $\Delta l/l$ ($\triangle$), induced by deuteron irradiation of copper[20].

small (but detectable) and continuum elasticity models are able to predict and estimate roughly the induced modifications of these parameters. In contrast, relevant changes are caused by point defects on the anelastic and plastic properties and therefore they are considered in some detail as useful detection and characterization techniques.

8.6 ANELASTICITY AND INTERNAL FRICTION

8.6.1 Basic phenomena

An ideal elastic body satisfies the Hooke law expressed as

$$\sigma = M\epsilon$$
$$\epsilon = \mathcal{J}\sigma$$
(8.10)

with $M = 1/\mathcal{J}$, relating the stress σ and the induced strain ϵ. M is the modulus of elasticity (or modulus) and $\mathcal{J}$ the modulus of compliance (or compliance).

Equations (8.10) imply a single-valued, instantaneous and linear response of the material to the applied stress. A more general behaviour can appear if some of these conditions are relaxed. In particular, the presence of defects introduces an additional coupling between stress and strain, involving atomic motion and/or migration, therefore leading to a finite response time. If linearity is preserved, we have anelastic behaviour described by

$$\mathcal{J}_R\sigma + \tau\mathcal{J}_N\dot{\sigma} = \epsilon + \tau\dot{\epsilon}$$
(8.11)

where $\mathcal{J}_R$ and $\mathcal{J}_N$ are the relaxed and unrelaxed compliances, respectively, and τ is the relaxation time. Detailed treatments of anelasticity, experimental methods and applications are covered elsewhere, e.g. by Nowick and Berry[22] and de Batist[23].

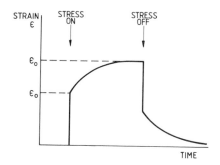

Figure 8.5.
Anelastic response (strain against time) of a material to a square stress pulse.

In response to a constant stress σ, the deformation ϵ increases with increasing time (Figure 8.5) according to

$$\epsilon(t) = \epsilon_0 + (\epsilon_\infty - \epsilon_0)\left[1 - \exp\left(-\frac{t}{\tau}\right)\right] \qquad (8.12)$$

where ϵ_0 is the instantaneous (elastic) strain and ϵ_∞ the long-term (relaxed) strain. ϵ_0 and ϵ_∞ obey

$$\epsilon_0 = \sigma \mathcal{J}_N$$
$$\epsilon_\infty = \sigma \mathcal{J}_R \qquad (8.13)$$

The amount of anelastic relaxation is characterized by the parameter Δ defined by

$$\Delta = \frac{\epsilon_\infty - \epsilon_0}{\epsilon_0} = \frac{\mathcal{J}_R - \mathcal{J}_N}{\mathcal{J}_N} \qquad (8.14)$$

On suddenly removing the stress ($\sigma = 0$), the strain follows the time exponential decay (Figure 8.5)

$$\epsilon(t) = (\epsilon_\infty - \epsilon_0) \exp\left(-\frac{t}{\tau}\right) \qquad (8.15)$$

Very often an oscillating stress $\sigma = \sigma_0 \exp(i\omega t)$, instead of a constant stress, is applied to the sample. The induced strain will also be oscillatory with the same frequency, i.e. $\epsilon = \epsilon_0 \exp[i(\omega t - \delta)]$, where a dephasing term δ has appeared with respect to σ. By introducing the complex compliance $\hat{\mathcal{J}} = \mathcal{J}_1 - i\mathcal{J}_2$, we can write

$$\epsilon = \hat{\mathcal{J}}\sigma \qquad (8.16)$$

By substituting equation (8.16) into equation (8.11), we find that

$$\mathcal{J}_1 = \mathcal{J}_N + \frac{\mathcal{J}_R - \mathcal{J}_N}{1 + \omega^2\tau^2}$$

$$\mathcal{J}_2 = (\mathcal{J}_R - \mathcal{J}_N)\frac{\omega\tau}{1 + \omega^2\tau^2}$$

(8.17)

i.e. the components of the complex compliance are both functions of the frequency ω of the stress (dispersive behaviour). In the particular case of instantaneous response $\tau = 0$, $\mathcal{J}_N = \mathcal{J}_R$ and the complex compliance $\hat{\mathcal{J}}$ becomes real: $\hat{\mathcal{J}} = \mathcal{J}_R = \mathcal{J}_N$.

The dephasing term δ is related to the imaginary part of $\hat{\mathcal{J}}$ through its tangent $\tan\delta = \mathcal{J}_2/\mathcal{J}_1$ so that

$$\tan\delta = \frac{\Delta}{(1 + \Delta)^{1/2}}\frac{\omega\tau}{1 + \omega^2\tau^2}$$

(8.18)

which for the common situation $\Delta \ll 1$ reduces to

$$\tan\delta = \Delta\frac{\omega\tau}{1 + \omega^2\tau^2} = \frac{\mathcal{J}_2}{\mathcal{J}_N}$$

(8.19)

This corresponds to a Debye-type dependence on ω. Debye peaks are typical of relaxation processes described by exponential decay laws with a single relaxation time (see section 8.10.5 on dielectric relaxation). In more general cases the analysis is complicated by the presence of several Debye relaxation processes and then

$$\tan\delta = \sum_n \Delta_n\frac{\omega_n\tau_n}{1 + \omega_n^2\tau_n^2}$$

(8.20)

$\tan\delta$ as well as $\mathcal{J}_2$ are related to the dissipation or loss of energy associated to the anelastic behaviour, so that $\tan\delta$ is often designated as an internal friction coefficient, and δ as the loss angle of the material. It can be shown that $\tan\delta = 1/Q$, where Q is the quality factor of the specimen, i.e. $Q = 2\pi W/\Delta W$, W being the stored energy and ΔW the energy loss per cycle.

The internal friction peak of $\tan\delta$ against ω in a forced-oscillation experiment as considered here has a maximum at a frequency $\omega_m = 1/\tau$. The width $\Delta\omega$ of the peak is related to the quality factor Q and internal friction coefficient by

$$\Delta\omega = 3_Q^{1/2}\frac{\omega_m}{Q} = 3^{1/2}\omega_m\tan\delta$$

(8.21)

Most often, the processes responsible for the relaxation time are thermally activated and so τ obeys the typical Arrhenius dependence $\tau = \tau_0\exp(-\Delta\epsilon/$

kT). Therefore, by carrying out experiments at various temperatures and then plotting log ω_m against $1/T$, the activation energy can be determined.

The decay of vibrations rather than a vibration driven at or near resonance can also be used as an experimental technique. This decay is usually characterized by the logarithmic decrement D_l, i.e. the logarithm of the ratio of the amplitudes of two successive vibrations. It can be verified that

$$\frac{1}{Q} = \tan\delta = \log\left(\frac{D_l}{\pi}\right) \tag{8.22}$$

relating the internal friction to the logarithmic decrement.

Finally, the propagation of a mechanical wave in an anelastic material presents attenuation associated with the energy dissipation. The attenuation coefficient is defined through the formula $I = I_0 \exp(-Ax)$ relating the intensity of the wave to the distance x travelled in the material.

For small damping ($\delta \ll 1$), the velocity and attenuation are given by:

$$v^2 = \frac{M_1}{\rho}$$

$$A = \frac{\delta\omega}{2v} = \frac{\pi\delta}{\lambda} \tag{8.23}$$

where M_1 is the real part of the appropriate modulus and ρ the density.

8.6.2   Experimental techniques

Several types of experimental technique[24] related to the various excitation and detection methods described above are used. In one of them, the decay or damping of the free vibration of a system is measured. From the logarithmic decrement D_l, the internal friction tan δ can be calculated through equation (8.22). In the so-called resonance methods, a resonant vibrating mode is established in the material and the parameters of the internal friction peak measured. The width of the peak yields the internal friction coefficient. Finally, wave propagation methods involve the measurement of the velocity and attenuation of a pulse or wave packet propagating through the specimen.

A very wide frequency interval has been used, extending from about 10^{-5} Hz to several gigahertz. The appropriate instrumentation depends on the frequency range and includes torsion pendulums[25] for very low and low frequencies (from about 10^{-5} Hz to 20 kHz), resonant methods in flexion or longitudinal compression (kilohertz range) and ultrasonic methods for the highest frequencies. The excitation and detection of the mechanical vibrations may use electrostatic, electromagnetic, electroacoustic, piezoelectric or magnetostrictive methods.

Measurements can be carried out as a function of temperature, frequency and amplitude of excitation. This latter method is particularly useful for investigating the liberation of dislocations from defects acting as pinning points.

8.6.3 Applications

Internal friction and anelastic methods have been extensively used in defect physics problems involving dislocations and/or point defects[24]. One of their advantages in comparison with other mechanical techniques is that they do not alter the microstructure of the samples.

A variety of relaxation processes which give rise to a number of internal friction peaks in the frequency or temperature spectra have been investigated. They are associated with dislocations, point defects (particularly interstitials) and point defect–dislocation interaction. Most studies have been performed on metallic systems for which an abundant body of information is available[24]. Relevant peaks connected with point defects are the Snoek peak associated with the relaxation of interstitial impurities (carbon, nitrogen and oxygen) in body-centred cubic metals or the Snoek–Koster peaks associated with the interaction of dislocations with hydrogen or other interstitial defects. The Snoek peaks for carbon in α-iron permit the determination of the jump frequency of the interstitial carbon between octahedral and tetrahedral sites and therefore the root-mean-square displacement per unit time. From this, the diffusion coefficient can be calculated, with the values obtained being in close agreement with those determined by radiochemical methods.

Internal friction methods have also been applied to semiconductor and insulating crystals, an illustrative example being the study[26] of substitutional sodium in CaF_2. Sometimes the reorientation behaviour of a centre under stress can be monitored by optical methods, e.g. the A centre in irradiated silicon[27], and/or EPR spectroscopy, e.g. the O_2^- molecule in KI[28] (see also Chapters 9 and 10).

8.7 PLASTIC DEFORMATION

8.7.1 Point defect hardening

Plastic deformation is an irreversible process involving long-range motion of dislocations. It occurs for applied stress larger than a critical stress σ_0, the yield stress. Several mechanisms account for σ_0: lattice friction, dislocation interactions and point defect–dislocation forces. This latter effect is responsible for the increase in σ_0 or hardening induced by chemical doping or irradiation. These effects are well documented for metallic systems[29] and for some non-metallic materials such as alkali halides[30,31]. From the available hardening models, the concentration of defects acting as efficient obstacles to dislocations could be inferred from yield stress data, if the system is well characterized. For a quantitative approach, a number of physical parameters have to be carefully controlled, e.g. temperature and strain rate. Even so, this cannot be considered as a precise method because of the role of a number of other variables, such as initial dislocation densities, defect distributions and aggregation state, which are very difficult to control.

The state of aggregation and precipitation of the impurities or defects can also be inferred from mechanical data, because the temperature and strain rate dependence of σ_0 are sensitive to the hardening mechanism and consequently to the type of active obstacle. For isolated point defects, we are dealing with solid-solution hardening. Models for solution hardening originally developed by Mott and Nabarro yield dependences of yield stress on defect concentration of the form $\sigma \propto c^n$ with n close to unity. For anisotropic (e.g. axial) defects such as the H-centre interstitial or dipolar complexes in alkali halides, the short-range defect—dislocation interactions lead to an efficient hardening with a marked temperature dependence, which should be fitted theoretically by a $\sigma_0 \propto T^{1/2}$ law (or more recently[32] to a $\sigma_0^{2/3} \propto T^{1/2}$ law). The associated dependence of the yield stress on concentration is of the type $\sigma_0 \propto c^{1/2}$. All theoretically predicted dependences have to be taken with some caution since they are quite sensitive to the distribution of obstacles and the statistical stress averaging which is used. Similarly these experimental data do not often distinguish between various possible laws.

For precipitate hardening, the data can often be analysed according to mechanisms involving the intersection of the precipitates by the moving dislocations[33]. A dependence of the type $\sigma_0 \propto T^n$ is generally observed, the exponent n being dependent on the energetics of the intersection, i.e. on the energy contribution of the various interphases involved during the process. The situation is simpler when the precipitate strength gives rise to an Orowan mechanism in which the precipitate pins the dislocation line. The dislocation segments consequently bow into loops which expand until they reach a critical radius. At this point the yield stress is given by

$$\sigma_0 \approx \frac{Gb}{\Lambda} \tag{8.24}$$

where G is the shear modulus, b the Burgers vector and Λ the average distance between obstacles. Therefore the average distance Λ can be estimated from a measurement of σ_0. If the total impurity concentration is known, the size of the precipitate can be determined.

Although the technique cannot compete, in general, with other more direct and precise methods, it could provide a complementary tool for elucidating some situations. For example, the temperature dependence of the yield stress is very marked in the initial stage near room temperature (stage I) of γ irradiation of NaCl[34] (Figure 8.6). The data are consistent with a Fleischer-type hardening model by axial disperse obstacles. This supports the formation of atomic-size defects formed by interstitial trapping at impurities. At higher irradiation doses (stage III), the yield stress becomes temperature independent in accordance with the Orowan mechanism associated with the presence of large interstitial clusters.

8.7.2 Creep

Creep experiments[35], i.e. straining at a constant stress and temperature, have been used, mostly in metals, for diffusion studies. The steady-state creep rate $\dot{\epsilon}$ has been shown to obey a general relationship

$$\dot{\epsilon} = C_d \frac{\sigma^n}{T} D \propto \frac{\sigma^n}{T} \exp\left(-\frac{h}{kT}\right) \tag{8.25}$$

where σ is the constant applied stress, n an exponent between 1 and 5, and D the appropriate self-diffusion coefficient (h is the activation enthalpy). This formula establishes a proportional relationship between steady-state

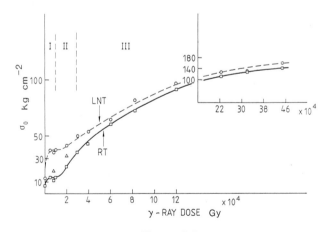

Figure 8.6.
Hardening of NaCl induced by room-temperature γ irradiation. The yield stress σ_0 data were taken at two temperatures, liquid-nitrogen temperature ($\bigcirc$, $---$) and room temperature ($\square$, ——). Note the different temperature behaviour of σ_0 for low and high doses.

creep and self-diffusion, both processes having the same activation enthalpy. Therefore, measurements of the creep rate at various temperatures have been used in metals to obtain h values, which are in good agreement with those determined by other techniques[21]. The coefficient C_d in equation (8.25), which depends on the initial dislocation contents, prevents the evaluation of an absolute value for D from creep data.

A number of theoretical models[36,37] have been developed to account for the observed creep rates and to justify equation (8.25). Depending on the operative mechanisms, different values of the exponent n are obtained.

Creep methods have been also applied to non-metallic materials, such as oxides[38] to investigate diffusion of point defects (vacancies). Although equation (8.25) is generally used, the situation appears more complicated and much work is still needed to obtain a satisfactory picture. In fact, the important deviations of stoichiometry in many oxides and the complexity of defects involved require confirmation of the validity of equation (8.25) for each particular case.

8.8 STORED ENERGY

As described in Chapter 2 on the thermodynamics of point defects, the presence of lattice disorder is associated with changes in the thermodynamic potentials with regard to the perfect material. In particular, the formation of one defect of a given type requires a certain amount of energy, i.e. the formation energy u. During heating of a sample containing defects, recovery processes set in which lead to partial restoration of the lattice perfection. Each recovery process is accompanied by a release of energy corresponding to the annealing (annihilation) of a certain group of defects. The defect formation energy is initially stored as lattice potential energy which is converted into lattice phonons during the annihilation stage. Thus a heating cycle gives a spectrum of energy released against temperature; this contains several recovery peaks whose magnitude and peak temperature give estimates of the number and type of defects in a particular annealing stage.

The measurement of the stored energy released is carried out by rather standard differential calorimeter techniques[39] and is also termed differential thermal analysis. A typical set-up consists of an isothermal chamber, containing two identical samples, one of them acting as a dummy or reference. The chamber is then heated at a constant rate and the temperature differences between the two samples, as well as between the dummy sample and the calorimeter walls, are recorded. The analysis of the data is usually made in accordance with the treatment developed by Overhauser[39]. If the heat transfer between the dummy and defective samples is neglected, the rate equations for the temperatures T_1 (dummy) and T_2 (sample) are

$$\frac{dT_1}{dt} = \alpha(T_w - T_1) \tag{8.26a}$$

$$\frac{dT_2}{dt} = \alpha(T_w - T_2) + \frac{1}{\rho C_h}\frac{dU}{dt} \tag{8.26b}$$

where T_w is the temperature of the calorimeter walls, C_h the specific heat, U the stored energy per unit volume and ρ the density. For a single type of defect of formation energy u, $U = Nu$, N being the defect concentration. α is a common heat transfer coefficient between samples and walls.

By eliminating α in equations (8.26), we obtain

$$\frac{1}{\rho C_h}\frac{dU}{dT_1} = \frac{T_2 - T_1}{T_w - T_1} + \frac{d(T_2 - T_1)}{dT_1} \tag{8.27}$$

In practice, since the two samples are not identical and do not have identical thermal coupling to the surroundings, it is necessary to introduce an unbalance coefficient β in equation (8.26b) which in general is a function of temperature. The heat transfer term from the sample to the walls is then given by $[\alpha/(1 - \beta)](T_w - T_2)$. Under these conditions, equation (8.27) transforms to:

$$\frac{1 - \beta}{\rho C_h}\frac{dV}{dT_1} + \beta = \frac{T_2 - T_1}{T_w - T_1} + (1 - \beta)\frac{d(T_2 - T_1)}{dT_1} \tag{8.28}$$

The parameter β is determined by repeating a heating run after the defects have been annealed. From a measurement of T_1, T_2 and T_w (assumed to be constant) as a function of time, we obtain dU/dT_1, i.e. the (derivative) spectrum of the stored energy.

The technique has a sensitivity of better than about 10^{-5} cal $°C^{-1}$ and has been successfully applied to irradiated metals[20] and insulators[40]. Figure 8.7 illustrates the type of signal that is obtained for an insulator.

Many of the early studies were concerned with energy storage in metals and materials used in nuclear reactors as the sudden release of energy during a thermal annealing cycle posed problems in early reactor technology. Indeed the release of stored energy contributed to one accident. Once the problem was understood, measurements of stored energy from radiation damage received less attention. Nevertheless the emission or absorption of stored energy is still a standard technique for recording phase changes. The technique applies to many systems and Figure 8.8 has the advantage of demonstrating energy changes in both antiferromagnetic and supercon-ducting materials. In this example the differential specific heat is plotted against temperature. The sample is a superconductor $YBa_2Cu_3O_7$ and the

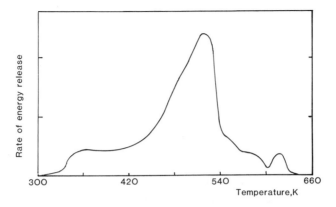

Figure 8.7.
Stored-energy spectrum of γ-irradiated NaCl[40].

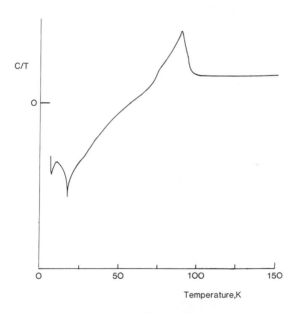

Figure 8.8.
An example of specific heat anomalies during phase transitions. The differential calorimeter contrasts features from an antiferromagnet, at a low temperature, and a superconductor, near 90 K[41].

reference standard is antiferromagnetic $YBa_2Cu_{2.7}Ni_{0.3}O_7$. The lowest-temperature anomaly results from the antiferromagnet and the 91 K feature from the superconductor[41].

8.9 THERMAL CONDUCTIVITY

Heat transport along the x axis of a material is governed by the equation $Q_h = K\,dT/dx$, where Q_h is the heat flux and K the thermal conductivity of the material.

Elementary kinetic arguments lead to the following expression for the lattice contribution to the thermal conductivity: $K_L = \frac{1}{3}C_h vl$, C_h being the heat capacity per unit volume, v the phonon velocity and l the phonon mean free path. In metallic systems, electrons also contribute to the thermal conductivity, their contribution being dominant for pure metals at high temperatures.

At high temperatures the lattice conductivity K_L is limited by phonon–phonon scattering and roughly follows a $1/T$ dependence because of the decrease in phonon mean free path with temperature. In contrast, at very low temperatures, geometrical scattering, e.g. by sample boundaries, becomes dominant and the phonon mean free path remains constant down to 0 K. Therefore the dependence of K_L on temperature is determined by $C_h(T)$ and a Debye law $K_L \propto T^3$ is usually expected. In addition to crystal boundaries, other phonon scattering obstacles such as dislocations, different isotopes, impurities and point defects are relevant in determining the phonon mean free path at low temperatures. At first sight a major change might not be expected in the thermal conductivity as a result of isotropic changes but in an early experiment[42] the peak phonon contribution near 20 K was reduced by a factor of 4 on changing from a 90% ^{74}Ge sample to a sample with a mixture of ^{70}Ge, ^{72}Ge, ^{73}Ge, ^{74}Ge and ^{76}Ge isotopes.

More spectacular changes occur with radiation damage as long-range order is destroyed. For example[43] irradiation of crystalline quartz leads to the formation of amorphous silica. The initially high thermal conductivity at the peak of the phonon-controlled conductivity in crystalline quartz (seen near 10 K) falls by a factor of 10^4 as the lattice order is destroyed.

Early examples of apparatus for measuring low-temperature thermal conductivity can be consulted in the literature[44,45].

8.10 ELECTRICAL TECHNIQUES

Charge transport measurements, and particularly electrical conductivity, provide information on the concentration and mobility of charge carriers. Point defects contribute to the current, when they act as sources, sinks, recombination centres or scattering centres for electrons or holes. The information that can be derived from the experiments depends on the types of carrier (electronic or ionic), and the experimental and sample conditions. Although most materials present a dominant type of conductivity, sizeable contributions of electronic and ionic charge transport can be operative for some compounds if a wide temperature range is considered.

The required instrumentation is, in general, simple and essentially depends on the value of the electric current and specimen characteristics. One particular technique, light-induced conductivity (photoconductivity), is described on dealing with optical spectroscopy methods.

8.10.1 Electronic conductivity

The situation is so different for metallic and semiconductor materials that the two cases will be considered separately.

(a) Metals. Since the concentration n of charge carriers is constant, point defects influence the conductivity σ through the mobility value μ, the relaxation time τ or the scattering mean free path l. A simple classical formula can be written as

$$\sigma = ne\mu = \frac{ne^2\tau}{m} \quad \left(l = \frac{1}{\sigma N} = v\tau\right) \tag{8.29}$$

N being atomic concentration and v the average thermal carrier velocity.

The scattering by point defects adds linearly to the scattering by phonons so that the Mathiessen rule is obeyed by the resistivity ρ_{tot}:

$$\rho_{tot} = \rho_d + \rho_T \tag{8.30}$$

where ρ_d is the defect resistivity and ρ_T the resistivity associated with the vibrational motion of the lattice. To enhance measurement sensitivity, it is customary to record resistivity near 4 K as this minimizes the phonon contributions to ρ_T. For defects in a low concentration, where defect interactions are negligible, it is expected that ρ_d is proportional to the concentration c. The absolute value of c cannot be determined from the measured ρ_d because of the uncertainty in the calculated value for scattering cross-sections associated with each defect. However, once the resistivity corresponding to a given type of defect has been calibrated by using another independent technique, we have a simple and reliable way to measure defect concentrations. Typically, resistivities associated with a vacancy or interstitial concentration of 1% are of the order of 1 $\mu\Omega$ cm.

The situation is much more favorable with regard to relative measurements. In this sense, resistivity provides a very useful method for studying variations in defect concentration brought about by any physical treatment such as irradiation or quenching. For example, by careful measurements of the resistivity of metals[20,21] quenched from different temperatures, it has been possible to obtain precise values for the formation energy of vacancies. Also the various stages usually appearing during the thermal recovery of irradiated metals can be conveniently monitored by the associated change in electrical resistivity.

(b) Semiconductors. Since both electrons and holes may contribute to the current in a semiconductor, the bulk conductivity σ is

$$\sigma = ne\mu_e + pe\mu_h \tag{8.31}$$

where n and p are the concentrations and μ_e and μ_h the mobilities for electrons and holes, respectively. For an intrinsic semiconductor, $n = p$ whereas, for extrinsic or doped material, n and p are generally different and vary over a very wide range. When $n \gg p$, we have a n-type semiconductor, whereas a p-type semiconductor corresponds to $p \gg n$. Although a given type of carrier is dominant, the sign of its charge cannot be determined from conductivity measurements alone and another complementary technique, such as the Seebeck or Hall effect, is required.

The dependence of σ on T arises through the mobility μ and the carrier density n (or p), although this latter factor is markedly dominant. In fact, the density of carriers is determined by a balance between the thermal generation of carriers from the lattice bands or from existing donor and acceptor impurities, and the electron–hole recombination process. Therefore, it is related to the location of the corresponding donor (or acceptor) level inside the gap and so conductivity measurements as a function of temperature are conventionally used to determine the levels of the impurities.

The dependence of μ on T is of the form $\mu \propto T^n$, with the exponent n depending on the dominant scattering mechanism (scattering by optical phonons, acoustic phonons, and charged or neutral defects).

Experimental transient methods which essentially rely on the lifetime of minority carriers because of the recombination processes will be described in section 8.11. They are used to investigate deep levels in semiconductor materials.

For some ionic materials the carriers self-trap in the lattice and move by hopping from one lattice site to another (e.g. some transition-metal oxides), in which case different conduction mechanisms are operative[46] depending on temperature. At very low temperatures, band-like motion should take place, which is associated to tunnelling of the small polaron between neighbouring sites without change in the phonon population. At higher temperatures, phonon-assisted tunnelling will become dominant. At sufficiently high temperatures a semiclassical approach will be in order and the mobility will be expressed as $\mu \propto \exp(-E_m/kT)$ where E_m is the activation energy for hopping ($E_m = E_f/2$, where E_f is the formation energy). Typically, small-polaron hopping mobilities are lower than 1 V cm^2 s^{-1}.

8.10.2 Hall effect

Hall effect measurements constitute a routine technique for measuring the concentration and sign of majority carriers in semiconductors. The experimental configuration has a magnetic field B applied in the z direction, perpendicular to the direction x of current flow. A voltage V_y (Hall voltage) and field E_y appear, given by

$$E_y = R_H \mathcal{J}_x B_z \qquad (8.32)$$

$\mathcal{J}_x$ being the current density and $R_H = 1/en$ the Hall constant. n is the carrier density and e the carrier charge, so that R_H is negative for n-type semiconductors and simple metals and positive for p-type semiconductors. By combining this Hall technique with conductivity measurements, both the carrier density n (and its sign) and the mobility can be determined.

For small-polaron transport, the behaviour is very different[46]. The Hall mobility may strongly differ from the drift mobility and have a completely different temperature dependence. Even the sign of the effect may be changed.

For materials having low mobilities, appropriate Hall methods have been developed[47].

8.10.3 Thermoelectric effects

Three coupled or crossed effects between thermal and electrical parameters are defined: the Seebeck, Peltier and Thomson effects. They are thermodynamically related and the same basic information can be obtained from any one of them. The Seebeck effect is more commonly used for metals and semiconductors. A potential difference appears across a junction if a temperature difference is established between the joint and free ends.

The Seebeck coefficient S_{AB} for the two materials is defined as $S_A - S_B = S_{AB} = dV_{AB}/dT$, where S_A and S_B are the appropriate Seebeck coefficients for each material. V_{AB} is the potential difference across the junction. For materials where band-like motion of the carriers takes place, the Seebeck coefficient is given by

$$S = \frac{k}{e}\left(\mathcal{A} + \frac{E_c - E_F}{kT}\right) \qquad (8.33)$$

which is valid under the assumption of a single type of carrier and parabolic bands. S values range from microvolts per Kelvin for metals up to millivolts per Kelvin for semiconductor systems. In equation (8.33), k is the Boltzmann constant, $\mathcal{A}$ is a constant parameter, and E_c and E_F are the conduction band bottom and Fermi energies, respectively.

The coefficient is negative for electrons and positive for holes, so that the Seebeck effect, as the Hall effect, may be used to determine unambiguously the sign of the current carriers.

For small-polaron transport conditions, the situation and the theoretical expressions for S markedly depend on the operative regime (i.e. temperature), as commented above for conductivity and Hall measurements. From a comparison between the temperature dependence of the conductivity and Seebeck coefficient, the electrical transport mechanism, either band-like or through small-polaron hopping, may sometimes be inferred[48].

8.10.4 Ionic conductivity

Good insulator materials with a wide electron energy gap show some conductivity associated with electric-field-induced ion transport[49,50], is generally mediated through lattice imperfections. For many non-metallic materials, the ionic conductivity may be comparable with and even higher than the electronic conductivity, particularly at high temperatures (near the melting point), where ion motion is markedly enhanced.

Ionic conductivity measurements have the advantage of being experimentally simple. For example, graphite or silver paint electrodes can be applied to alkali halide specimens of convenient size and standard direct-current or low-frequency alternating-current measurements made with straightforward electronic measuring devices. Some caution has to be taken, however, to guarantee good ground connections and to avoid spurious surface currents, because of the low value of the detected currents, when measurements are performed at or below room temperature. Guard-ring arrangements are a common practice.

The main difficulty comes in analysing the data. Even if we restrict ourselves to predominantly Schottky defects in perfectly pure crystals of the simplest A^+X^- stoichiometry, we need at least six parameters to discuss the conductivity in terms of point defects. We need the enthalpy and entropy associated with Schottky defects (equal number of positive and negative vacancies) and with the motion of both positive and negative ion vacancies. In any real crystal, we must also know the parameters for association between vacancies and aliovalent impurities and we must consider Coulombic interactions (the Debye−Hückel clouds) and the formation of complex aggregates of defects and even precipitates of a definite crystallographic structure.

Fortunately, measurements over a range of temperature may enable the separation of different contributions to be made. For the illustrative case of alkali halides A^+X^-, for which a vast amount of work has been performed, the plot of log σT against $1/T$ (Figure 8.9) includes a number of linear (or close to linear) regions, each one corresponding to a different temperature

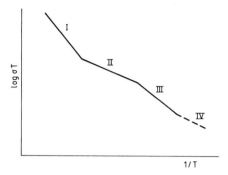

Figure 8.9.
Schematic diagram illustrating the stages usually appearing in the ionic conductivity of alkali halides.

range. It should be pointed out that the transition from one region to the next is not always well defined. Stage I at the highest temperatures is determined by intrinsic or thermally generated defects, i.e. Schottky pairs for alkali halides. On the assumption, as is usually the case, that cation vacancies make the dominant contribution to the ionic current, the conductivity is given by

$$\sigma = ne\mu \qquad (8.34)$$

where n is the concentration of cation vacancies (equal to that of Schottky pairs), e the electron charge and μ the mobility. Now, by recalling the simple discussion in Chapter 2, we can easily write

$$n = N \exp\left(-\frac{g_f}{2kT}\right) \qquad (8.35)$$

and

$$\mu = \frac{4a^2ev}{kT} \exp\left(-\frac{g_m}{kT}\right) \qquad (8.36)$$

where g_f and g_m are the Gibbs free energy for Schottky pair formation and cation vacancy migration, respectively. Substituting equations (8.35) and (8.36) into equation (8.34), we obtain

$$\sigma = N\frac{4a^2e^2v}{kT} \exp\left(-\frac{g_m + g_f/2}{kT}\right) \qquad (8.37)$$

which is verified from the linear dependence of log σT against $1/T$. Obviously, this formula should be modified to include the anion contribution to the current in cases where this is relevant (e.g. NaF). The fitting of the experimental data to the theoretical prediction, including several parameters, is a very appropriate task for a computer and this is now a standard analysis procedure for obtaining the best-fit values for the relevant thermodynamic parameters.

Stage II corresponds to a lower-temperature range, where the concentration of the mobile cation vacancies is dominated by the amount of aliovalent impurities present in the crystal. However, the temperature is high enough to assure that any vacancies are not bound to the impurity partners. Under these conditions, the concentration of cation vacancies is fixed at n_I and the conductivity given by

$$\sigma = n_I \frac{4a^2 e^2 v}{kT} \exp\left(-\frac{g_m}{kT}\right) \tag{8.38}$$

which also yields a straight line in the plot of log σT against $1/T$, allowing for the determination of the activation energy g_m for cation vacancy migration.

At lower temperatures, the situation becomes more complex. A stage III is often detected, where vacancy–impurity association is occurring. The theoretical analysis leads to a formula similar to equation (8.38) but including the binding energy g_b between vacancies and impurities in the activation exponent. At even lower temperatures (room-temperature range), clustering and precipitation occur. The reproducibility of the data is poor and the interpretation becomes rather difficult.

It should be pointed out that the ionic conductivity technique has been very successfully used to obtain values of energies and entropies of formation and migration in alkali halides and other ionic materials.

8.10.5   Dielectric relaxation techniques

Any material presents electronic, ionic and reorientational contributions to its electric polarization and consequently to its dielectric response. The presence of lattice defects modifies this response and constitutes a possible method for their study. Depending on the response time (or frequency) range for the defect-induced polarization, different instrumental techniques are used, although the basic physical methods are the same. They are briefly discussed next.

(a)   Debye relaxation (or the tan δ) method. This method[51] closely resembles that previously considered in discussing internal friction and anelastic losses. The most relevant situation corresponds to defects having an associated electrical moment which can adopt several equivalent orientations in the crystal lattice with a finite jump probability between them. A typical example is provided by the divalent cation impurity–vacancy complexes in alkali halides described in Chapter 5. The preferential alignment of the dipoles under an applied field will give rise to an induced polarization P (dipole moment per unit volume). Because of the non-instantaneous character of the response, i.e. finite relaxation time, the processes of polarization and depolarization are accompanied by energy losses in the material.

The situation can be conveniently analysed for a sinusoidal applied field $E = E_0 \exp(i\omega t)$, which is moreover a practical case. The polarization P will also be sinusoidal with the same frequency ω and will include two contributions, one in phase with E and the other $\pi/2$ out of phase. This latter component is responsible for the energy loss (dielectric losses) in the material. Under the assumption commonly approached in practice that the dielectric relaxation follows an exponential decay law $P = P_0 \exp^{(-t/\tau)}$, τ being the relaxation time, the complex dielectric constant ϵ is given by the Debye formula

$$\epsilon = 1 + \frac{4\pi\alpha_s N}{1 + i\omega\tau} \tag{8.39}$$

where α_s is the static polarizability associated with each dipole and N is the concentration of dipoles. α_s depends on temperature and elementary dipole moment p according to the law $\alpha_s \propto p^2/kT$ (at high temperatures), the detailed expression being a function of the orientational properties of each dipole system. For a molecular dipole in an isotropic medium $\alpha_s = p^2/3kT$.

From equation (8.39), the real part ϵ_1 (in phase with E) and the imaginary part ϵ_2 ($\pi/2$ out of phase with E) can be obtained:

$$\epsilon_1 = 1 + \frac{4\pi\alpha_s N}{1 + \omega^2\tau^2}$$
$$\epsilon_2 = \frac{4\pi\alpha_s \omega\tau N}{1 + \omega^2\tau^2} \tag{8.40}$$

The dependence of ϵ_1 and ϵ_2 on ω is illustrated in Figure 8.10. The dielectric power loss in the material is then given by

$$\mathscr{P} = \frac{E^2}{4\pi} \omega\epsilon_2 = \frac{\epsilon_1 E^2}{4\pi} \omega \tan \delta \tag{8.41}$$

where $\tan \delta = \epsilon_2/\epsilon_1$, δ being the so-called loss angle.

The Debye method for dielectric relaxation usually involves the measurement of ϵ_2, $\mathscr{P}$ or $\tan \delta$ as a function of ω at fixed temperature. The maximum height of the $\epsilon_2(\omega)$ curve measures the dipole concentration, whereas the associated frequency $\omega_m = 1/\tau$ yields the relaxation time τ. By performing experiments of this type at several temperatures, $\tau(T)$ may be determined; this is often of the form $\tau = \tau_0 \exp(-E/kT)$, E being the activation energy for dipole reorientation. A useful way of representing the

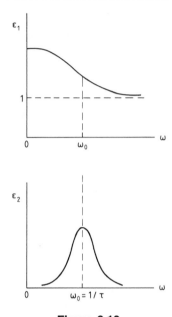

Figure 8.10.
Dependence of the real part ϵ_1 and imaginary part ϵ_2 of the dielectric constant ϵ on frequency ω, for dipoles inside a solid.

Debye-type behaviour is by means of the so-called Cole–Cole plot showing the relation between the real and imaginary parts of the dielectric constant. It is easy to show that the (ϵ_1, ϵ_2) points should be on a half-circle centred at $\epsilon_1 = 1 + 4n\alpha_s N/2$ (or, in general, $\epsilon_1 = \epsilon_0 + 4n\alpha_s N/2$) and $\epsilon_2 = 0$. The Cole–Cole plot is well suited to analysis of situations involving several relaxation processes[52], as illustrated in Figure 8.11. The conventional set-up consists of an impedance bridge which measures the capacitance and conductance of a solid-plate capacitor containing the sample as the dielectric material. Two- and three-terminal holders are in use. Generally, the bridge works in the audiofrequency range and the typical sensitivity limit is 10^{-5} for the atomic dipole concentration.

(b) Dielectric methods at high frequencies. Although capacitance bridges can be used over a wide frequency range (e.g. from 10^{-2} Hz to 100 MHz), depending on sample holders, other techniques are available for very-high-frequency and microwave dielectric measurements[53]. They include the transmission line and cavity methods, which use the influence of a dielectric material on line impedance or on the resonance parameters of a

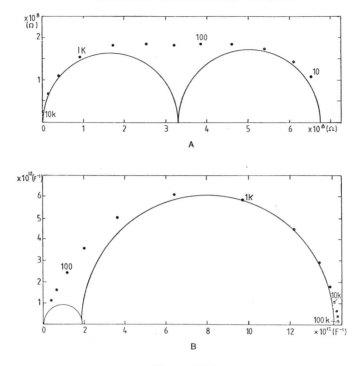

Figure 8.11.

A variant of the Cole–Cole plot for a mixed-phase structure (mostly Suzuki phase) of NaCl–CdCl$_2$. ϵ_1 and ϵ_2 are derived[52] from these (A) complex impedance and (B) permittivity modulus plots. Data were taken at frequencies from 10 Hz to 100 kHz.

resonant cavity. In particular, a rather simple method consists of a comparison between the resonant frequency and Q factor of an empty cavity and the same parameters for the dielectric-filled cavity.

Some very specific methods have also been used for particular applications. We may quote[54] the broadening of the EPR lines for a dipolar impurity–vacancy complex at high temperatures or the use of double electric–magnetic modulation of EPR lines. The latter technique makes it possible to associate a peak in the dielectric relaxation spectrum with a particular structure determined by EPR.

(c) Ionic thermocurrents. In 1964, Bucci and Fieschi[55] introduced a new technique for studying dielectric relaxation which involves the measurement of the thermally assisted depolarization current during heating of the polarized sample. The working principle of the technique can be understood in terms of four main experimental steps in the procedure, which are discussed below and illustrated in Figure 8.12. In the initial step 1, no electric field is applied to the sample and dipoles are oriented at random. In step 2, an electric field E_p is applied during a time τ_p with the sample at a constant temperature T_p. This induces a preferential orientation of the dipoles, i.e. a macroscopic polarization. The sample is then cooled without removing the field during step 3, so that dipole orientation is frozen. Then the field is removed so that the electronic and ionic (not the dipolar) polarization is relaxed. Finally, in step 4, the sample is heated (preferably at a constant rate) and the depolarization current associated with the dipole orientation is detected. This occurs at around a temperature where the relaxation time reaches values comparable with the time scale of the experiment.

One of the main advantages of this technique lies in the fact that the detection of the depolarization current takes place at zero field, resulting in low noise and high sensitivity. In alkali halides, dipole concentrations of 1 ppm or less are routinely detected.

The analysis of the thermally assisted depolarization current is most often carried out by assuming a single Arrhenius-type behaviour for the relaxation process, i.e. having a unique relaxation time τ given by

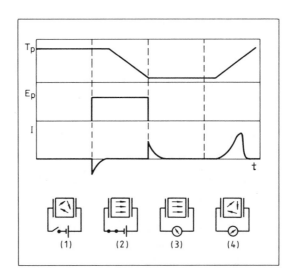

Figure 8.12.
Schematic diagram showing the sequence of operations in a typical ITC experiment (see text). In steps 1 and 4, dipoles are at random whereas, in steps 2 and 3 after application of an electric field, they are preferentially oriented.

$$\tau(T) = \tau_0 \exp\left(-\frac{E}{kT}\right) \tag{8.42}$$

The depolarization current density j is expressed as a function of time by

$$j(t) = -\frac{dP(t)}{dT} = \frac{P}{\tau} \tag{8.43}$$

where P is the instantaneous polarization of the material. The differential equation (8.43) can be easily solved and leads to

$$j(t) = \frac{P(0)}{\tau} \exp\left(-\int_0^t \frac{dt'}{\tau}\right) \tag{8.44}$$

For a linear heating rate $\beta = dT/dt$, equation (8.44) becomes

$$j(T) = \frac{P(0)}{\tau_0} \exp\left(-\frac{E}{kT}\right) \exp\left[\int_{T_0}^T (\beta\tau_0)^{-1} \exp\left(-\frac{E}{kT'}\right) dT'\right] \tag{8.45}$$

This expression corresponds to an asymmetric current peak such as that illustrated in Figure 8.13(a) for Sr^{2+} dipoles in RbCl. Some interesting properties of this ionic thermoconductivity (ITC) peak are as follows.

(1) It presents a maximum at a temperature

$$T_M = \left(\frac{\beta E \tau(T_M)}{k}\right)^{1/2} \tag{8.46}$$

independent of T_p and E_p (temperature and field during polarization of the material).

(2) In the low-temperature limit,

$$\log j = \text{constant} - \frac{E}{kT} \tag{8.47}$$

From equations (8.46) and (8.47), both the activation energy E and the pre-exponential factor τ_0 can be obtained.

Therefore, at variance with the Debye method, a single experiment is required to characterize the dielectric relaxation of the material. Equation (8.47) is only approximate but an exact method can be used to determine τ_0 and E. The area under the ITC peak from the instant t up to ∞ is given by

$$A_t^\infty = \int_t^\infty I(t')\,dt' = S_s\int_t^\infty j(t')\,dt' = S_s P(t) = S_s j(t)\tau \tag{8.48}$$

S_s being the surface of the sample. From equation (8.48), we immediately obtain $\tau = A_t^\infty/I(t)$ and, taking into account the Arrhenius expression for τ, we obtain

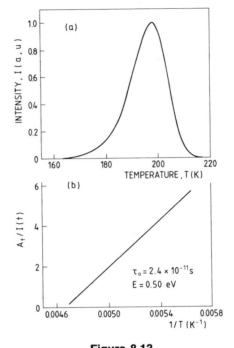

Figure 8.13.
(a) ITC peak of Sr^{2+} dipoles in RbCl; (b) analysis of this peak by using the area method as discussed in the text (after P. Aceituno, Thesis Universidad Autonoma Madrid, 1985).

$$\ln\left(\frac{A_t^{\infty}}{I(t)}\right) = \ln \tau_0 - \frac{E}{kT} \qquad (8.49)$$

giving τ_0 and E from the values of A_t^{∞} and $I(t)$ at various times. The use of this analysis is illustrated in Figure 8.13(b).

The initial polarization $P(0)$ of the material and consequently the total concentration of dipoles can also be obtained from the ITC peak. In fact, the total area under the peak is given by

$$A_0^{\infty} = S_s P(0) \qquad (8.50)$$

Assuming N dipoles per unit volume with dipole moment μ, we may obtain (for a cubic field)

$$P(0) = \frac{A_0}{S} = \frac{N\mu^2 E_p}{3kT_p} \qquad (8.51)$$

which determines N, if μ is known.

Another very useful advantage of the ITC method is the possibility of separating heavily overlapping peaks by recourse to spectral cleaning techniques. These involve the partial erasure of the lower-temperature peaks during a previous run over a limited temperature range.

In the theoretical analysis for the ITC peaks, non-interacting dipoles have been assumed. The role of dipolar interactions on the shape of the ITC peak has been considered by several researchers[56,57].

In a typical experimental set-up the sample is sandwiched between two electrodes in a chamber which is either evacuated (for direct thermal contact between sample and cryostat) or under a low-pressure atmosphere, e.g. nitrogen. In most experiments the heating cycle extends from about liquid-nitrogen temperature to room temperature which covers the ITC peak corresponding to impurity—vacancy dipoles in alkali-halide crystals. The control of temperature and heating rate is quite critical for analysis of the data. The depolarization currents are often in the range $10^{-14}-10^{-15}$ A, thus requiring very good isolation and screening.

As a final comment on dielectric relaxation methods, we would like to remark that some workers have measured the depolarization current as a function of time at various temperatures. This method permits the determination of the relaxation parameters, although in principle the ITC method appears preferable.

8.11 TRANSIENT-CAPACITANCE AND TRANSIENT-CURRENT METHODS FOR SEMICONDUCTORS

8.11.1 Basic phenomena

These methods now play a prominent role in the field of defects in semiconductors. They have been developed to a point where they constitute a true spectroscopic tool to locate impurity and defect levels inside the gap of a semiconductor material. They present some similarities with the dielectric relaxation methods previously discussed for insulators. A variety of transient methods exist which are well documented[58,59]. All of them essentially use the change in capacitance (or current) for a Schottky or p—n junction induced by the relaxation of the electronic system following an initial perturbation of the thermal equilibrium situation. Before entering into the discussion of the most relevant methods, some general concepts referring to semiconductor junctions will be briefly summarized.

Let us consider, as a working example, a step (abrupt) p—n junction determined by joining a p-type material with an N_A acceptor concentration and a n-type material with a donor concentration N_D. At thermal equilibrium, the energy level diagram is as illustrated in Figure 8.14, as a consequence of the matching of the Fermi levels on both sides of the junction. Majority carriers (electrons from the n side and holes from the p side) have crossed to the opposite side, forming a double charged layer at the junction. The potential difference between the p and n sides is given by

$$V_n - V_p = \frac{D_e}{\mu_e}\frac{n_n}{n_p} \qquad (8.52)$$

where $n_n = N_D$ and $n_p = n_i^2/N_A$ are the majority carrier concentrations at the n and p sides, respectively, of the junction, n_i being the intrinsic carrier density of the semiconductor. One important feature is the much reduced concentration of free carriers in a depletion region around the junction plane. The depths x_n and x_p of the depletion region in the n and p sides obey the simple relationship $x_n N_D = x_p N_A$, indicating that the depletion region will increase in thickness as the doping level decreases. The total potential difference V across the junction is the sum of the voltages across the two parts:

$$V = \frac{e}{2\epsilon}(N_D x_n^2 + N_A x_p^2) \qquad (8.53)$$

The electronic behaviour of the junction allows us to define a so-called differential or small-voltage capacitance as $C = dQ/dV$, where Q is the charge on each side of the junction and V is the corresponding voltage

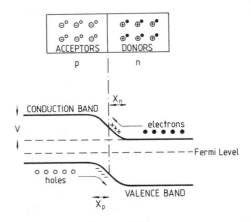

Figure 8.14.
Electronic structure of an abrupt p–n junction, showing band bending and majority-carrier populations.

difference. It can be easily shown that C is the capacitance corresponding to an ordinary capacitor with the thickness $x = x_n + x_p$ and the permittivity ϵ of the depletion layer:

$$C = A\frac{\epsilon}{x} \qquad (8.54)$$

A being the area of the junction.

By changing the voltage across the junction, the length x of the depletion region is modified and so is the junction capacitance. The relationship between capacitance and voltage is $C \propto V^{-1/2}$, indicating that, as the capacitance decreases, the voltage increases, i.e. on going to increasing reverse-bias polarization.

The influence of electron or hole traps on the junction capacitance can be analysed by considering their capture and emission behaviour for free carriers. Let us consider an electron trap with concentration N in the depleted region of a n^+-p junction (i.e. $x_p \gg x_n$). Its capture parameters c_1 and c_2 and emission parameters e_1 and e_2 for electron and holes, respectively, are indicated in the schematic diagram of Figure 8.15. The two emission rates are thermally activated with activation energies equal to the energy separation from the conduction and valence bands. In the depleted p-type region, where the free-carrier density is very low, the electron occupation of the trap level at thermal equilibrium (steady state) is given by $n/N = e_2/(e_1 + e_2)$, which for a proper electron trap is close to zero. This situation applies to reverse-bias polarization. When a forward-bias pulse is applied, minority carriers (electrons) are injected in the depleted p-type region, leading to an increased occupation ratio for the traps. This additional negative charge at the depletion region will lead to a higher junction capacitance. For a saturating injecting pulse, all the traps will be filled and the capacitance transient can be easily calculated. From equation (8.53), for an n^+-p junction, $V = (e/2\epsilon)N_A x_p^2$; the decrease in the depth x_p which is

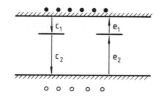

Figure 8.15.
Diagram illustrating the various emission and trapping processes in a doped semiconductor.

induced by filling the traps at constant voltage is $dx_p/x_p = \frac{1}{2}N/N_A$. Therefore, equation (8.54) immediately leads to

$$\frac{dC}{C} = \frac{1}{2}\frac{N}{N_a} \tag{8.55}$$

resulting in a positive change in capacity. A similar analysis applies if a majority-carrier pulse instead of a minority-carrier pulse is used in the experiment.

After the completion of the pulse, the junction is again under reverse-bias conditions and the electron trap will be emptied exponentially with time with a decay constant $e_1 + e_2$ ($e_1 \gg e_2$ for electron traps). This will appear as a decrease in the associated capacitance which can be measured by suitable bridge techniques. From this transient-capacitance experiment, e_1 can be obtained and by repeating the experiments at various temperatures the thermal depth of the level in the gap can be determined.

Further versions of this technique are to be discussed next. It should be pointed out that all transient-capacitance methods can be replaced by essentially equivalent current methods. In fact, the relaxation of the capacitance following an injection pulse is associated with electron (or hole) emission from the filled traps and therefore with a current transient which can be measured instead of the capacitance transient.

8.11.2   Thermally stimulated capacitance and current

An interesting version of the transient-capacitance methods involves the measurement of the thermally stimulated capacitance change induced by a linear-heating run at constant-bias voltage following a carrier injection pulse. It is one of the various thermally stimulated methods that can be used to study transient electronic or ionic phenomena in solids. The thermally stimulated capacitance technique operates as follows. The bias voltage causing the minority-carrier injection considered in the above section to fill the electron traps is maintained whereas the material is cooled to an appropriate low temperature. After removal of the bias field at this low temperature, the carrier emission rate e_2 from the filled traps is kept frozen. Finally, the reverse-polarized junction is heated at a given rate. When a temperature is reached where the emission rate e_1 becomes appropriate to induce appreciable carrier emission, a transient-capacitance peak is observed. The thermal ionization energy of the level can then be determined from the temperature T_m of the maximum and the rate β of temperature change.

The transcription of this technique to measure the thermally stimulated current instead of capacitance does not involve any new physical concepts.

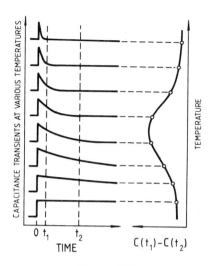

Figure 8.16.
Illustration of the working principle of DLTS (see text).

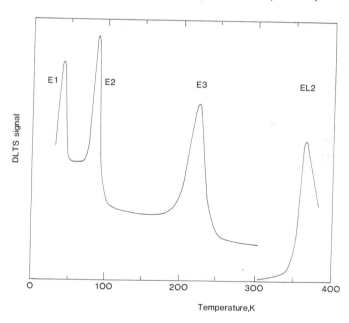

Figure 8.17.
Examples of the E1, E2 and E3 DLTS signals induced by electron irradiation of n-type GaAs at 80 K and the EL2 signal found in material grown by vapour-phase epitaxy[63].

8.11.3 Deep-level transient spectroscopy

This technique is now most widely used to investigate deep traps in Schottky or p–n junctions. Initially developed by Lang and co-workers[60,61] and improved by Lefevre and Schulz[62], it is based on the same physical principles as the transient methods previously discussed. The peculiarity of the technique is that the measurement of the capacitance transient in every run is made through a time (or rate) window which can be selected. Let us assume that the window is defined by the time interval $t_1 - t_2$. The deep-level transient spectroscopy (DLTS) signal S_{DLTS} is the normalized difference between the capacitance values at t_1 and t_2, i.e.

$$S_{\text{DLTS}} = \frac{C(t_1) - C(t_2)}{\Delta C(0)} \tag{8.56}$$

where $\Delta C(0) = C(0) - C(\infty)$. A time window is selected which is appropriate for a selected lifetime and as shown in Figure 8.16 it records the signal as a function of temperature during the heating cycle. In general, the spectrum obtained by DLTS will consist of several maxima, each corresponding to a different active trap. For each maximum, the value τ_m for the relaxation time is related to the "window" times through the relationship

$$\tau_m = (t_1 - t_2)\left[\ln\left(\frac{t_1}{t_2}\right)\right]^{-1} \tag{8.57}$$

i.e. one window selects a number of DLTS peaks all having the same rate (equation (8.57)) at the peak temperature. The effect of changing the window is to move the peak position, in temperature, and hence the relaxation time at the peak value. A plot of log τ_m against $1/T$ can be constructed and therefore the activation energy determined from $\tau_m = \tau_0 \exp(-E/kT)$.

A wide range of trap depths can be investigated by this technique, which lends itself conveniently to automatic operation. Its advantages over the alternative transient techniques are discussed in the original paper by Lang[60]. Figure 8.17 shows spectra obtained by DLTS for traps in n-type GaAs[63] in material grown by vapour phase epitaxy.

REFERENCES

1 J. C. H. Spence, *Experimental High Resolution Microscopy*, Clarendon, Oxford (1981).

2 J. M. Cowley, High resolution microscopy of crystal defects and surfaces, *Annu. Rev. Phys. Chem.* **29**, 251 (1978).

3 L. Reimer, *Scanning Electron Microscopy*, Springer, Berlin (1984).

4 G. Thomas and M. J. Goringe, *Transmission Electron Microscopy of Materials*, Wiley, New York (1979).

5 L. Hobbs, *J. Phys. (Paris), Colloq.* C7 **37**, 3 (1976).

6 D. K. Bowen and C. R. Hall, *Microscopy of Materials*, Wiley, New York (1975).

7 B. G. Yacobi and D. B. Holt, *J. Appl. Phys.* **59**, 1 (1986).

8 E. W. Müller and T. T. Tsong, *Field-Ion Microscopy*, Elsevier, New York (1969).

9 T. M. Hayes and J. B. Boyce, *Solid State Phys.* **37**, 173 (1982).

10 D. R. Sandstrom and F. W. Lytle, Development in extended X-ray absorption fine structure applied to chemical systems, *Annu. Rev. Phys. Chem.* **30**, 215 (1979).

11 T. M. Hayes, J. B. Boyce and J. L. Beeby, *J. Phys. C: Solid State Phys.* **11**, 2931 (1978).

12 J. Jaklevic, J. A. Kirby, M. P. Klein and A. S. Robertson, *Solid State Commun.* **23**, 679 (1977).

13 G. Binnig, H. Rohrer, Ch. Gerber and E. Weibel, *Appl. Phys. Lett.* **40**, 178 (1982); *Phys. Rev. Lett.* **49**, 57 (1982).

14 P. K. Hansma and J. Tersoff, *J. Appl. Phys.* **61**, R1 (1987).

15 A. M. Baro, G. Binnig, H. Rohrer, Ch. Gerber, E. Stoll, A. Baratoff and F. Salvan, *Phys. Rev. Lett.* **52**, 1304 (1984).

16 A. Guinier and G. Fournet, *Small Angle Scattering of X-Rays*, Wiley, New York (1955).

17 P. Haasen, *Physical Metallurgy*, Cambridge University Press, Cambridge (1978).

18 A. E. Hughes and S. C. Jain, *Adv. Phys.* **28**, 717 (1979).

19 M. Lambert, C. Mazieres and A. Guinier, *J. Phys. Chem. Solids* **18**, 129 (1968).

20 M. W. Thompson, *Defects and Radiation Damage in Metals*, Cambridge University Press, Cambridge (1969).

21 Y. Queré, *Defauts Ponctuels dans les Métaux*, Masson, Paris (1967).

22 A. S. Nowick and B. S. Berry, *Anelastic Relaxation in Crystalline Solids*, Academic Press, New York (1972).

23 R. de Batist, *Internal Friction of Structural Defects in Crystalline Solids*, North-Holland, Amsterdam (1972).

24 G. Fantozzi, C. Esrouf, V. Benoci and I. G. Ritchie, *Prog. Mater. Sci.* **27**, 311–451 (1982).

25 J. Woirgard, Ph. Mazot and A. Riviere, *J. Phys. (Paris), Colloq. C5* **42**, 1135 (1981).

26 H. B. Johnson, N. J. Tolar, C. R. Miller and I. B. Cutler, *J. Phys. Chem. Solids* **30**, 31 (1969).

27 J. W. Corbett, G. D. Watkins, R. M. Chrenko and R. S. McDonald, *Phys. Rev.* **121**, 1015 (1961).

28 R. H. Silsbee, *J. Phys. Chem. Solids* **28**, 2525 (1967).

29 U. F. Kocks, in P. Haasen, V. Gerold and G. Kostorz (eds), *Proceedings of the Fifth International Conference on the Strength of Metals and Alloys, Aachen, 1979*, Pergamon, Oxford (1980).

30 T. E. Mitchell and A. H. Heuer, *Mater. Sci. Engn.* **28**, 81 (1977).

31 W. Skrotzki and P. Haasen, *J. Phys. (Paris), Colloq. C3* **42**, 119 (1981).

32 J. Soullard and P. Veyssiere, *Acta Metall.* **31**, 1177 (1983).

33 U. F. Kocks, A. S. Argon and M. F. Ashby, *Prog. Mater. Sci.* **52**, 196 (1975).

34 I. S. Lerma and F. Agulló-Lopez, *J. Appl. Phys.* **41**, 4628 (1970).

35 J. P. Poirier, *Plasticité à Haute Température des Solides Cristallins*, Eyrolles, Paris (1976).

36 J. Weertman, in J. C. M. Li and A. K. Mukherjee (eds), *Rate Processes in Plastic Deformation of Materials*, American Society for Metals, Metals Park, OH (1978).

37 J. Bretheau, J. Castaing, J. Rabier and P. Vegsiere, *Adv. Phys.* **28**, 835 (1979).
38 W. R. Cannon and T. G. Langdon, *J. Mater. Sci.* **18**, 1 (1983).
39 A. W. Overhauser, *Phys. Rev.* **94**, 1551 (1954).
40 L. Delgado and J. L. Alvarez Rivas, *J. Phys. C: Solid State Phys.* **12**, 3159 (1979).
41 J. Loram, to be published.
42 T. H. Geballe and G. W. Hall, *Phys. Rev.* **110**, 773 (1958).
43 R. Berman, P. G. Klemens, F. E. Simon and T. M. Fry, *Nature (London)* **166**, 864 (1950).
44 D. S. Billington and J. H. Crawford, *Radiation Damage in Solids*, Princeton University Press, Princeton, NJ (1961).
45 Ph.D. Tracher, *Phys. Rev.* **156**, 975 (1967).
46 D. Emin, *J. Solid State Chem.* **12**, 246 (1975); *Adv. Phys.* **24**, 305 (1975).
47 W. E. Spear, *Proc. Phys. Soc.*, London, **76**, 826 (1960); *J. Phys. Chem. Solids* **21**, 110 (1961).
48 J. M. Honig, in A. Dominguez-Rodriguez, J. Castaign and R. Marque (eds), *Basic Properties of Binary Oxides*, Universidad de Sevilla, Seville (1984).
49 P. Suptitz and J. Teltow, *Phys. Status Solidi* **23**, 9 (1967).
50 A. V. Chadwick, *Radiat. Eff.* **74**, 17 (1983).
51 V. V. Daniel, *Dielectric Relaxation*, Academic Press, New York (1967).
52 P. J. Chandler and E. J. Lilley, *Phys. Status Solidi (a)* **66**, 183 (1981).
53 H. E. Bussey, *Proc. IEEE* **55**, 1046 (1967).
54 A. Edgar, *J. Phys. E: Sci. Instrum.* **10**, 1261 (1977).
55 C. Bucci and R. Fieschi, *Phys. Rev. Lett.* **12**, 16 (1964).
56 W. Von Weperen, B. P. M. Lenting, E. J. Bijvank and H. W. der Hartog, *Phys. Rev. B* **18**, 2857 (1978).
57 F. Cusso, R. Aceituno, H. Murrieta and F. J. Lopez, *Phys. Rev. B* **31**, 8119 (1985).
58 J. L. Pautrat, B. Katiraoglu, N. Magnea, D. Benschel, J. C. Pfister and L. Revoil, *Solid-State Electron.* **23**, 1159 (1980).
59 D. Stievenard, M. Lannoo and J. C. Bourgoin, *Solid-State Electron.* **28**, 485 (1985).
60 D. V. Lang, *J. Appl. Phys.* **45**, 2033 (1974).
61 G. L. Miller, D. V. Lang and L. C. Kimmerling, *Annu. Rev. Sci.* **7**, 377 (1977).
62 H. Lefevre and M. Schulz, *Appl. Phys.* **12**, 45 (1977).
63 A. A. Rezazadeh and D. W. Palmer, *J. Phys. C: Solid State Phys.* **18**, 43 (1985); *Proceedings of the Eleventh International Conference on Defects and Radiation Effects in Semiconductors, Oiso, 1980, Inst. Phys Conf. Ser.* **59**, 317 (1981).

9
Experimental Techniques II—Optical

9.1 INTRODUCTION

In this chapter a second group of experimental techniques, mostly optical, used to investigate the structure and properties of point defects is described. Essentially all spectroscopic techniques involve a beam of photons or other particles (electrons, neutrons, etc.) incident on a defective solid and acting as structural probes. The solid may be appropriately perturbed by subjecting it to electric, magnetic or stress fields.

Electromagnetic spectroscopy techniques are particularly suitable for semiconductor and insulating materials, because of the window offered by their electronic structure to electromagnetic radiation. Moreover, major advantages of optical techniques are their simplicity, sensitivity and resolution. Depending on the techniques and the operative wavelength range, they are efficient tools for investigating the geometrical, rotational, vibrational and electronic structure of point defects in materials.

9.2 OPTICAL ABSORPTION AND LUMINESCENCE

9.2.1 Spectra

In non-metallic materials, defects and impurities can be considered as localized electron or hole centres that may have electronic (as well as vibrational or rotational) levels within the energy gap. The transitions between these levels give rise to absorption and luminescence in spectral

255

regions where the host material is transparent. Consequently, each impurity or defect centre is characterized by its absorption or luminescence spectrum containing one or several bands.[1-5] The absorption and luminescence spectra are respectively specified through the $\mu(E)$ or $I(E)$ curves, where μ is the absorption coefficient, I is the intensity of the emitted light and $E = h\upsilon = hc/\lambda$ is the photon energy (υ is the frequency and λ the wavelength of the light).

The absorption coefficient μ is defined by $\mu = (1/d) \log(I_0/I)$ where I is the light intensity transmitted through a parallel plane sample of thickness d and I_0 is the intensity of the beam incident normally on the sample. (Reflectivity corrections will be mentioned later.) This formula immediately follows from the differential equation giving the intensity dI absorbed by an infinitesimal thin plate of material of thickness dx at a distance x from the first surface: $dI = -\mu I(x)\, dx$. Simple integration yields $I(x) = I_0/\exp(-\mu x)$, from which μ is easily obtained.

A simple relationship exists between concentration N of absorbing centres and the area under the absorption band. This is the Smakula formula (as modified by Dexter):

$$Nf = \frac{9}{2} \frac{cm^\star}{\pi^2 e^2 \hbar} \frac{n}{(n^2 + 2)^2} \int \mu(E)\, dE \qquad (9.1)$$

where f is the oscillator strength of the optical transition and n is the refractive index at the spectral region of the band. f is related to the matrix element of the transition and is a measure of the allowed character of the transition ($0 \leqslant f \leqslant 1$). $m^\star$ is the effective mass of the electron. For optical bands of known shapes, equation (9.1) can be written as an equivalent relationship between N, the height μ_m and the full width W at half-maximum of the band. As an example, for Gaussian bands, which are most often found in practice,

$$Nf = 0.87 \times 10^{17} \frac{n}{(n^2 + 2)^2} \mu_m W \qquad (9.2)$$

whereas, for Lorentzian bands,

$$Nf = 1.29 \times 10^{17} \frac{n}{(n^2 + 2)^2} \mu_m W \qquad (9.3)$$

if μ_m is expressed in reciprocal centimetres and W in electronvolts.

In this way, absorption spectroscopy provides a simple relative method for measuring the concentration of absorbing centres. Absolute determinations are very difficult as N must be separated from the product term Nf.

The measurement of μ is carried out with a spectrophotometer. The basic instrument consists of a light source and a monochromator. The intensity of the transmitted beam is detected and compared with a reference beam. Most instruments yield either the transmittance $T = I/I_0$ or the absorbance (i.e. the optical density) $D = \lg (1/T) = \mu d/2.303$. In principle, commercial spectrophotometers are capable of giving a reliable measurement of transmittance within the approximate range 10^{-6} to 0.99 or in absorbance 0.01– 6. The lower limit in transmittance is essentially set by the stray light of the monochromator rather than electronic noise. On the assumption that the oscillator strength is unity, the minimum signals correspond to some 10^{14} defects in the optical path; that is, for a typical sample of 10 mm × 10 mm × 1 mm dimensions a defect concentration of more than 10^{15} defects cm^{-3} is required. In order to achieve high sensitivity and low noise over a wide dynamic range, or to correct for reflectivity losses, many modern spectrophotometers incorporate computer systems which record data and present a signal-enhanced output. Whilst these are generally excellent, there are some well-known examples of these automatic data-processing packages which generate false features in the absorption curves. A note of caution is therefore appropriate.

On interpreting the spectrophotometer readings, we must take into account that a fraction R_i of the incident beam is reflected at each sample–air interface, so that the total transmitted light is given by

$$I = I_0(1 - R_i)^2 \exp(-\mu d) + I_0(1 - R_i)^2 \exp(-\mu d) R_i^2 \exp(-2\mu d) + \ldots$$

$$= \frac{(1 - R_i)^2 \exp(-\mu d)}{1 - R_i^2 \exp(-2\mu d)} I_0 \tag{9.4}$$

For low-reflectivity materials, such as alkali halides, $R_i \approx 5\%$ and the approximation $I = I_0 \exp(-\mu d)$ is adequate. In contrast, for high-reflectivity materials such as rutile, $R_i \approx 20\%$, and the full equation (9.4) must be used for absorption measurements. As a first-order correction for the reflected light contribution, it is a common practice to use a dummy sample in the reference beam of the spectrophotometer.

All the above expressions are valid for light normally incident on plane parallel specimens. For rough material, it is possible to use an integrating sphere with perfectly reflecting walls. Under illumination a uniform power density is reached inside the sphere, which is then monitored by a photodetector. If a specimen is inserted into the sphere, the power level is reduced and this gives a measure of the strength of the optical absorption band. Finally it should be noted that false absorption values are recorded if the sample is luminescent and the light emitted is recorded by the detector.

The absorption and luminescence bands, when monitored with polarized light, provide information on the symmetry, ordering and energy separation of electronic levels of defects in the material. Theoretical models for defect centres can readily be tested against the experimental data and values of the Hamiltonian parameters ascertained. For example, for transition-metal or rare-earth impurities, where a crystal or ligand field scheme is in order, the comparison between optical data and theoretical predictions provides estimates of the crystal field strength and degree of covalency with the ligand ions.

In a defective material, not only do the observed optical bands (absorption and luminescence) correspond to transitions between discrete levels of an impurity or defect, but they may also be associated with transitions to or from one of the levels in the conduction or valence band[6]. From the fitting of these spectra to theoretical predictions, donor and acceptor binding energies have been determined for a large number of semiconductor systems.

The familiar example of the absorption spectra of ruby demonstrates several of these features. Ruby contains chromium atoms on aluminium sites in Al_2O_3. This is a rhombohedral crystal and thus there is an anisotropic crystal field acting on the impurity. Optical absorption measurements with light polarized parallel or perpendicular to the c axis of the crystal, which in this case is also the symmetry axis of the defect site, show distinct differences, as presented in Figure 9.1.

Defect sites may equally perturb the energy states or transition paths of other lattice features. For example excitons can preferentially be created or annihilated near a defect and this is apparent in changes in the exciton lifetime or in the absorption or emission spectra. Figure 9.2 shows luminescence spectra associated with exciton decay in NaCl for nominally "pure" crystals obtained from different manufacturers in which both the precise emission energies and the balance in decay routes are shifted[7]. A large variety of impurity and defect-bound excitons have been identified through their optical bands in insulator and semiconductor materials.

Finally, free-carrier (intraband) absorption can take place in a material if the presence of defects acting as donors promotes electrons to the conduction band. This happens, indeed, in most extrinsic semiconductors and in many reduced oxides, e.g. $SrTiO_3$, where oxygen vacancies induce donor states close to the conduction band. Free-carrier absorption is characterized by a monotonic, generally structureless spectrum, growing as λ^p, with p often close to 2 as expected from the Drude theory. The value of exponent p depends on the dominant electron-scattering mechanism. Therefore, this mechanism as well as the concentration of carriers can be inferred from the measured absorption.

Infrared absorption and luminescence spectra can provide information on the vibrational properties of impurities and defects. Point defects destroy

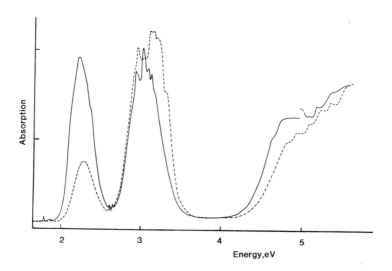

Figure 9.1.
Optical absorption of ruby (Al_2O_3:Cr) for light polarized parallel (– – –) and perpendicular (——) to the c axis.

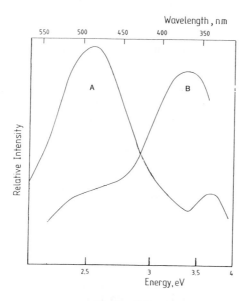

Figure 9.2.
Ion-beam-induced luminescence of exciton bands in NaCl showing differences for "pure" crystals from different manufacturers: curve A, Harshaw; curve B, Rank.

the translational symmetry of the crystal and modify the normal modes of vibration. Two cases are possible.

(1) The perturbed modes lie within the frequency bands of the perfect lattice (band modes) and so they correspond to travelling waves.

(2) New modes appear at frequencies higher than the maximum for the perfect crystal (localized modes) or between the bands of allowed frequencies (gap modes).

In case (2) the vibrational mode is localized at and around the defect. A localized mode appears if the impurity atom is lighter than the host atoms (provided that binding forces are equal).

The theory of vibrations in defective crystals and a number of experimental results for impurities and impurity clusters in ionic and covalent crystals can be read for example in review papers by Newman[8]. Infrared spectroscopy has been applied to many systems ranging from hydrogen ions in alkali halides and alkaline-earth fluorides, interstitial oxygen in silicon and germanium to impurities in GaAs.

The GaAs example[8] has been monitored at high resolution and the lattice location of the impurity can be identified, the isotopic effects sensed and the consequences of coupling between pairs of ions noted. Figure 9.3 is particularly interesting as it shows that in silicon-doped GaAs the silicon can occupy the gallium or the arsenic lattice sites and in some cases the impurity ions pair together or with other defects.

9.2.2   Band shapes and bandwidths—configuration coordinate diagrams

Depending on the electron−lattice coupling, different band shapes are observed in absorption and emission spectra, as discussed in sections 5.5.1 and 5.5.4. For strong linear coupling, the minima of the adiabatic potential energy surfaces for the ground and excited states are well separated in the q (normal coordinate) space, as exemplified by the configuration coordinate diagram (Figure 5.7). This is the situation found for F centres and most electronic states in transition-metal ions. The optical bands have, then, a Gaussian shape and their width obeys the following law as a function of temperature:

$$W(T) = W(0)\left[\coth\left(\frac{\hbar\omega}{2kT}\right)\right]^{1/2} \tag{9.5}$$

where $\hbar\omega$ refers to the ground ($\hbar\omega_g$) or excited ($\hbar\omega_e$) states for absorption or luminescence bands, respectively. From a measurement of the bandwidth as a function of temperature, the frequency ω of the phonon coupled to the ground (absorption) or excited (luminescence) states can be obtained. Moreover, the adiabatic potential energy surfaces for both the ground and the

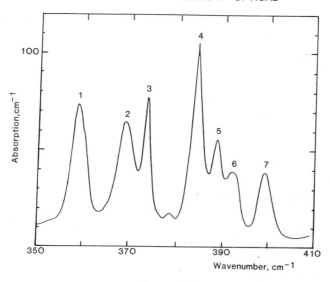

Figure 9.3.
Localized vibrational modes for silicon-doped GaAs: peak 1, ^{30}Si$-$X; peak 2, ^{28}Si$-$X; peak 3, ^{30}Si$_{Ga}$; peak 4, ^{30}Si$_{Ga}-^{30}$Si$_{As}$, ^{28}Si$_{Ga}$; peak 5, ^{28}Si$_{Ga}-^{30}$Si$_{As}$, ^{30}Si$_{As}$, ^{30}Si$_{Ga}-^{28}$Si$_{As}$; peak 6, ^{28}Si$_{Ga}-^{28}$Si$_{As}$; peak 7, ^{28}Si$_{As}$.

excited states as well as their relative position can be fully determined from the low- and high-temperature values of the peak position and width of both absorption and emission bands.

For weak linear coupling ($S \approx 1$), the shape of the optical bands is quite different. In this case, at low temperatures the $0 \rightarrow 0$ vibronic transition (or zero-phonon line) or a low-order vibronic transition are dominant, depending on the S value, with a smaller contribution of other close vibronic lines (e.g. Figure 6.2). If higher-order coupling is neglected, the energy separation between the various vibronic components is constant and equal to the quantum energy of the vibrational mode coupled to the final level of the transition.

One relevant advantage of the zero-phonon line (as well as other vibronic lines) is its small width (about 1 Å). Moreover, this spectral purity is very convenient for studies of the effect of external perturbations. However, the width of these vibronic lines is much larger than it should be from intrinsic (homogeneous) mechanisms since it is essentially dominated by inhomo-geneous broadening effects.

9.2.3 Perturbed optical spectra

The information gathered with the conventional optical spectra can be extended by studying the effect of external perturbations, such as mechanical stress and electric or magnetic fields. These effects are, indeed, much more conveniently studied in narrow-line (vibronic) spectra, although the availability of highly sensitive modulation and phase detection techniques has made possible their extension to broad-band spectra.

(a) Mechanical stress.   Hydrostatic as well as uniaxial stress can be applied to a sample. Hydrostatic stress does not reduce the point group symmetry of the centre and therefore it causes a shift of the zero-phonon line without inducing any splitting. This provides information on the volume dependence of the energy levels of the centre or the electron coupling to the totally symmetric (breathing) mode.

The pressure dependence of zero-phonon peaks are now so well documented that they can be used as a monitor of pressure in work in the kilobar pressure range. As an example the R_1 line of ruby moves about 4 nm with a pressure change[9] of 100 kbar.

For broader absorption bands such as those caused by transitions to the excited states of the F centre in alkali halides (the K, L_1, L_2 and L_3 bands), it was noted that the energy shifts for the bands differed in both magnitude and sign. The explanation for this was that the excited states lie within different conduction bands and the changes in dE/dP are a function of energy band curvature[10,11].

In contrast, uniaxial stress experiments may increase the electronic and orientational degeneracy of the centre. The stress-splitting pattern and any stress-induced dichroism then indicate the character of the electronic degeneracy and whether it occurs in the ground or excited states or both. The stress-splitting pattern directly derives from the application of group theory and first-order perturbation theory[12].

(b) Electric-field-induced (Stark) spectra.   Externally applied electric fields may induce a number of effects in the electronic structure of defects, such as splitting of degenerate levels and mixing of states sufficiently close in energy. Consequences of the electric field are shifting and splitting of line positions, broadening of bands, changes in radiative lifetimes, polarization and quenching of luminescence. Examples of all these effects are reported in the literature[1−4].

(c) Magnetic-field-induced spectra.   The application of a magnetic field to a sample influences the optical spectra[1,13,14] in several related ways, which constitute the so-called Zeeman, magnetic circular dichroism and Faraday rotation effects. It is easy to understand these effects if the spin and the electron−lattice coupling are ignored. Let us consider an optical transition s → p from the ground s state to an excited orbital triplet of a given centre. The threefold degeneracy is resolved by the application of a magnetic field H, as indicated in Figure 9.4. The energy separation between consecutive split levels is $g\beta H$ and consequently the three possible transitions have different energies (Zeeman effect). Moreover, the three magnetic sublevels, $M_S = +1$, 0 and -1, can be reached from the ground state by transitions which are respectively allowed for right-handed circular (σ^+), left-handed circular (σ^-) and linear (π) polarized light. Consequently, different spectra will be obtained depending on the polarization state of the incident light. In particular, the difference $\alpha^+ - \alpha^-$ between the absorption for right- and left-handed circularly polarized light constitutes the magnetic circular dichroism (MCD). The MCD spectra or curves of $\alpha^+ - \alpha^-$ against ω appear typically as illustrated in Figure 9.5(a).

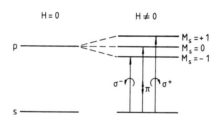

Figure 9.4.
Effect of a magnetic field on s and p levels. The σ and π transitions between split levels are shown.

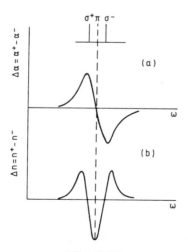

Figure 9.5.
(a) MCD spectrum of $\alpha^+ - \alpha^-$ against ω; and (b) Faraday rotation spectrum of $n^+ - n^-$ against ω. The spectra correspond to the transitions in Figure 9.4.

The Faraday rotation is associated with the difference in the refraction indexes n^+ and n^- for σ^+ and σ^- light. The difference causes a rotation of the polarization axis of linearly polarized light traversing the sample. The Faraday rotation spectra of $n^+ - n^-$ against ω are as illustrated in Figure 9.5 (b). Of course, these spectra are related to the MCD spectra by the Kramers–Kronig relations and therefore contain essentially the same information.

A more thorough discussion of the magnetic-field-induced phenomena, including spin–orbit and electron–lattice interactions can be consulted in the appropriate literature.

The basic information to be inferred from those magnetic-field-induced spectra refers to the degeneracy of electronic levels, g values and spin–orbit coupling strength.

9.2.4 Complementary techniques

More information about the geometrical and electronic structure of defect centres can be obtained by using some complementary optical tools, such as modulation methods, polarized light and lifetime measurements.

(a) Modulation spectroscopy. This has been extensively used in semi-conductor physics[15] to learn about the Van Hove singularities in interband transitions and can be applied to enhance weak features in the spectra. Two alternative experimental methods are in order. One of these involves the direct modulation of the light wavelength at a certain frequency, and the derivative of the spectrum with respect to wavelength is then recorded by means of phase-sensitive detection synchronized with the modulation. In the other method, the light wavelength is kept fixed, whereas the transition energy is modulated by applying an oscillating electric or stress field. The derivative spectrum, as for other spectroscopy techniques such as electron paramagnetic resonance (EPR) improves resolution and enhances signal-to-noise ratios.

(b) Polarized light measurements. Absorption and emission spectra in-volving polarized light are particularly useful when dealing with anisotropic defects. Linearly polarized light will only induce transitions whose transition matrix vector has a component parallel to the light vector. Conse-quently, absorption or emission spectra obtained using linearly polarized light will depend on the relative orientation of the polarization with regard to the centre axis (dichroism). A simple example for ruby optical absorption has already been given (Figure 9.1) in which the crystal and defect axes are aligned. However, the crystal and defect symmetries may differ. Further, the transition matrix vector is not simply related to the structure of the defect; hence in practice the analysis is non-trivial.

The situation of non-cubic defects in a cubic host has been thoroughly analysed by Kaplyanskii[12]. Experimental methods and analysis of data are also examined in detail in the book by Feofilov[16].

The dichroic absorption of anisotropic centres has been used to induce preferential orientation by selective bleaching from energetically equivalent spatial configurations (e.g. to enhance $\langle 110 \rangle$ defects from the family of [110] centres). Much work along these lines has been performed on F_2 and V_k centres in alkali halides.

(c) Lifetime measurements. The data obtained by means of static absorption and luminescence experiments can be complemented by measuring the lifetime τ of a given transition. Radiative lifetimes together with their temperature dependence may yield interesting information on the character and space extension of electronic wavefunctions, electron–phonon coupling and separation of energy levels from the conduction or valence bands of the perfect material. As an example, let us consider the emission from the excited to the ground level of a given centre. If the upper level is close in energy (ΔE) to the conduction band, the lifetime τ will depend on temperature according to the simple formula

$$\frac{1}{\tau} = \frac{1}{\tau_R} + \frac{1}{\tau_0} \exp\left(-\frac{\Delta E}{kT}\right) \tag{9.6}$$

where τ_R is the radiative lifetime which is measured by performing the experiments at a very low temperature. From the temperature dependence of τ, we then determine ΔE (see the F-centre example in section 5.5.1).

More detailed information on the radiative decay can be inferred from time-resolved spectroscopy experiments where the spectra are recorded at various times after excitation. This technique is particularly useful when complex decay schemes are operative, including participation of more than one centre.

9.2.5   Spectra of concentrated systems

For simplicity, many defects are discussed as though they exist in isolation in a perfect lattice. However, as defect concentrations increase, there will be interactions which will be apparent as modifications of the energy spectra or as formation of totally new defects. The more sensitive measurements such as EPR can detect defect interactions as far away as some ten atomic shells but normally such long-range coupling between defects could be neglected. More direct defect association is expected either on a purely statistical basis or if there is an additional driving force which favours it. Several such driving forces exist. For charged defects in insulators, local charge equilibrium can be restored by pairing impurities, or by associating an intrinsic and extrinsic defect. For large impurities, which strain the lattice, the total strain energy of the structure may be minimized by pairing or clustering several defects. Similarly, charge and size considerations can favour the movement of point defects towards (or away from) the extended defects such as dislocation lines.

In concentrated systems, the presence of impurities or defects close by may give rise to observable changes in the energy levels and spectral bands because of the electronic interaction between the two centres. Many examples of these changes associated with pairs or clusters of impurities or defects are available for insulator and semiconductor materials. As mentioned in Chapter 1, the vacancy centre in alkali halides, the F centre, can cluster into F_2, F_3 and F_4 groupings which are readily identified by new absorption bands. The number of point defects in the aggregate can be inferred by the change in signal levels of the original and new optical absorption bands (assuming that both defects exhibit similar oscillator strengths).

In addition to broad-band spectra, line absorption or emission features may be observed. This is particularly relevant for pairs coupled by exchange interaction as is the case for d (transition-metal) ions[17,18]. This coupling gives rise to a number of interesting new features in the spectra by comparison with the isolated centres, of which the following should be mentioned.

(1) Increase in the intensity of spin-forbidden bands.

(2) Strong temperature dependences.

(3) Appearance of additional fine structure in these forbidden bands.

(4) Observation of absorption bands at energies corresponding to the sum of two single excitations, i.e. double excitations.

The ruby structure with Cr^{3+} ions substituting for Al^{3+} in Al_2O_3 shows a clear example of impurity ions occupying adjacent sites. The emission lines at 700.4 and 703.3 nm are associated with an exchange interaction of the Cr^{3+} ions and an increase in intensity with increasing concentration which is more rapid than the increase in the signals from isolated ions. Hence it suggests there is preferential pairing of the Cr^{3+} ions, which may be justified as this would reduce the strain energy from the large Cr^{3+} ions in the lattice. The new spectral feature can be understood starting from the spin-forbidden transition between the ^4A (ground) and ^2E (excited) levels of a monomer centre. For an exchange-coupled pair (dimer), the pair ground state $^4A-^4A$ gives rise, after exchange splitting, to the four levels illustrated in Figure 9.6 having a well-defined total spin. The first excitation state $^4A-^2E$ splits into two levels with spins 2 and 1. Typical splittings are in the

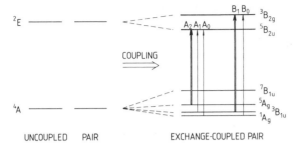

Figure 9.6.
Effect of exchange coupling on the levels of a pair of close centres with levels ^4A (ground) and ^2E (excited), e.g. Cr^{3+}.

range of 10 cm^{-1}. Now, two spin-allowed transitions marked B_1 and A_2 in the figure are possible and will give rise to prominent absorption bands in the spectrum. However, the intensities of these bands will very abruptly decrease on lowering the temperature because of the Boltzmann-type depopulation of the ^5A_g and $^3B_{1u}$ levels to the benefit of the lowest level ^1A_g. This is the situation corresponding to the Cr^{3+} pairs as occurs in heavily doped ruby crystals.

The enhancement of spin-forbidden transitions of Mn^{2+} by exchange coupling to radiation-induced F centres close by has been very clearly revealed in a number of fluoroperovskite systems[19]. This is a consequence of the fact that the coupling enhances the f value.

The occurrence of double excitations involving the absorption of one photon of high energy to cause the simultaneous excitation of two ions close by is a relevant consequence of the coupling and many examples have been reported. An observation of double Mn^{2+} transitions in manganese-doped NaCl has been considered[20] as evidence for the formation of a manganese-concentrated phase, the so-called Suzuki phase.

A very interesting example of pair transitions is provided by donor–acceptor pairs in semiconductor materials[21]. The exciton-like luminescence emission from the donor to the acceptor consists of sharp peaks whose energy hv obeys the well-known formula

$$hv = E_G - E_D - E_A + \frac{e^2}{\epsilon r} - f(r) \qquad (9.7)$$

where E_G is the band gap energy, ϵ is the dielectric constant, and E_D and E_A are the optical ionization energies for the donor and acceptor, respectively. It represents the energy separation between isolated donor and acceptor levels, corrected by the Coulomb interaction of the pair. The correction term $f(r)$ is discussed by Stoneham and Harker and is useful at small r values[22]. Very rich spectra containing as many as 300 well-resolved lines corresponding to pairs separated up to about 40 Å have been obtained and identified in a variety of semiconductors. An important characteristic of these spectra is that lifetimes of the spectral lines increase on going to higher separations, i.e. lower energies. This is a consequence of the exponential dependence of the radiative transition probability with the pair distance. In other words, distant pairs have a longer life than close pairs do. This feature can be very nicely revealed by time-resolved spectroscopy.

Halide crystals provide several well-documented examples of defect clustering which are of importance in technological applications. For LiF doped with magnesium to form a thermoluminescence dosimeter (see Chapter 14) the Mg^{2+} ion is charge compensated by an adjacent Li^+ vacancy; however, this dipole-like defect is mobile at about 100°C and readily precipitates into a cluster of three units, most probably in the form of an $(Mg^{2+} - Li^+_{vacancy})_3$ six-sided arrangement in the (111) plane. In the dosimeter the magnesium dopant level is about 100 ppm. The simple dipole-type defect controls the lower temperature thermoluminescence peaks, and the trimers the higher peaks. Differences may also be sensed in the near-ultraviolet absorption spectra. At higher concentrations (about 1 000 ppm) the magnesium is incorporated in quite a different fashion as a Suzuki precipitate phase. This structure is a vacancy-rich system[23] of $6LiF.MgF_2$.

The second, and familiar, example of impurity precipitation is the clustering of photolytically released silver from silver halides. The metallic speck of $[Ag^0]_4$ forms the latent image both in the photographic process and in reversible photochromic glasses which incorporate silver halides.

In many systems with the fluorite structure the simple concept of isolated point defects is inadequate as the observations and calculations suggest a local clustering of the imperfections. For example in UO_2 the production of oxygen interstitials leads to a defect cluster involving six components. The notation[5] 2:2:2 refers to two oxygen interstitials relaxed along the $\langle 110 \rangle$ axes, two oxygen ion on normal sites but perturbed along $\langle 111 \rangle$ axes and a pair of oxygen vacancies. An extension of the labelling scheme allows impurity ions, etc., to be incorporated in the description of the cluster. As an example, from neutron diffraction studies of La^{3+}- and Er^{3+}-doped CaF_2, we see larger clusters of the type 4:3:2 (i.e. four interstitials, three vacancies and two impurity interstitials)[24].

The preceding examples demonstrate that point defects are frequently not isolated and even at low concentrations may aggregate into more complex structures. At high intrinsic or extrinsic defect concentrations the more major rearrangements of colloidal precipitates, dislocation structures around defect planes or new crystallographic phases are almost inevitable and must be considered during analyses of the crystal properties.

9.2.6 Excitation transfer spectroscopy

This topic is intimately related to that previously discussed on the spectroscopy of concentrated systems. The excitations of a given atom or defect group in a solid can be transferred to other defects close by which are some distance from the original excitation site[25]. This transfer may cause quenching of the luminescence or the emission of light by a different species from that initially excited. The study of these excitation transfer processes has been shown to be relevant for the investigation of close defect pairs and concentrated systems.

Let us consider that a sensitizer centre A is excited from the ground level 0 to a short-lifed excited state H from which it decays to a long-lived level labelled 1, with energy ΔE above 0 (Figure 9.7). If an activator centre B is at a close distance R from A and the excited level $1'$ lies at an energy $\Delta E' \approx \Delta E$ above the ground level $0'$, the excitation energy can be transferred from A to B (resonant transfer). The transfer probability per unit time is

$$P = \frac{2\pi}{\hbar} |\langle \psi_i | \mathcal{H}_{AB} | \psi_f \rangle|^2 \rho_E \qquad (9.8)$$

where $\psi_i = \psi_{A1} \psi_{B,0'}$ and $\psi_f = \psi_{A,0} \psi_{B,1'}$ are the initial and final states, respectively, of the system (pair), ρ_E is the density of final states and $\mathcal{H}_{AB}$ is the coupling Hamiltonian between both centres. This coupling Hamiltonian includes electrostatic multipolar as well as exchange interaction terms. If the sensitizer and activator, $0 \to 1$ and $0' \to 1'$, transitions are dipole allowed, we can derive from equation (9.8) a dipole–dipole transfer probability P_{d-d}:

$$P_{d-d} = \frac{3\hbar^4 c^4}{4\pi n^4 \tau_A} \frac{1}{R^6} Q_A \int f_A(E) \frac{f_B(E)\,dE}{E^4} \qquad (9.9)$$

where n is the refractive index of the material, τ_A the lifetime of the excited level 1 of the sensitizer and $Q_A = \int \sigma_A(E)\,dE$, σ_A being the absorption cross-section of the sensitizer. The integral factor includes the overlapping of the normalized shape functions $f_A(E)$ and $f_B(E)$ for the absorption band of the sensitizer and emission band of the activator respectively.

It should be noted that the energy decays via an intermediate level 1 or $1'$ so that the luminescence spectra of A and B defects are distinguishable. Thus the changes in luminescence signals A_L and B_L monitor the energy transfer.

A critical distance R_0 can be defined such as $P_{d-d}\tau_A = 1$; that is, at closer distances, excitation transfer is dominant over sensitizer de-excitation and emission from the activator is mostly observed. At longer distances, the

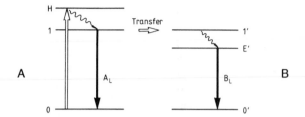

Figure 9.7.
Schematic diagram showing the transfer of excitation from excited centre A to emitting centre B.

emission from the sensitizer is preferentially observed. For favourable dipole–dipole transfer conditions, $R_0 \approx 25$ Å.

Higher-order multipole coupling terms yield even more abrupt distance dependences. For example, the dipole–quadrupole transfer probability varies as R^{-8}, and the quadrupole-quadrupole probability as R^{-10}. These terms may become dominant at short separation distances when at least one of the transitions is strongly dipole forbidden. In contrast, the probability for exchange-induced transfer presents a decreasing exponential dependence on distance, the critical value R_0 being of the order of first-neighbour atomic distances in solids.

When the mismatch between the excitation energies ΔE and $\Delta E'$ of sensitizer and activator greatly exceeds the sum of widths of the two transitions, phonon-assisted energy transfer may take place. An analysis of this case can be consulted in the literature[26].

In view of the abrupt dependence on distance the efficiency of transfer provides a very sensitive measure of the separation between activator and sensitizer. This technique finds very interesting applications ranging from solid-state lasers to biological systems. As a recent example, the excitation transfer from Pb^{2+} to Mn^{2+} has been used[27] to follow the clustering of these impurities in NaCl, from the initial dispersed stated obtained after fast quenching and storage at room temperature, where the dipoles are mobile.

9.3 PHOTOACOUSTIC SPECTROSCOPY

At variance with the more conventional optical spectroscopy (absorption, emission and scattering), photoacoustic spectroscopy[28] measures the energy liberated as heat (lattice phonons) during the excitation–de-excitation cycle. The production of these phonons occurs because of a non-radiative de-excitation or because of the relaxation following absorption and emission for the radiative transitions.

The basic scheme of a photoacoustic spectrometer is quite similar to that of a spectrophotometer, except that the sample cell and detector system are rather specific. A light source, a chopper (about $10-1000\ Hz$) and a monochromator are standard elements. Two main detector systems are in use, although some new developments have recently appeared[29,30]. The most classical system uses a gas-filled cell containing the sample and a microphone for the detection of the photoacoustic signal. The cell has a window to allow the entrance of the incident light beam. Various cell types have been designed and built. The second main detector system uses a lead zirconate titanate transducer glued to the sample. Because of the much better acoustic impedance matching in the solid–solid interphase, this method is much more sensitive than the microphonic method. It has also additional advantages, particularly the possibility of performing measurements over a very wide range of temperature (including low-temperature work). Typically, heat generation rates of $10^{-6}\ cal\ cm^{-3}\ s^{-1}$ can be detected.

The information extracted from the photoacoustic spectra relies on the theoretical model used to describe the heat generation processes and the subsequent transfer to acoustic waves. For local absorbing centres such as impurities and colour centres, the photoacoustic signal $P(\omega)$ is selected to give the absorption spectrum $\alpha(\omega)$ by the following simple law, valid under some rather general assumptions:

$$P(\omega) = A\alpha(\omega)\left(1 - \eta\frac{\omega_f}{\omega}\right) \qquad (9.10)$$

where A is a coefficient depending on the geometry and detection system, η the quantum efficiency of the fluorescence emission and ω_f the peak fluorescence frequency. From a comparison between $P(\omega)$ and $\alpha(\omega)$, the value of η can be determined accurately. This is a very important parameter in characterizing laser applications.

Illustrative examples are provided by measurements of the quantum efficiency of Nd^{3+} in several hosts[31] and Eu^{2+} in $NaCl$[32]. For this latter case, the crystal, after a rapid quenching, was excited at the t_{2g} absorption band of Eu^{2+} and the luminescence band at 427 nm from free dipoles was used for photoacoustic detection. By plotting $P(\omega)/\alpha(\omega)$ against λ, a η value of 0.59 was obtained at room temperature.

9.4 LASER SPECTROSCOPY

The properties of laser light, monochromaticity, coherence and high spectral intensity, have prompted the development of a number of techniques, which are being increasingly used for point defect problems. We give here a brief sketch of some of the most relevant methods and the reader is referred elsewhere[33,34] for a more thorough coverage.

A first application that takes advantage of the high spectral intensity of the laser light constitutes the so-called excited-state spectroscopy. Absorption of an intense laser light beam (generally a short giant pulse with a duration of nanoseconds and a power of kilowatts to megawatts) promotes an appreciable number of the absorbing centres to an excited state. Therefore, this state can reach a high enough concentration to permit optical probing with a second light source. This complements the spectroscopic data obtained from the ground level; in some cases this allows observation of levels that cannot be directly reached from the ground level.

Another application that also profits from the high spectral intensity of laser light involves multiphoton excitation of a centre[35]. At a sufficiently high incident photon flux, there is an appreciable probability for the simultaneous absorption of more than one photon. With multiple-photon absorption, we can access energy states in the interior of a crystal which would have been masked by high absorption coefficients in a simple one-photon step. The most practical approach is to use just two photons per transition. At variance with one-photon transitions where the Laporte rules require opposite parity between the involved states, two-photon transitions connect states with the same parity. This makes two-photon spectroscopy a complementary tool to reach states forbidden to one-photon spectroscopy. For example exciton transitions overlap in energy with the band-to-band absorption edge in conventional one-photon spectroscopy. An early example of this was given using CsI[36]. Thus, the two-photon spectra only map the band gap transitions. The difference between the one- and two-photon spectra gives the exciton spectrum.

It should also be mentioned that the availability of Q-switched and mode-locked lasers, yielding very intense nanosecond, picosecond and even femtosecond (about 10^{-15} s) pulses, is fostering the study of rapid relaxation phenomena in solids.

The monochromaticity and coherence capabilities of laser light has been used to develop a variety of high-resolution methods. These are classified in two broad categories according to whether they operate in the frequency or the time domain. One of the interesting frequency domain techniques is the so-called site selection spectroscopy that involves selective excitation of a given set of centres out of those contributing to an inhomogeneously broadened line. The luminescence emission of those centres is then quite narrow because of the removal of the inhomogeneous broadening effect (fluorescence line narrowing) (Figure 9.8). In the ideal limit, the width of the emission line will be determined by intrinsic (homogeneous) broadening mechanisms. Another related technique uses an intense laser beam to saturate the absorption corresponding to a given set of centres whose spectrum

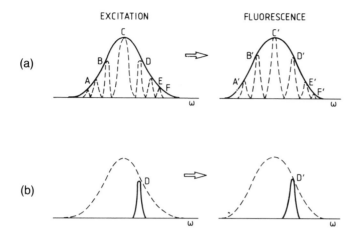

Figure 9.8.
Illustration of the narrowing of the emission, when (a) a broad inhomogeneous band is excited with (b) a narrow laser line.

lies within an inhomogeneously broadened line[37]. Under these conditions, a "hole" can be burnt in the inhomogeneous line and it can be probed by a second laser of sufficiently high resolution. The processes involved are schematically drawn in Figure 9.9.

These frequency domain methods rely on the high-resolution ($\Delta v \approx 1$ MHz) and stability of present-day lasers. Resolutions down to about megahertz can be obtained.

Time domain methods[38] include two main types of experiment.

(1) Coherent transients.

(2) Photon echoes.

Ultrahigh-resolution (about kilohertz) studies are possible and these are now described. The basic idea is to produce a coherent excitation of the active centres and to measure the relaxation times T_1 (transverse) and T_2 (longitudinal) directly for the time decay of the coherence. This is essentially equivalent to the experiments performed in the frequency domain because of the Fourier transformation properties between decay times and linewidths.

In the coherent transient technique, the decay of the coherent emission (free-induction decay) is often recorded by monitoring the beats against the excitation laser frequency after shifting it out of resonance.

The photon echo experiments (which are a transcription to the optical domain of the original nuclear magnetic resonance experiments) are somewhat more complicated, although essentially equivalent[39].

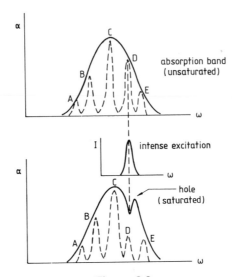

Figure 9.9.
Formation of a hole in the shape of a broad inhomogeneous band when excited with an intense laser line.

Not only static but also dynamic spectroscopic data can be very conveniently obtained from these high-resolution techniques. One such method, particularly relevant, is time-resolved fluorescence line narrowing. If the excitation of a given set of centres in a fluorescence-line-narrowing experiment is followed by excitation transfer to nearby centres at different physical sites, then the initially narrowed fluorescence line will become broadened because of the contribution of these new centres to the luminescence. This behaviour is schematically illustrated in Figure 9.10. By monitoring the time evolution of this broadening, after the end of the excitation pulse the dynamics of transfer can be elucidated.

9.5   RAMAN SPECTROSCOPY

The Raman effect[40] involves the inelastic scattering of an incident photon beam with a material so that elementary excitations (phonon, polaritons, plasmons, etc.) are created or annihilated, therefore causing a change in the photon frequency. Most work concerns the study of phonons or vibrational modes in the material and we refer to it throughout this section.

A typical experimental set-up is shown in Figure 9.11. The laser beam with wavevector k_i and polarization unit vector e_i is directed onto the sample. The scattered light with selected wavevector k_s and polarization

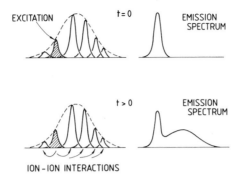

Figure 9.10.
Illustration of the change in the emission spectrum with time after excitation of a broad inhomogeneous band with a narrow laser line.

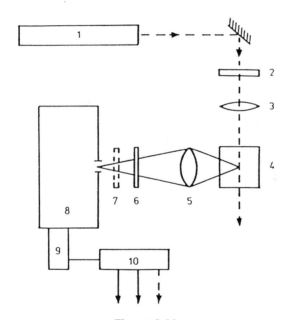

Figure 9.11.
Raman spectroscopy set-up (schematic): 1, light source; 2, polarization rotator; 3, focusing lens; 4, sample; 5, collecting lens; 6, polarizer; 7, polarization scrambler; 8, double-grating monochromator; 9, photomultiplier; 10, photon counting.

vector e_s is detected with a photomultiplier tube, which is usually coupled to a photon-counting system for optimum sensitivity.

The first-order spectrum of the scattered light includes bands at frequencies $\omega_s = \omega_i \pm \Omega$, where ω_i is the frequency of incident light and Ω

the frequency of the phonon. The equation indicates that the energy $\hbar\omega_i$ of the incoming photon has been respectively increased or decreased by the amount $\hbar\Omega$, corresponding to the annihilation of creation of a phonon with frequency Ω during the scattering event. The Raman peaks with $\omega_s < \omega_i$ (creation of phonons) are designated as Stokes bands, whereas those with $\omega_s < \omega_i$ (annihilation of phonons) as anti-Stokes bands. The intensity of the Stokes peaks is larger than the corresponding anti-Stokes peaks. The shift Ω of the Raman peaks with regard to ω_i is characteristic of the material and obviously independent of ω_i. In this respect, Raman peaks are distinguished from luminescence emissions, for which the frequency itself and not the shift is independent of the frequency of the incident radiation. However, the intensity of the Raman spectrum is proportional to ω_i^4 because of the frequency dependence of the Raman cross-section.

For a given phonon to be observed in the first-order Raman spectrum of a crystal, two conditions have to be simultaneously fulfilled.

(1) The phonon must produce a net change in the electronic polarizability of the unit cell. In terms of group theory, this means that the phonon-associated irreducible representation of the crystal point group must coincide with the irreducible representation of some element of the polarizability tensor. Depending on the phonon symmetry, only certain experimental polarization and scattering configurations designed as $k_i(e_i, e_s)k_s$ in the Porto notation will allow for the observation of a given Raman peak. Therefore the symmetry of the active phonons can be determined.

(2) The polarizability change must be simultaneous in all unit cells of the crystal. This is a consequence of the momentum conservation rule, $k_i = k_s + k_{ph}$, k_{ph} being the phonon wavevector which, in turn, originates from the crystal periodicity. Since the momentum of the light is about 10^{-3} of that for the Brillouin zone edge, i.e. $|k_i| \approx |k_s| \approx 0$, only those phonons with $k_{ph} \approx 0$ (centre of the Brillouin zone) can participate in the first-order Raman process.

For crystals with inversion symmetry, the first condition implies that Raman-active phonons must be even under inversion, since all elements of the polarizability tensor are even. In this case, only second-order Raman spectra are observed. When a defect or impurity is present in such a crystal, the inversion symmetry is removed and first-order Raman scattering generally appears. This makes Raman scattering very appropriate for studying point defects in crystals with inversion symmetry, such as alkali halides, alkaline-earth oxides, etc. Some defects (e.g. Tl^+ in KCl) do not significantly change the force constants or the phonon spectrum and are called isotopic impurities. They modify the distribution of scattered light intensity between the various modes of the perfect crystal.

When there is a change in the force constants between the substitutional impurity and its neighbours, the Raman spectrum becomes more complicated and generally new frequencies appear. If the new frequencies lie above the upper limit of the phonon density of states of the pure crystal, or inside the phonon gap, the result is a local mode. A clear example of this type is the OH^- impurity, whose stretching frequency (about 3200 cm^{-1}) is well above any normal mode of the host.

Raman scattering can also be applied to precipitate particles with a different crystallographic structure from that of the host. Precipitates should show their own active Raman modes in the spectra. The intensity of the corresponding peaks will be proportional to the concentration of precipitated impurity, whereas their frequencies can be slightly affected by the strains produced when a lattice mismatch exists between the precipitate and the host crystal. Selection rules cannot be readily applied to the precipitates, unless they are coherently oriented with respect to the host crystal. For instance, Suzuki phase precipitates in alkali halides are well oriented and their Raman spectra show very clear selection rules[41]. By contrast, precipitates of an incoherent phase such as the dihalide structure MX_2 are generally oriented at random in the host and no selection rules can be observed.

In summary, the Raman bands associated with the precipitates can yield information on their nature, orientation, mismatch with the host lattice, etc. Sometimes, we can also study[42] the change in precipitate size when thermal treatments are applied to the crystal, by measuring the changes in phonon frequencies and in bandwidth. In principle, it would be also possible to observe surface modes due to vibration of atoms at (or close to) the precipitate–host interphase. This would be a special case of localized or quasilocalized modes similar to those recently found in semiconductor superlattices[43].

9.6 PHOTOCONDUCTIVITY

9.6.1 Basic phenomena

All insulator or semiconductor materials show photon-induced conductivity for appropriate wavelength ranges[44]. This photoconductivity adds to the "dark" value which is due to electronic or ionic carriers of thermal origin. Ilumination with photons of energy greater than the band gap excites electrons across the gap, creating electron and hole pairs which contribute to the conductivity. Photoconductivity can also be produced for longer-wavelength radiation, if electrons are promoted from or to impurity levels which exist in the gap of the material. Experiments can be performed under continuous or pulsed illumination and they are described next.

(a) Continuous illumination. The steady-state photoconductivity is

$$\sigma_{ph} = \rho_e \mu_e e + \rho_h \mu_h e \tag{9.11}$$

where ρ_e and ρ_h are the steady density of electrons and holes, respectively, and μ_e and μ_h the corresponding mobilities. The electron density ρ_e can be written as

$$\rho_e = \eta_e I_e \tau_e \tag{9.12}$$

where η_e is the quantum efficiency for the production of a free electron, I_e the density of photons absorbed by donors and τ_e the free-electron lifetime. An analogous expression applies for ρ_h. Substitution of these densities into equation (9.11) yields

$$\sigma_{ph} = \eta_e \mu_e I_e \tau_e e + \eta_h \mu_h I_h \tau_h e = \eta_e I_e \omega_e e + \eta_h I_h \omega_h e \tag{9.13}$$

with $I_e + I_h = I$, and ω is the so-called schubweg or average carrier displacement per unit electric field.

The carrier mobility μ is an important parameter that critically depends on the scattering mechanisms. Each mechanism contributes differently to the mobility value, which ranges from a few square centimetres per volt per second for ionic compounds such as alkali halides to values such as 65 000 cm^2 V^{-1} s^{-1} for InSb at room temperature. Lower mobilities occur when carriers are self-trapped and move by hopping. Mobility increases markedly with decreasing temperature, the dependence being usually written in the form of $\mu \propto T^n$, with an exponent n different for each particular scattering mechanism.

The lifetime of carriers is determined by the recombination probability of electrons and holes at recombination centres. If v is the thermal carrier velocity in the appropriate band and N is the concentration of recombination centres, the carrier lifetime is

$$\tau = \frac{1}{v S_c N} \tag{9.14}$$

where S_c is the capture cross-section. For charged centres, which interact in a Coulombic manner with the carriers, S_c can be estimated from the distance r of approach at which the binding energy is equal to kT, i.e.

$$kT = \frac{e^2}{\epsilon r} \tag{9.15}$$

where ϵ is the dielectric constant of the material. The capture cross-section is then

$$S_c = \pi r^2 = \frac{\pi e^2}{k^2 T^2 e^2} \tag{9.16}$$

which for $\epsilon = 10$ and $T = 300$ K gives

$$S_c \approx 10^{-12} \text{ cm}^2$$

Coulomb repulsion can reduce the cross-section to as low as 10^{-22} cm^2. An uncharged centre has a typical cross-section of about 10^{-15} cm^2.

In real photoconductors, electrons and holes may be captured by imperfections with energy levels in the band gap. If the level is shallow, the trapped carriers may be subsequently released by thermal energy, before capturing an oppositely charged carrier and be annihilated. These traps prolong the decay of conductivity when the light is switched off and hence transient-photoconductivity measurements enable us to study the nature of these imperfections.

Experimental photoconductivity data can yield two main pieces of information. First, the critical wavelength for the onset of photocurrent from a given donor or acceptor level gives the optical separation from the conduction band (for donor) or from the valence band (for acceptor), i.e. the appropriate optical ionization energies. The measured values of the photoconductivity σ_{ph} give the schubweg ω, if the quantum efficiency η is known. By varying the temperature, the dependence $\omega(T)$ or equivalent $\mu(T)$, if τ is assumed to be constant, can be measured and therefore information on the main scattering mechanism for the mobility can be obtained. From a practical view and with reference to device applications, the photoconductive response of a material is usually characterized by the so-called quantum or photoconductive gain G. It is defined as the number of charges collected at the electrodes per photon absorbed. It is easy to show that

$$G = \frac{\sigma_{ph}E}{Ied} = \frac{V}{Id^2}(\tau_e\eta_e\mu_eI_e + \tau_h\eta_h\mu_hI_h) \tag{9.17}$$

V being the applied potential and E the corresponding field.

(b) Pulsed illumination. The sample of thickness d constitutes a dielectric which fills the interelectrode space of a capacitor charged to a potential V with charge $\pm Q$. The stored electrostatic energy is $W = \frac{1}{2}CV^2$, C being the capacity. Let us assume that during a light pulse, q ($= I\eta$) charges (electrons or holes), are liberated as a consequence of the fact that I photons are absorbed in the material, with quantum efficiency η. Each free carrier can move an average distance $l = \omega E$ (where $E = V/d$ is the applied field) in the direction of the field before being annihilated. The work carried out by the field is $Eqel$, which induces a decrease in the stored energy given by

$$\delta W = \frac{1}{2}C(2V\,\delta V) = CV\,\delta V \tag{9.18}$$

Therefore,

$$\delta V = \frac{Eqel}{CV} = \frac{qel}{Cd} \tag{9.19}$$

and the charge collected and measured at the electrometer is

$$\delta Q = C \, \delta V = \frac{qel}{d} \tag{9.20}$$

which can be used to determine l or the schubweg ω as for the continuous-illumination case.

9.6.2   Space-charge-limited currents

Experimentally, photoconductivity measurements are complicated by the development of a space charge in the crystal which is influenced by the presence of defect states. Therefore the apparent signal strength can be a non-linear function of the light intensity and voltage applied to the crystal. This non-Ohmic and non-linear behaviour is most apparent in steady-state measurements and the problem is often reduced by using pulsed illumination and/or alternating-current/electric fields[44].

The problem of space charge within the crystal is exacerbated by electrodes which are non-Ohmic.

9.7   PHOTOEMISSION

In this technique[45], monochromatic light is shone on a material and the induced emission of electrons studied. Photoemission data include the photoelectric yield, i.e. number of electrons emitted per absorbed photon and the energy distribution. The energy ϵ_{ph} of the photoelectron escaping the material is simply given by $\epsilon_{ph} = h\nu - E_b - \phi$, if no change in the electron energy has taken place since the photoelectric event. In the formula, ν is the light frequency, E_b is the binding energy of the electron (referred to the Fermi level) and ϕ is the work function of the material. Photoemission spectroscopy has been extensively used to study electronic properties of the material–vacuum interphase, such as the work function, electron affinity, ionization energy, band bending near the surface, distribution of energy levels inside bands and surface state levels. The availability of synchrotron sources has fostered the development of angle-resolved photoemission spectroscopy as a powerful tool for studying the electronic structure of solids, particularly energy dispersion curves[46].

Defect problems have also been investigated. Illustrative examples are the study of colour centres in alkali halides[47] and Ti^{3+} centre in reduced $SrTiO_3$[48]. For alkali halides, the technique used has involved the measure-

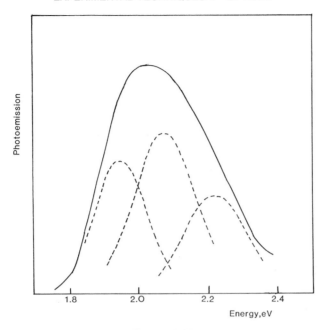

Figure 9.12.
Photoemission spectrum associated with the F band in CsCl. The three components bands match optical absorption features.

ment of the photoemission yield as a function of light wavelength. The spectra obtained show bands which can be very well correlated in shape with the colour centre bands in optical absorption. The method appears to have good resolution and high sensitivity, if only a moderate or low vacuum is used or if previous alkali-metal (e.g. caesium) deposition on the sample is made in order to reduce the electron affinity and to facilitate electron emission. As an example[47] the F band photoemission of CsCl at room temperature is shown in Figure 9.12.

9.8 THERMOLUMINESCENCE

9.8.1 Basic theory

Insulators which contain electrons trapped at defect sites are in a metastable condition. The trapped charge can be liberated by heat and will cross the potential barrier and move to a lower-energy state with the emission of light. This is called thermoluminescence. The rate at which electrons, or holes, escape from the trap is governed by the vibrational frequency S_v of the charge within the trap and the height E of the potential barrier, so that the overall rate of escape is proportional to $S_v \exp(-E/kT)$. Since we are considering a very sensitive technique which is capable of detecting as few as 10^7 defects in a sample, we are faced with a complex problem in a real system where there will be a total concentration of 10^{16} or more defects. Many approximations have been proposed to analyse the form of the thermoluminescence signal. The earliest was the simple model of Randall and Wilkins[49] in which the electrons are thermally excited from a single defect level to the conduction band and radiatively decay to a luminescent site, as is shown in Fig. 9.13. It is assumed that no direct transitions take place from the defect to the recombination centre, that the number of defect sites is small compared with the number of luminescent centres and also that the recombination lifetime in the conduction band is small. The rate of depopulation of the defect is

$$\frac{dn}{dt} = -n(N_c S_1 v) \exp\left(-\frac{E}{kT}\right) + S_d n_c (N - n) \qquad (9.21)$$

In the first term, n is the electron concentration in the defect traps, N_c is the density of states in the conduction band, S_1 is the electron capture cross-section in the conduction band and v is the electron velocity. The second term allows for the possibility of a back reaction with S_d as the electron capture cross-section of the defect, N is the total concentration of the shallow traps and n_c the electron concentration in the conduction band. By comparison with the simple model

$$S_v = N_c S_1 v \qquad (9.22)$$

As a first simplification, we assume that retrapping is negligible. This is a valid assumption if the concentration of traps is small compared with the concentration of recombination centres. Further, if the recombination lifetime τ is very short, the electron concentration n_c in the conduction band will remain essentially constant. Loss takes place to the recombination centres or the trapping levels; so, in general,

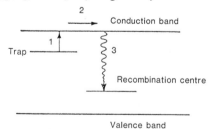

Figure 9.13.
The idealized model for thermoluminescence: 1, thermal charge release; 2, migration; 3, radiative recombination.

$$\frac{dn_c}{dt} = \frac{n_c}{\tau} + \frac{dn}{dt} \qquad (9.23)$$

When τ is short, dn_c/dt is negligible compared with the other terms. We may also relate the time and temperature by $T = \beta t$, if we raise the temperature at a constant rate so that equation (9.23) becomes

$$\frac{n_c}{\tau} = \beta \frac{dn}{dT} \qquad (9.24)$$

For no retrapping ($S_d = 0$), integration of equation (9.21) provides

$$n = n_0 \exp\left(-\int_{T_0}^{T} \frac{N_c S_1 v \exp(-E/kT)\,dT}{\beta}\right) \qquad (9.25)$$

It is the movement of electrons from the conduction band to the deeper levels which provides the luminescence; so n_c/τ is proportional to the light intensity I. This gives the light intensity

$$I = \beta \frac{dn}{dT}$$

$$= n_0 N_c S_v v \exp\left(-\frac{E}{kT} - \int_{T_0}^{T} \frac{N_c S_1 v \exp(-E/kT)\,dT}{\beta}\right) \qquad (9.26)$$

which is a glow curve of the form shown in Figure 9.14 and is termed first-order annealing.

If the shallow energy levels also compete for the conduction electrons, then we can derive a second-order expression for the glow curve which is of the form

$$I = n_0 N_c S_1 v \exp\left(-\frac{E}{kT}\right)\left(1 + \int_{T_0}^{T} \frac{N_c S_v v \exp(-E/kT)\,dT}{\beta}\right)^{-2} \qquad (9.27)$$

if we assume an equal probability of recombination or retrapping. Figure 9.14 shows such a glow curve.

For simplicity, most analyses assume that the product $N_c S_1 v$ can be taken as a single parameter which is temperature independent. Keating[50] pointed out that $N_c \propto T^{3/2}$, $v \propto T^{1/2}$ and $S_1 \propto f(T)$ might be expected. Then it is only for $S_1 \propto T^{-2}$ that the above estimate is valid.

In this simple first-order model there are only two variables, E and S_1, and so it must be assumed that these kinetics match both the peak temperature and the shape of the curve. In practice, we expect the factor S_v to be $10^{12}-10^{13}$ s^{-1}, i.e. it reflects the vibrational frequency of the lattice. In this situation a simple measurement of the peak temperature T_m gives a reasonable estimate of the activation by assuming $E \approx 25kT_m$. Normally the

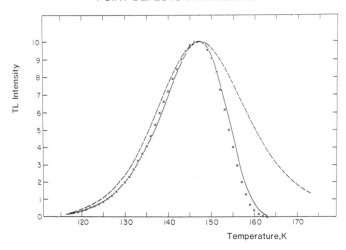

Figure 9.14.
Simple first-order and second-order (– – –) thermoluminescence (TL) kinetics compared with data(x) from LiF (E = 0.26 eV; S_v = 1.5 × 10^6 s^{-1}).

activation energies increase with increasing glow peak temperature but occasionally extreme variations in S_v offset this order. For example Table 9.1 gives an early LiF example[51] (see also Figure 9.15).

Despite the very extensive literature[52,53] on thermoluminescence in which methods of extracting E and S_v values from thermoluminescence data are discussed, this simplistic estimate is of considerable value. The high precision in the determination of E and S_v allows us to distinguish between defect centres but does not of itself offer any understanding of the defect sites producing the thermoluminescence.

9.8.2   Computer analysis of glow curves

The equations governing the luminescence process do not readily lead to analytically tractable solutions even for simple cases. It is therefore more convenient to ignore the simplifications and simply to compute the resultant glow curve. One of the earliest treatments was performed by Kemmey et al.[54]. He assumed a set of defect levels and assigned to each a frequency factor, trap depth, concentration, retrapping cross-section and a state of excitation. The last specifies whether the defect is empty, electron filled or hole filled and whether it is in a ground or excited state. Rate equations are used to calculate the rearrangement of the charges which take place during a short time interval. The new population conditions are then used in a reiteration process.

Table 9.1.
Examples of thermoluminescence analysis for LiF[51]

T_m (K)	E (eV)	S_v (s^{-1})	$25kT_m$ (eV)
115	0.32	1.3×10^{12}	0.25
133	0.26	1.0×10^{8}	0.29
147	0.28	1.5×10^{6}	0.31
158	0.25	1.5×10^{6}	0.34
194	0.53	4.3×10^{12}	0.42
202	0.46	3.9×10^{12}	0.44

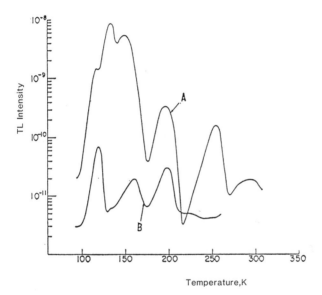

Temperature,K

Figure 9.15.
Thermoluminescence (TL) of LiF: Mn after X-ray irradiation (curve A) and regenerated thermoluminescence after exposure to F band light (curve B).

A computational method is valuable when attempting to determine the effect of different concentrations of shallow traps and recombination centres as shown in Figure 9.16.

9.8.3 Influence of heating rate

The glow peak can be displaced by varying the heating rate, and the peak temperature is related to the rate of rise of temperature by

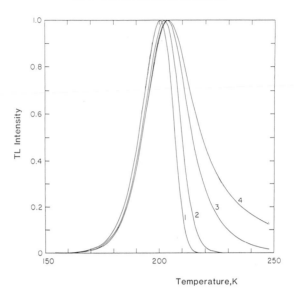

Figure 9.16.
Theoretical thermoluminescence (TL) curve shapes for 10^{16} trapped charges and 10^{18} recombination centres (curve 1), 5×10^{16} recombination centres (curve 2), 2×10^{16} recombination centres (curve 3) and 1×10^{16} recombination centres (curve 4) with $E = 0.5$ eV and $S_v = 5.8 \times 10^{10}$ s^{-1}.

$$\frac{S_1 N_c v k T_{max}^{2}}{\beta E} = \exp\left(\frac{E}{k T_{max}}\right) \tag{9.28}$$

In practice, this allows the peak position to be moved by 20 or 30 K and over this limited range temperature-dependent parameters to be sought. Equation (9.28) has also been used as a means of estimating the trap depth, and an example of this type of analysis is shown in Figure 9.17 for glow peaks in LiF.

Figure 9.17 also demonstrates that repopulation of defect centres may perturb the centres under study. In this example the peak near 194 K was first produced by X-ray irradiation and remeasured after charge repopulation by ultraviolet excitation, which was assumed to transfer electrons from the F centre into the 194 K defect. Whilst the glow peak appears in both cases, there is a small but finite change in the activation energy, suggesting that a modified form of the defect is being viewed.

In other cases, changes in the defect structure may be directly attributable to the analysis technique. There are many examples of defect clustering which proceed slowly by thermal diffusion. In the LiF thermoluminescence

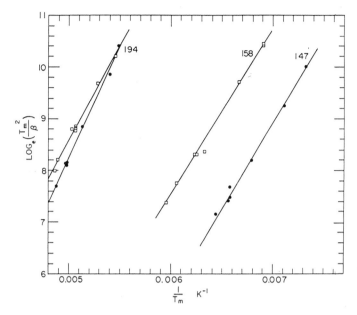

Figure 9.17.
Shifts of LiF:Mn thermoluminescence peaks with heating rate for the T_m values specified on the curves: •, X-ray irradiated; □, ultraviolet light irradiated.

dosimeter material (see Chapter 14), $Mg^{2+}-Li^+_{vac}$ dipoles appear to cluster into groups of three during the thermoluminescence measurement. Very rapid heating produces fewer glow peaks as the electronic processes of charge release and recombination proceed on a faster time scale than the diffusion steps of the cluster formation.

A similar difficulty exists in thermal "cleaning" or "T_{stop}" analyses[52] in which lower-temperature glow peaks are removed from a set of samples by a series of heating to successively higher temperatures. The apparent result may be a "clean" high-temperature peak which can be well characterized. This, in turn, can be subtracted from the total glow curve to analyse progressively the lower-temperature features. In the T_{stop} method the position of component peaks appear as plateau regions on the plot of signal against T_{stop}. In both cases, we must be confident that the observed peaks do not result from the thermal treatments of the samples.

9.8.4 Emission spectra

Charge release which proceeds via a lattice energy band will produce the same luminescence spectrum for all defects which release the same charge and activate the same set of recombination centres. Therefore, we might expect different emission spectra for electron and hole release. Analysis is rarely this simple as even for a single type of charge carrier the temperature dependence of luminescence efficiency, changes in Fermi level or concentrations of recombination centres will modify the spectra. An interesting example of modifications of the emission spectrum with temperature is provided by natural calcite[55] in which manganese impurity ions dominate the recombination sites of the $CaCO_3$. This shows evidence of broad-line spectra from manganese transitions. Over the temperature range $20-400°C$ all glow peaks give similar line spectra for dilute manganese doping levels. At higher dopant levels, only the higher-temperature peaks have resolved lines, as shown in Figure 9.18. One possible explanation for this is that the manganese is precipitated in clusters at lower temperatures but is atomically dissolved at the higher temperatures. The change-over from broad emission bands at lower temperatures to 'line' spectra at higher values is somewhat unexpected if we are accustomed to consider solely the phonon broadening effects discussed earlier in section 9.2.2.

Conventional thermoluminescence curves obtained by the use of a broad-band filter over a restricted region of the blue end of the spectrum can be misleading. For example Figure 9.19 shows the "classical" thermoluminescence curve from a meteorite sample. Although the classical thermoluminescence data suggest that the 200 and 400°C peaks are in ratios as low as 8:1, more complete spectral information suggests[56] this is 200:1.

Similar complexity in spectra are recorded for alkali halides, LiF[57], quartz, fluorite, calcite, feldspars, zircon, obsidian, etc.[58] Figure 9.20 presents a contour map of part of the $I(\lambda, T)$ data obtained during the thermoluminescence of an obsidian crystal. In this example the major line features at the lower temperature are associated with rare-earth impurity ions.

It is evident that failure to account for changes in spectra undermines any kinetic analysis for E and S_v values as well as the obvious loss of key information on the various electron and hole recombination centres.

To summarize, analysis of the wavelength spectrum for each glow peak can give the following information.

(1)   The sign of the captured charge.
(2)   The temperature dependence of the trapping cross-section.
(3)   The position of the trapping level with respect to the Fermi level.

9.8.5   Electron and hole features

The determination of the sign of the released charge is often facilitated by performing other optical, electrical or EPR experiments in addition to the

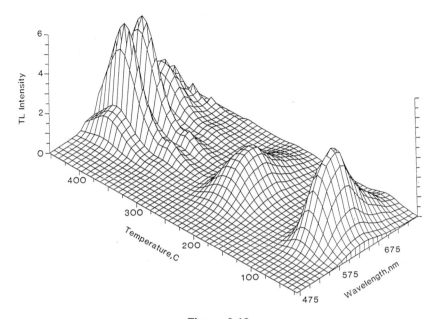

Figure 9.18.
TL emission spectrum of X-ray-irradiated natural calcite (CaCO$_3$:Mn$^+$).

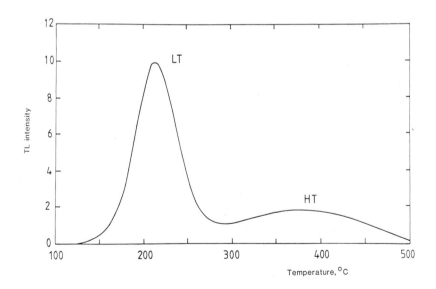

Figure 9.19.
Thermoluminescence (TL) of a meteorite sample recorded via a blue filter.

POINT DEFECTS IN MATERIALS

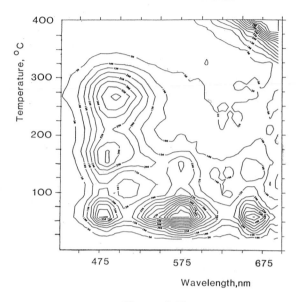

Figure 9.20.

A contour map of thermoluminescence emission from a sample of obsidian.

thermoluminescence spectra. Electrically the currents are rarely sufficient to determine the sign of the carrier. Optically there are several choices, and the first of these is to measure the emission spectra for each glow peak. It can be seen that, if the charge moves from the trap to the luminescent centre via a band, then all electron traps will have one spectrum and this will differ from that produced by hole release. A totally different spectrum may also imply that there is direct charge transfer between a pair of localized levels. When a band is involved, it is only necessary to determine the sign of the charge for one defect and all the others are automatically identified. Trivial solutions exist if we can directly associate a glow peak with the annealing, or formation of an EPR signal or a known optical absorption band. Unless this can be done for several defects, there is still the possibility of a mistake if charge is moved from a complex defect or defect which contains both a bound electron and a hole, since in this instance the release of an electron will allow a hole-type defect to appear. (This is precisely the problem observed in the LiF:Mg:Ti dosimeter material.) An alternative method is to provide a known charge from another defect and to observe the changes in the glow curve and the spectra. As already mentioned, in alkali halides it is possible to provide a source of electrons by bleaching the F centres with F band light. Glow curves produced as a result of this are much simpler to interpret than those following

X-ray or γ-ray irradiation where the more energetic ionization liberates both electrons and holes. The method of analysis is demonstrated by the results shown earlier in Figure 9.15 and 9.17 and Table 9.1. In these we see that all the glow curves appeared after X-ray irradiation, whereas charge transfer of electrons from the F centres only stimulated the glow peaks at 115, 158 and 194 K. These three defects are thus electron traps.

Non-appearance of the other glow peaks could occur because they are hole traps or because they represent electron-trapping defects which are formed during X-ray irradiation but completely anneal out after the loss of the charge. In the example quoted, it was also possible to irradiate with X-rays and to bleach the F centre before heating the crystal. The additional electrons increased the fraction of electron traps which were populated, even if they had been formed by the X-ray irradiation; secondly, electron–hole recombination took place at hole traps with a reduction in the intensity of the corresponding glow peaks. The result of this experiment indicated that the peaks at 133 and 147 K were hole traps. Further confirmation was possible as the light intensity was sufficient to measure the emission spectra of the individual peaks. This showed the two basic emission spectra predicted in section 9.8.4.

9.8.6 Diffusion-controlled thermoluminescence

Most thermoluminescence spectra are interpreted in terms of processes such as those considered in section 9.8.1 involving untrapping and luminescent recombination of electrons or holes. However, it is to be expected that in some cases the rate-determining step in the process is not untrapping but diffusion of carriers when they are self-trapped (e.g. small polarons). This is well documented from the thermoluminescence of irradiated NaCl, for which a prominent glow peak is attributed[59] to the onset of V_K-centre migration through a hopping process. The same situation may apply to materials in which the existence of self-trapped carriers is well ascertained. The analysis of diffusion-controlled thermoluminescence requires a specific theoretical framework involving the diffusion coefficient of the carrier. However, the shapes of the glow peaks do not differ in practice from those resulting from first- or second-order kinetics for electron release but the activation energy now refers to the migration of carriers and very low pre-exponential factors are usually found.

The fact that impurity ions may contribute to thermally activated reactions which cause thermoluminescence emission should be taken into account. In such a case, the kinetics of the glow peaks may be determined by the rate at which the active ions escape from certain defect traps or their diffusion rate to the luminescent sinks. An illustrative example appears to be provided by the thermoluminescence of alkali halides irradiated at or near liquid-helium temperatures[60,61]. The escape and motion of H (or I) centres and the recombination with F centres (or other electron centres such as F') could be responsible for the light emission at some low-temperature glow peaks. All these comments reveal that a correct analysis of thermoluminescence curves can be very complex and requires extensive knowledge of the physical processes, both electronic and ionic, operating during heating.

REFERENCES

1 W. Beall Fowler (ed.), *Physics of Color Centers*, Academic Press, New York (1968).
2 J. I. Pankove, *Optical Processes in Semiconductors*, Prentice-Hall, Princeton, NJ (1971).
3 P. D. Townsend and J. C. Kelly, *Colour Centres and Imperfections in Insulators and Semiconductors*, Chatto and Windus, London (1973).
4 D. Curie, *Luminescence in Crystals*, Methuen, London (1963).
5 W. Hayes and A. M. Stoneham, *Defects and Defect Processes in Non-metallic Solids*, Wiley, New York (1985).
6 Y. Farge and M. Fontana, *Perturbations Electroniques et Vibrationelles Localisées dans les Solides Ioniques*, Masson, Paris (1974).
7 M. Aguilar, P. J. Chandler and P. D. Townsend, *Radiat. Eff.* **40**, 1 (1979).
8 R. C. Newman, *Adv. Phys.* **18**, 545 (1969); in P. A. Glasow, Y. I. Nissim, J. P. Noblanc and J. Speight (eds), *Advanced Materials for Telecommunication 1986*, Les Editions de Physique, Les Ulis, pp. 99–110 (1986).
9 B. A. Weinstein, *Rev. Sci. Instrum.* **57** 910 (1986).
10 A. D. Brothers and D. W. Lynch, *Phys. Rev.* **164**, 1124 (1967).
11 P. G. Dawber and I. M. Parker, *J. Phys. C: Solid State Phys.* **3**, 2186 (1970).
12 A. A. Kaplyanskii, *Opt. Spectros.* **16**, 329, 557 (1964).
13 Y. Merle d'Aubigne, in B. Henderson and A. E. Hughes (eds), *Defects and Their Structure in Non-metallic Solids*, Plenum, New York (1976).
14 F. Luty and J. Mort, *Phys. Rev. Lett.* **12**, 45 (1964).
15 M. Cardona, Modulation spectroscopy, *Solid State Phys.*, Suppl. 11 (1969).
16 P. P. Feofilov, *Polarized Luminescence of Atoms, Molecules and Crystals*, State Publishers of Physical–Mathematical Literature, Moscow (1959).
17 J. Owen and E. A. Harris, in S Geschwind (ed.), *Electron Paramagnetic Resonance*, Plenum, New York, pp. 472–492 (1972).
18 H. U. Gudel, in R.D. Willett *et al.* (eds), *Magneto-structural Correlations in Exchange Coupled Systems*, D. Reidel, Dordrecht, pp. 297–327 (1985).
19 W. A. Sibley and N. Koumvakalis, *Phys. Rev. B* **14**, 35 (1976).
20 F. Rodriguez, M. Moreno, F. Jaque and F. J. Lopez, *J. Chem. Phys.* **78**, 73 (1983).
21 F. Williams, *Phys. Status Solidi* **25**, 493 (1968).
22 A. M. Stoneham and A. M. Harker, *J. Phys. C: Solid State Phys.* **8**, 1109 (1975).
23 K. Suzuki, *J. Phys. Soc. Jpn.* **16** 67 (1961).
24 C. R. A. Catlow, A. V. Chadwick and J. Corish, *Radiat. Eff.* **75**, 61 (1983).

25 B. di Bartolo (ed.), *Radiationless Processes*, Plenum, New York (1980).
26 R. Orbach, in B. di Bartolo (ed.), *Optical Properties of Ions in Solids*, Plenum, New York (1975).
27 P. Aceituno, C. Zaldo, F. Cusso and F. Jaque, *J. Phys. Chem. Solids* **45**, 637 (1984).
28 A. Rosencwaig, *Photoacoustics and Photoacoustic Spectroscopy*, Wiley, New York (1980).
29 A. C. Tam, Applications of photoacoustic sensing techniques, *Rev. Mod. Phys.* **58**, 381 (1986).
30 J. B. Kinney and R. H. Staley, *Annu. Rev. Mater. Sci.* **12**, 295 (1982).
31 A. Rosencwaig and E. A. Hildum, *Phys. Rev. B* **23**, 3301 (1981).
32 J. Etxebarria, J. Zubillaga and J. Fernandez, *Proceedings of the Fourth International Topical Meeting on Photoacoustic, Thermal and Related Sciences, Montreal, 1985, Canadian J. Phys.* NRC, Ottawa (1985).
33 W. M. Yen and P. M. Selzer, (eds), *Laser Spectroscopy of Solids, Topics in Applied Physics*, Vol. 49, Springer, Berlin (1981).
34 R. L. Swofford and A. C. Albrecht, Non linear spectroscopy, *Annu. Rev. Phys. Chem.* **29**, 421 (1978).
35 H. Mahr, *Quantum Electronics: A Treatise*, Vol. 1, p. 283 (1975).
36 J. J. Hopfield and J. M. Worlock, *Phys. Rev.* **137**, 1455 (1965).
37 R. G. Brewer and R. G. De Voe, in *Coherence and Energy Transfer in Glasses, Proceedings of a NATO Workshop*, Plenum, New York (1984).
38 A. L. Schoemaker, *Annu. Rev. Phys. Chem.* **30**, 239 (1979).
39 R. M. MacFarlane, R. T. Hartley and R. M. Shelby, *Radiat. Eff.* **72**, 1 (1983).
40 G. B. Wright (ed.), *Light Scattering Spectra of Solids*, Springer, Berlin (1964).
41 J. M. Calleja, F. Flores, V. R. Velasco and A. Ibarra, in J. T. Devreese, L. F. Leemens, V. E. Van Doren and J. Van Royen (eds), *Recent Developments in Condensed Matter Physics*, Plenum, New York (1981).
42 A. de Andrés and J. M. Calleja, *Solid State Commun.* **48**, 949 (1983).
43 A. K. Soud, J. Menendez, M. Cardona and K. Ploog, *Phys. Rev. Lett.* **54**, 2111 (1985).
44 R. Bube, *Photoconductivity of Solids*, Wiley, New York (1960).
45 F. J. Himpsel, *Adv. Phys.* **32**, 1 (1983).
46 Y. Petroff and J. Lecante, *J. Microsc. Spectrosc. Electron.* **1**, 319 (1976).
47 A. Soler, L. Galan and F. Rueda-Sanchez, *J. Phys. C, Solid State Phys* **7**, 820 (1974).
48 S. Ferrer and G. A. Samorjai, *J. Appl. Phys.* **52**, 4792 (1981).
49 J. T. Randall and M. H. F. Wilkins, *Proc. R. Soc. London, Ser. A* **184**, 366 (1945).
50 P. N. Keating, *Proc. Phys. Soc., London* **78**, 1408 (1961).
51 P. D. Townsend, C. D. Clark and P. W. Levy, *Phys. Rev.* **155**, 908 (1967).
52 S. W. S. McKeever, *Thermoluminescence of Solids*, Cambridge University Press, Cambridge (1985).
53 R. Chen and Y. Kirsh, *Analysis of Thermally Stimulated Processes*, Pergamon, Oxford (1981).
54 P. J. Kemmey, P. D. Townsend and P. W. Levy, *Phys. Rev.* **155**, 917 (1967).
55 J. S. Down, R. Flower, J. A. Strain and P. D. Townsend, *Nucl. Tracks* **10**, 581 (1985).
56 J. A. Strain, P. D. Townsend, B. Jassemnejad and S. W. S. McKeever, *Earth Planet Sci. Lett.* **77**, 14 (1985).

57 P. L. Mattern, K. Lengweiler, P. W. Levy and P. D. Esser, *Phys. Rev. Lett.* **24**, 1287 (1970).
58 Y. Kirsh and P. D. Townsend, *Contemp. Phys* to be published.
59 F. J. Lopez, M. Aguilar and F. Agulló-Lopez, *Phys. Rev. B* **23**, 3041 (1981).
60 F. J. Lopez, M. Aguilar, F. Jaque and F. Agulló-Lopez, *Solid State Commun.* **34**, 869 (1980).
61 K. Tanimura, M. Fujiwara, T. Okada and T. Suita, *Phys. Lett. A* **50**, 301 (1976).

10
Experimental Techniques III—Electron and Nuclear

10.1 INTRODUCTION

In this final section on experimental techniques the powerful spectroscopic methods of electron paramagnetic resonance (EPR) and nuclear magnetic resonance (NMR) will be described, together with a number of hyperfine methods such as Mössbauer and spin precession methods. The various approaches allow us to map out the electron density in the vicinity of a defect, to record distortion of neighbouring atoms from lattice sites and to probe the internal electric and magnetic fields at defect sites.

Brief mention is made of some ion and electron beam methods which record surface, or near-surface, composition and structure.

10.2 ELECTRON PARAMAGNETIC RESONANCE SPECTROSCOPY

10.2.1 Introduction

A large number of point defects in solids have unpaired electrons and/or non-zero orbital angular momentum and therefore are paramagnetic. The measurement of the paramagnetic susceptibility can then provide information on the nature and concentration of the paramagnetic species. In fact, this static method has sometimes been used but the main drawback is its low sensitivity requiring rather high defect concentrations. Much higher sensitivity can be attained with the dynamic method, the so-called electron spin resonance

(ESR) or more properly EPR[1-5]. This can easily handle defect concentrations in the parts per million range. A resonant absorption will be detected when microwave photon energy $\hbar\omega$ coincides with the magnetic-field-induced Zeeman splitting ΔE of the ground-state electronic level of the centre. This resonance condition is $\Delta E = \mu H_0 = g\mu_B H_0 = \hbar\omega_0$ where $\mu_B = e\hbar/2mc$ is the Bohr magneton and g the Landé factor, i.e. $\omega_0 = g\mu_B H_0/\hbar$.

Inside a solid, a variety of interactions with the surrounding lattice occur which induce a splitting of the electron energy levels of the centre. Then the allowed transitions between levels give rise to the various lines in the EPR spectra.

As will be shown below, the EPR spectrum gives information on the charge state of the defect, the type of binding, the amount of interaction with nearby nuclei and the symmetry of the environment. This is sufficient information to propose a fairly detailed model of the centre.

10.2.2   The effective spin Hamiltonian and electron paramagnetic resonance spectrum

The more advanced quantum treatment uses an effective spin Hamiltonian, which tries to describe the effect of the external magnetic field, crystalline field and hyperfine interactions on the lowest-energy manifold of the paramagnetic centre[1-5]. The degeneracy $2S + 1$ of this manifold yields the dimension of the space of states appropriate for constructing the effective spin Hamiltonian matrix. At the same time, S is considered as the effective spin of the centre. The diagonalization of the Hamiltonian matrix yields the eigenvalues and eigenstates of the manifold under all interactions present. From the eigenvalues, we can then easily obtain the allowed transitions (magnetic dipole type) that are responsible for the resonance lines in the microwave spectrum. When eigenstates can be characterized by the quantum number m_S corresponding to the projection of the effective spin on the magnetic field direction, the appropriate selection rule is $\Delta m_S = \pm 1$. This condition approximately applies as long as the Zeeman interaction is dominant.

The effective spin Hamiltonian can be written in the form

$$\mathcal{H} = \mu_B HgS + SDS \quad + SAI \quad + \mu_N Hg_N I + IQI \quad (10.1)$$

Total energy =	Zeeman	Stark fine structure	magnetic hyperfine	nuclear Zeeman	electric quadrupole

where g, D, A, g_N and Q are the relevant coupling tensors for the various contributing interactions. Figure 10.1 shows how the addition of each term splits the energy levels. The coupling tensors should be determined from

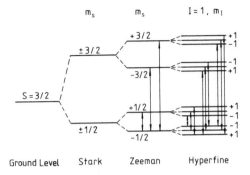

Figure 10.1.
Diagram illustrating the effect of successive terms of the Hamiltonian on energy levels (more detailed nuclear Zeeman and quadrupole splittings are not shown). Allowed EPR transitions are indicated.

the experimental spectra and they contain all available information on the symmetry and structure of the centre.

(a) Zeeman term. For a free electron, g is isotropic (i.e. a scalar) and equal to $g_e = 2.0023$. Inside a solid, g differs from the free-electron value. If the ground state of the centre is an orbital singlet, the system will essentially behave as having a "pure" spin S. Some difference $\Delta g = g - g_e$ will appear because of the mixing of the orbital states by spin−orbit interaction. On the assumption of a small spin−orbit energy, it can be shown that

$$g = g_e - C\, \frac{\lambda_{s-0}}{\Delta E} \qquad (10.2)$$

where ΔE is a typical energy separation of the excited levels, C is a constant in the range $1-10$ and λ_{s-0} is the spin−orbit coupling constant. For crystal fields with symmetry lower than cubic, g becomes anisotropic and the structure of the tensor reflects the symmetry of the crystal field. In fact, the principal axes of the g tensor determine the symmetry axes of the paramagnetic centre. A direct consequence of this anisotropy is the variation in the position of the resonance lines with the orientation of the magnetic field seen as the angular dependence of the spectrum.

(b) Fine-structure term. The fine-structure term SDS in the effective spin Hamiltonian (10.1) is due to the second-order interaction of the effective spin with the non-cubic crystal field, together with the spin−orbit interaction. It is responsible for the zero-field splitting of the ground electronic level and consequently for the occurrence of multiple-line EPR spectra, except for effective spin $S = \frac{1}{2}$ where no fine structure exists (the Kramer theorem). For the common physical case of an axial crystal field (z axis), the (quadratic) fine-structure term can be written $\mathcal{H}_D = DS_z^2(3 \cos^2 \theta - 1)$, where θ is the angle between the defect axis and the direction of the magnetic field. The value of the parameter D can then be readily derived from the angular dependence of the spectrum. The Zeeman levels and EPR transitions for an axial centre with $S = 1$ are illustrated in Figure 10.2 for H_0 parallel and perpendicular to the axis of the defect. An important consequence of the fine structure is the occurrence of forbidden transitions, violating the selection rule $\Delta m_S = \pm 1$. This results from the fact that m_S is not a "good" quantum number when the crystal field axis does not coincide with the direction of the applied magnetic field.

In practice, the symmetry of the defect can be sensed by noting the angular dependence of the line pattern. The axis of the defect is inferred from the maxima and minima in the spread of the pattern.

(c) Hyperfine term. The magnetic hyperfine term SAI is associated with the magnetic interaction between the effective electronic spin S of the paramagnetic centre and the nucleus with spins I with which it interacts. The A coupling tensor includes a dipolar contribution referring to the interaction of the magnetic moments of the electron and nucleus and an isotropic contribution. This so-called Fermi or contact term is associated with the interaction of the nuclear moment with the electronic spin density at the nucleus site. The (scalar) coupling constant A_F for this Fermi term is

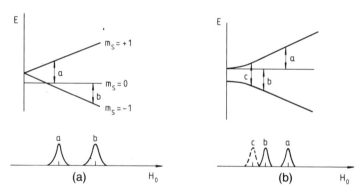

Figure 10.2.
Energy levels and EPR spectra for two different orientations of H_0 with regard to the axis of an axial defect with $S = 1$: (a) H_0 parallel to defect axis; (b) H_0 perpendicular to defect axis. In (b), transition 3 is forbidden.

$|\psi(0)|^2$, i.e. proportional to the electron density at the nucleus, so that it can only be different from zero for orbital states having some s character.

The magnetic hyperfine interaction includes a splitting of the fine-structure levels with a constant spacing between sublevels (independent of the orientation). This leads to the occurrence of a set of $2I + 1$ equally spaced hyperfine lines for each fine-structure line. This is exemplified in Figure 10.3, for a system with $S = \frac{1}{2}$ interacting with the nucleus with $I = \frac{3}{2}$. The electron spin may interact with the various nuclei, constituting a paramagnetic defect, as in colour centres, or with the ligand nuclei surrounding the central ion in the case of paramagnetic impurities (superhyperfine interaction). This effect gives rise to additional splitting of the levels. To deal with this situation, it is useful to consider two cases.

(1) Interactions with N non-equivalent nuclei.
(2) Interactions with N equivalent nuclei.

For case (1) the interaction with each particular nucleus i with spin I_i will yield a different coupling constant A_i so that we can consider in order the successive hyperfine splittings induced by the decreasing coupling terms SA_iI. As a general rule, a total number of $\Pi_{i=1}^{N} (2I_i + 1)$ equally intense lines will be observed (although accidental degeneracies may appear).

For equivalent nuclei (case 2) with spin I, the various split levels are associated with the $2NI + 1$ different values (from $- NI$ to $+ NI$) of the component M_I of the total nuclear spin along the magnetic field axis. Each M_I value and therefore the corresponding split line will have a configurational

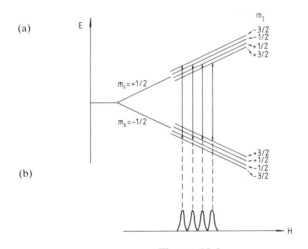

Figure 10.3.
(a) Hyperfine levels and (b) spectra corresponding to a centre with electronic spin $S = \frac{1}{2}$ and nuclear spin $I = \frac{3}{2}$.

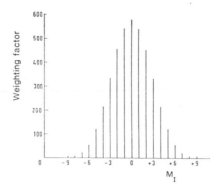

Figure 10.4.
The statistical weights of hyperfine resonance lines produced by interaction with six equivalent nuclei of $I = \frac{3}{2}$.

degeneracy given by the number of arrangements of the individual M_I values consistent with the total M_I. This degeneracy is proportional to the intensity of the corresponding hyperfine line. For N nuclei with spin $S = \frac{1}{2}$ (e.g. hydrogen nuclei) the degeneracy of the M_I component is given by the I coefficient of the binomial expansion $(1 + x)^N$. For nuclei with $I > \frac{1}{2}$ the degeneracies can be found through a general formula that can be found in the literature. An example of the weighting factor for the hyperfine lines is shown in Figure 10.4.

We can also deal with a situation where the electron spin interacts with p groups of nuclei, each one consisting of N_i equivalent nuclei with spin I_i. The total number of hyperfine lines will then be:

$$\prod_{i=1}^{p} (2N_i I_i + 1) \tag{10.3}$$

Through the Fermi term, the hyperfine structure of the EPR spectra provides very interesting information on the distribution of the electronic charge density on the nuclei forming or surrounding a given defect. In principle, a good mapping of this electron density can be accomplished, although this is very often impeded by the lack of appropriate resolution to separate the various hyperfine lines. As an example, for F centres in most alkali halides, the hyperfine structure is not resolved, although for some crystals, e.g. NaF, a distinct structure with a number of components is observed. To resolve the hyperfine structure and to obtain very detailed maps of the electron wavefunction, a electron–nuclear double-resonance (ENDOR) technique can be used (see section 10.5). Other possible contributions to the hyperfine structure of the EPR spectra are the nuclear Zeeman interaction and the electric quadrupole interaction, included in equation

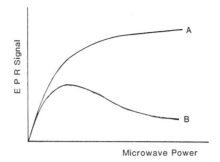

Figure 10.5.
The behaviour of the EPR signal as a function of microwave power for an inhomogeneously broadened line (curve A) and a homogeneously broadened line (curve B).

(10.1). The quadrupole interaction is associated with the coupling between the quadrupole moment of the nucleus and the gradient of electric field at the nucleus site. It only appears for nuclei with $I > \frac{1}{2}$ and crystal fields with a symmetry lower than cubic.

(d) Relaxation processes.   Energy is dissipated from the paramagnetic defect after it has been stimulated to the higher-spin state. This relaxation, leading to the inversion of spins, is due to interaction with the lattice via time-dependent electric or magnetic fields acting on the spin site. It is characterized by a spin–relaxation time T_1, which is temperature dependent. For example, for F centres in KCl, T_1 varies from 10^{-4} s at 300 K to 100 s at 2 K. This finite lifetime T_1 of the excited spin level causes broadening of the resonance line and leads to saturation of it when the rate of microwave absorption matches the relaxation rate.

In addition to energy relaxation, the coherence of individual spin precessions is destroyed by the interaction with nearby spins. This effect leads to a spin–spin relaxation time T_2, which is mostly independent of temperature and usually shorter than T_1. For F centres in KCl it remains close to 10^{-6} s. This type of relaxation does not induce saturation of the line, but it is most often responsible for the linewidth.

It should be noted that the saturation behaviour of a resonance line depends on whether or not inhomogeneous broadening effects become dominant, as illustrated in Figure 10.5. Inhomogeneous broadening is caused by anisotropic Zeeman or hyperfine interactions as well as by inhomogeneities in applied magnetic field or in the environments of the defect.

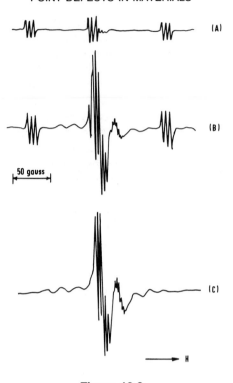

Figure 10.6.
Effect of microwave power on an EPR signal in KN_3 monitored at 77 K. The power levels are (A) 0.002 mW, (B) 0.2 mW and (C) 20 mW.

As shown in Figure 10.5, the inhomogeneously broadened lines steadily saturate at high power levels whereas homogeneous broadening results in a population inversion at high power levels. Hence the signals decrease in intensity at the highest microwave powers. An example of this is shown in Figure 10.6 for EPR of KN_3 after γ-ray irradiation.

10.2.3　EPR in glassy materials

EPR spectra in glasses present a broad-band structure[6]. To interpret these spectra, the Hamiltonian matrix must be averaged over all angles, as for a polycrystalline sample. In principle, the spectrum can be calculated in the limit of zero linewidth and then convoluted with suitable broadening functions (Gaussian or Lorentzian). As an illustrative example, Figure 10.7 shows the appearance of the integral and derivative spectra corresponding

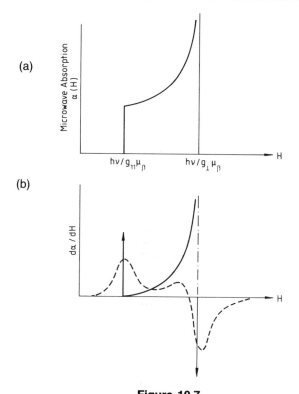

Figure 10.7.
(a) Integral and (b) derivative EPR spectra for an axial centre in a glassy material.

to axial and rhombic symmetry centres in an amorphous material. The role of a finite linewidth on the differential spectra is also shown as a dashed line in 10.7(b).

The analysis of the EPR spectra in amorphous material, at variance with the case of powder crystalline samples, has to make allowance for statistical distribution of the spin-Hamiltonian parameters. This is an obvious consequence of the random variation in crystal fields and bond angles in a glassy material.

10.2.4   Experimental techniques

In the EPR spectrometer, samples are kept in a resonant microwave cavity, which is placed inside the gap of a magnet providing the static applied field. At the sample location, the oscillating magnetic field of the cavity mode is maximized and is perpendicular to the applied field, whereas the microwave electric field is essentially zero. The microwave source is connected to the cavity through a waveguide consisting of a hollow metal pipe with a rectangular cross-section and gold plated inside surfaces. The common type of spectrometer contains a "magic tee" bridge, where the reflected radiation from the cavity is balanced with the other arm having a resistive load. At or near resonance, energy absorption by the sample unbalances the bridge and then microwave radiation falls onto a semiconducting diode detector crystal and yields the EPR signal. Normally, spectrometers work at slowly varying magnetic field H and fixed microwave frequency in one of the X, Q or K bands.

To improve the detection system, a small alternating-current-modulated magnetic field is superimposed on the slowly changing H and analysis is made with phase-sensitive detection. It is then customary to display the detection signal as the first derivative of the absorption curve. The typical minimum detectable concentration of paramagnetic defects is about 10^{15} spins cm^{-3}.

EPR detection can be extended to the submillimetre wavelength region so that a higher sensitivity is, in principle, possible. A submillimetre gas laser, as a radiation source, together with a high-field Bitter solenoidal magnet, has been used to detect the arsenic antisite in GaAs[7].

A major problem with EPR is the richness of detail and the overlap of different spectra. One way to simplify the problem is to change the microwave power level to separate homogeneously and inhomogeneously broadened lines. Figure 10.6 gives an example of the resolution achieved in potassium azide.

10.2.5   Classical examples

Some examples of the identification of intrinsic defects by EPR, which have constituted landmarks in the field, are now briefly discussed.

(a) V_K centre in alkali halides[8].   The spectrum of the V_K centre in KCl (molecular Cl_2^- ion along the $\langle 110 \rangle$ direction) is quite rich in hyperfine lines, as illustrated in Figure 10.8. Because of the two isotopes of chlorine, ^{35}Cl (75%) and ^{37}Cl (25%), both with $I = \frac{3}{2}$, there will be three superimposed spectra corresponding to three types of V_K centre: $^{35}Cl-^{35}Cl$ (56%), $^{35}Cl-^{37}Cl$ (38%) and $^{37}Cl-^{37}Cl$ (6%).

(b)   E$'$ centre in α-quartz[9] and silica glass[10].   As already discussed in Chapter 6, the structure of the E$'$ centre consists of an electron trapped in a non-bonding sp^3 hybrid orbital on a silicon atom at the site of an oxygen vacancy. The spectrum in α-quartz consists of a single line with the following principal g values: $g_x = 2.0017$, $g_y = g_z = 2.003$; these give rise to an

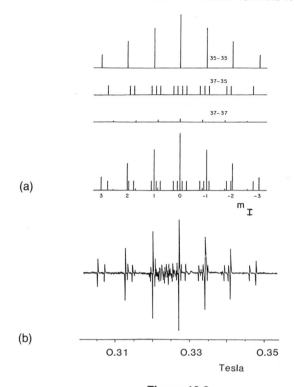

Figure 10.8.
(a) The relative sizes of absorption signals for the various ^{35}Cl and ^{37}Cl isotopic interactions as a function of m_I; (b) the conventional differential EPR spectrum of V_K centres in KCl with applied field.

angular dependence of the line location. For those centres located at ^{29}Si nuclei (4.7% abundance and $I = \frac{1}{2}$), an isotopic hyperfine splitting with $A \approx 410$ G appears. The spectra of the E′ centre in SiO_2 glass can be very accurately simulated from that of α-quartz as previously discussed. The real and computer-simulated spectra[6] are shown in Figure 10.9.

(c) **The V^+ vacancy centre in silicon[11].** The spectra of this defect consists of three equally intense, and sharp (1.5 G at 4.2 K) lines. Each line contains four pairs of weak satellites symmetrically disposed on either side. As expected, the g tensor is axially symmetric around $\langle 100 \rangle$: $g_\parallel = 2.0087$ and $g_\perp = 1.9989$. These features of the spectra are consistent with the model proposed for the centre and discussed in section 7.4, i.e. an unpaired electron on the t_2 electronic level originating from the four sp^3 bonds tetrahedrally surrounding a silicon vacancy. The three main lines arise from

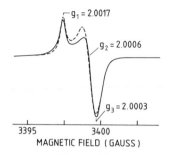

Figure 10.9.

EPR spectrum of E' centre in fused silica (Corning 7943): the dashed line is a computer simulation[6].

the three possible orientations of the centre. The four satellite lines are associated with the hyperfine interaction of the unpaired spin with the four neighbour ^{29}Si nuclei ($I = \frac{1}{2}$). The hyperfine tensor A is therefore axially symmetric around the $\langle 111 \rangle$ direction connecting the vacancy and the active silicon nucleus. The Jahn–Teller relaxation effect operating on the t_2 electronic level induces a small tilting (-7%) off the axis of the hyperfine tensor with regard to the $\langle 111 \rangle$ direction, because of the associated D_{2d} (tetragonal) distortion.

10.2.6   Transition-metal impurities

The cations of the three transition series have incomplete d shells and are consequently paramagnetic. They have often been used as paramagnetic probes to investigate the local structure in a large variety of material hosts. Table 10.1 taken from the book by Low[1] gives a list of the electron configuration, valency, ground term and spin–orbit coupling for the iron group ions. Since most of them crystallize with octahedral coordination, the lowest orbital state for this symmetry is also included in the table. Those ions having ground-state singlets, such as d^3 (Cr^{3+} and V^{1+}), d^5 (Fe^{3+} and Mn^{2+}) and d^8 (Ni^{2+}), are particularly suitable as paramagnetic probes, since the long relaxation times permit the observation of the EPR spectra at, and even above, room temperature whereas this is precluded by short relaxation times. Details about the spectra of transition-metal ions (as well as rare-earth ions) can be found in the appropriate references[1-5].

10.3   OPTICAL DETECTION OF PARAMAGNETIC RESONANCE

The transitions between close electronic levels induced by microwave radiation can be detected through their influence on optical transitions from or to

Table 10.1
The iron group[1]

Electron configuration	d^1	d^2	d^3	d^4	d^5	d^6	d^7	d^8	d^9
Valencies of ions	Ti^{3+} Mn^{6+}	V^{3+} Ti^{2+}	Cr^{3+} V^{2+}	Mn^{3+} Cr^{2+}	Fe^{3+} Mn^{2+}	Fe^{2+}	Ni^{3+} Co^{2+}	Ni^{2+}	Cu^{2+}
Ground state of free ion	$^2D_{3/2}$	$^3F_2$	$^4F_{3/2}$	$^5D_0$	$^6S_{5/2}$	$^5D_4$	$^4F_{9/2}$	$^3F_4$	$^5D_{5/2}$
Spin−orbit coupling constants (cm^{-1}) for the free ion	154	104	87 55	85 57		−100	−180	−335	−852
Lowest orbital state in octahedral symmetry	Γ_5 Triplet	Γ_4 Triplet	Γ_2 Singlet	Γ_3 Doublet		Γ_5 Triplet	Γ_4 Triplet	Γ_2 Singlet	Γ_3 Doublet
Spin degeneracy	$\frac{1}{2}$	1	$\frac{3}{2}$	2	$\frac{5}{2}$	2	$\frac{3}{2}$	1	$\frac{1}{2}$
Total degeneracy of lowest state	4	3	4	1	Either 4 or 2	3	2	3	2
Spin−lattice relaxation time	Short	Short	Long	Long	Long	Short	Short	Long	Long

these levels[12,13]. This provides an alternative detection route to that used in conventional EPR spectroscopy, i.e. absorption of microwaves. One of the main advantages of optical detection is that it is particularly suited to investigating paramagnetic resonance in excited electronic levels, where conventional EPR studies are not easy to perform. Additionally, much higher sensitivities can be reached with defect concentrations as low as 10^{12} ions cm^{-3}.

The operative principle of the method can be easily understood by reference to Figure 10.10. The defect centre presents an orbital triplet which is split under a magnetic field into three sublevels corresponding to the azimuthal quantum numbers $m_S = +1$, 0 and -1. The fluorescence transitions from these levels to the ground singlet are respectively σ^+, π and σ^- polarized. If the relaxation of the excited state is much faster than the radiative lifetime, a Boltzman equilibrium will be reached prior to the emission. Therefore the $\sigma-$ emission will be dominant. However, if the sample is then subjected to a microwave beam at the resonance condition between magnetic sublevels, transitions $\Delta m_S = \pm 1$ will be produced. The effect of these transitions will be to perturb the thermodynamic equilibrium and decrease the intensity of the σ^- emission to the benefit of the π transition (Figure 10.10). Two detection schemes are possible: measuring either the intensity of one of the polarized emissions (e.g. the σ^-) or the state of polarization of the emitted light.

10.4 NUCLEAR MAGNETIC RESONANCE SPECTROSCOPY

10.4.1 Introduction

Many nuclei possess an intrinsic angular momentum I and consequently a magnetic moment $\mu = \gamma I = g_I(\mu_N/\hbar)I$, where γ is the nuclear gyromagnetic

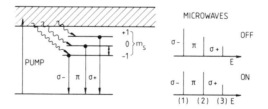

Figure 10.10.
Illustration of the operating principle of the optically detected EPR. Changes in the relative intensity of the π and σ luminescence emissions are induced by the application of a microwave field.

ratio, g_I is the nuclear splitting factor and $\mu_N = eh/2m_p$ is the nuclear magneton (m_p is the proton mass).

In a typical NMR experiment[14–16], a homogeneous magnetic field H_0 is applied to the sample containing the active nuclei. This field induces a Zeeman splitting of the nuclear level in $2I + 1$ equally spread sublevels. Then a radiofrequency field with amplitude H and frequency ω is applied along a direction perpendicular to H_0. If the absorption of radiofrequency by the sample is recorded as a function of ω, a resonant line will appear in the spectrum when the resonance condition is achieved:

$$\hbar\omega_0 = \mu H_0 = \gamma I H_0 = g_I \frac{\mu N}{\hbar} I H_0 \qquad (10.4)$$

The experimental situation is essentially similar to that formed for EPR spectroscopy. Modern systems frequently include superconducting magnets, pulsed fields and Fourier transform spectroscopy[15].

The position of the resonance line is, in principle, characteristic of each nucleus and allows for its identification.

It is important to remark that nuclei with $I > \frac{1}{2}$ are not spherical and present an electric quadrupole momentum Q which interacts with a gradient of electric field. This gradient appears whenever a nucleus is located at a site whose symmetry is less than cubic. In such a case, the quadrupole coupling also causes a splitting of nuclear levels and its effect has to be taken into account together with the Zeeman interaction for a correct interpretation of the spectra.

An advantage of the NMR technique is its applicability to all types of solids, either crystalline or amorphous. Moreover, it provides very detailed and local information about the lattice site where the nucleus sits. Its disadvantages are the low sensitivity and minimum concentrations (about $0.01-0.1\%$) usually required.

10.4.2 Wide-line nuclear magnetic resonance in solids

Resonance lines in NMR present a finite width because of the broadening mechanisms, which are either homogeneous or inhomogeneous. One of these relevant mechanisms is the electric and magnetic interaction of each nucleus with other nuclei in the sample. For liquid samples, these interactions average out to zero, and typical linewidths of about 0.3 Hz or 10^{-4} G are obtained, which implies a relative definition of the line of about 10^{-8}. Under these conditions, high-resolution studies can be performed and, in particular, lines of the same nucleus in different chemical environments can be resolved. In fact, values of this chemical shift are typically in the range $0-100$ ppm, requiring a resolution of 10^{-5} or less.

For solid samples, the situation is much less satisfactory since widths larger than about 1 G are common. This is even worse when paramagnetic impurities are present because of their much higher electronic magnetic moment. The linewidth is related to the spin−spin relaxation time T_2 and spin−lattice relaxation time T_1. T_2 is typically about $10^{-3}-10^3$ s.

In the application of NMR to solid-state physics, we take advantage of these relaxation mechanisms to gain information on the dynamics of nuclei. Some typical applications are briefly discussed next.

10.4.3   Application to diffusion

A classic application of wide-line NMR is to the study of diffusion in solids. One of the main advantages of NMR spectroscopy is the wide range of jump frequencies that can be studied by comparison with alternative techniques.[16].

Several techniques have been developed, although all of them rely on the influence of diffusion on relaxation times and consequently on linewidths.

The simplest and oldest method involves direct measurement of the decrease in linewidth induced by the diffusion of the resonant nuclei (motional narrowing). Let us consider that the operating nuclei jump among different lattice sites at a frequency $1/\tau$ which is higher than the $1/T_2$ associated to the transverse relaxation time for a given site. Then the effective relaxation time will increase and consequently lead to a narrowing of the resonance line. The method can be applied over the temperature range where $\tau = \tau_0 \exp(-E/kT) < T_2$, i.e. to rather high temperatures. In fact, it has often been used to study diffusion in both metallic and insulating solids. As an early example[17], there is an Arrhenius-type dependence of $1/\tau$ (obtained from the linewidth) with temperature for ^{27}Al ($I = 5/2$; 100% abundance) in aluminium metal. From these data, the activation energy ΔE_m for diffusion has been obtained as 1.4 eV.

Alternative methods which follow diffusion processes and their effect on T_2 include free-induction decay and spin echo[5,15,18,19].

Another method involves the determination of the minimum in the dependence of the longitudinal relaxation time T_1 on temperature. During diffusion, a Fourier component of the fluctuating dipolar interaction appears, whose frequency equals the jump frequency of the active nuclei: $\omega \approx 1/\tau$. Therefore, for a temperature where $\omega \approx \omega_0$ (resonance frequency at the applied magnetic field), an optimum dipolar coupling will be obtained and consequently a minimum in the relaxation time will be observed.

The above method can be extended to nuclei with a weak magnetic moment but having an electric quadrupole moment Q. The diffusion of such a nucleus is accompanied by fluctuations in the electric field gradient;

this contributes to the relaxation time T_1 and eventually leads to a minimum in this time at an appropriate temperature.

Another interaction also contributing to T_1 which can be used to study nuclear motion, as discussed for dipolar interaction, is the chemical shift anisotropy.

Some sophisticated version of the relaxation techniques involving measurements in rotating frames can be found in the literature[15,18].

10.4.4 Bound diffusion—molecular reorientation

The above NMR methods can also be applied to translational diffusion and to bound atomic motion, such as that involved in molecular reorientation. An interesting example is provided by the rotation of the off-centre Ag^+ in $RbCl^{20}$ or that of the CN^- group in $NaCN^{15}$. In both cases, the fluctuating field responsible for the reduction in T_1 is the electric field gradient at the site of the active nuclear magnet (silver or ^{23}Na neighbour to the CN^- ion). Figure 10.11 shows the data cited by Ailion in the book by Kaufmann and Shenoy[15] corresponding to ^{23}Na nuclei in NaCN with a minimum relaxation time T_1 near 200 K.

10.4.5 High-resolution nuclear magnetic resonance in solids

The high-resolution NMR methods routinely used in chemical research are acquiring great relevance in the study of solid materials. In fact, some

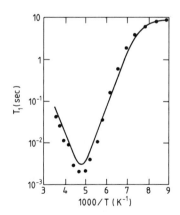

Figure 10.11.
Minimum in the relaxation time T_1 of ^{23}Na nuclei in NaCN as a function of temperature at 24 MHz.

techniques have been developed[19,21] which cause a total or partial annihilation of the dipolar interactions and open up the possibility of high-resolution studies in the solid state.

The dipolar interaction between two nuclei with magnetic moments μ_i and μ_j separated by a distance r_{ij} is given by the Hamiltonian

$$\mathcal{H}_D{}^{ij} = \frac{\mu_i \cdot \mu_j}{r_{ij}{}^3} - 3\frac{(\mu_i \cdot r_{ij})(\mu_j \cdot r_{ij})}{r_{ij}{}^5} \tag{10.5}$$

If we take into account that $\mathcal{H}_D \ll \mathcal{H}_{Zeeman}$, equation (10.5) can be written in the form

$$\mathcal{H}_D{}^{ij} = \frac{\gamma_i\gamma_j\hbar^2}{2r_{ij}{}^3} (1-3\cos^2\theta)(I_i \cdot I_j - 3I_{zi}I_{zj}) \tag{10.6}$$

where θ is the angle formed by the vector r_{ij} and the external magnetic field H_0 (along the z axis).

The high-resolution technique developed so far involves a temporal averaging of the above Hamiltonian, through either the angular term $1-3\cos^2\theta$ or the spin term $I_i \cdot I_j - 3I_{zi}I_{zj}$. In the first case, which was developed earlier, the temporal variation in θ is accomplished by a rapid rotation of the sample. It can be shown that by rotating the sample around a direction forming an angle θ of 54°44' (magic angle when $1-3\cos^2\theta = 0$) with H_0, a complete cancellation of $\mathcal{H}_D$ is obtained if the rotation velocity ω_r is higher than the relaxation rate $1/T_2$. The effect of this rotation procedure is illustrated in Figure 10.12.

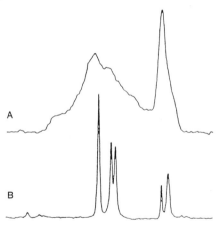

Figure 10.12.
The NMR spectra for cross-polarization of ^{13}C in a polycarbonate sample without (curve A) and with sample spinning at the "magic" angle (curve B).

The second possibility for cancelling $\mathcal{H}_D$ involves the averaging out of the spin term. High-resolution NMR provides detailed local information that cannot be obtained by X-ray diffraction techniques but it is limited to heavily disordered systems. One such example is the lattice arrangement of mica. Micas are layer silicates containing two-dimensional tetrahedral sheets of composition M_2O_5 (normally $M \equiv$ Si or Al). Every tetrahedron is linked with three neighbouring tetrahedra by sharing corners to form a hexagonal arrangement. The high-resolution NMR spectra of ^{29}Si consist of several resolved components which correspond to different possible environments of silicon in the tetrahedral sheet. Figure 10.13 shows the spectra[22] for four types of mica: muscovite, phlogopite, vermiculite and margarite, exhibiting components numbered from 0 to 3. They have been respectively assigned to the following: one silicon atom surrounded by three silicon atoms; one silicon atom surrounded by two silicon and one aluminium atoms; one silicon atom surrounded by one silicon and two aluminium atoms; one silicon atom surrounded by three aluminium atoms. This type of study throws light on the thermodynamic stability and physicochemical behaviour of these minerals.

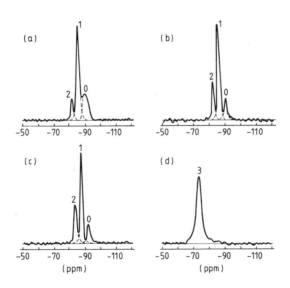

Figure 10.13.
High resolution ^{29}Si NMR spectra of (a) muscovite, (b) phlogopite, (c) vermiculite and (d) margarite. Chemical shifts are on the abscissa axis. The number of aluminium atoms around the silicon atom are indicated over the peaks.

10.5   ELECTRON–NUCLEAR DOUBLE-RESONANCE SPECTROSCOPY

The hyperfine structure of the EPR spectra can be studied by a technique which allows much higher resolution, namely ENDOR[2,23]. It involves the detection of the NMR transitions associated with the hyperfine and super-hyperfine levels of the centre through their effect on the EPR lines. Compared with EPR the resolution enhancement of superhyperfine and ligand quad-rupole interactions is about three orders of magnitude for typical solid-state defects. In ENDOR the NMR sensitivity is enhanced by up to about six orders of magnitude by a quantum transformation from the NMR quanta in the megahertz region to EPR quanta in the 10–20 GHz region. The amount of amplification depends on the details of the electron and nuclear spin relaxation mechanisms, which determine the ENDOR signal intensity.

The basis of ENDOR is conveniently discussed by referring to the simplest situation consisting of an electron spin $S = \frac{1}{2}$ and a nuclear spin $I = \frac{1}{2}$, whose level scheme is shown in Figure 10.14. If the microwave beam is tuned at the resonance frequency v_0 corresponding to the $(m_S = -\frac{1}{2}; m_I = +\frac{1}{2}) \rightarrow (m_S = +\frac{1}{2}; m_I = +\frac{1}{2})$ transition, the EPR line can be saturated if enough beam power is used. The saturation effect leads to a very small microwave absorption and is equivalent to "burning" a hole in the resonance line. However, if a radiofrequency radiation at frequency v_+ is then applied to stimulate the transition resonantly between the hyperfine levels $(m_S = +\frac{1}{2}; m_I = +\frac{1}{2})$ and $(m_S = +\frac{1}{2}; m_I = -\frac{1}{2})$, the microwave absorption line will reappear.

This effect can be observed in a steady state if a cross-relaxation (e.g. between the levels $(m_S = -\frac{1}{2}; m_I = -\frac{1}{2})$ and $(m_S = -\frac{1}{2}; m_I = +\frac{1}{2})$ operates with a rate comparable with the spin–lattice relaxation rate T_1^{-1}. The stationary effect is usually 1% or less of the EPR line intensity for typical solid-state defects. Thus the EPR line has to be strong in order that ENDOR signals can be seen.

To understand more quantitatively the structure of the ENDOR spectrum corresponding to this example, let us assume isotropic hyperfine coupling. In first-order perturbation theory the hyperfine energy levels are given by

$$E m_S m_I = A m_S m_I - g_I \mu H m_I \qquad (10.7)$$

Two NMR transitions ($\Delta m_S = 0$, $\Delta m_I = \pm 1$), inducing nuclear spin reversal, will appear at frequencies

$$v(m_S) = \frac{1}{h} \left| m_S A - g_I \mu_n H \right| \text{ for } g_I > 0 \qquad (10.8)$$

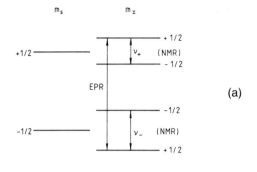

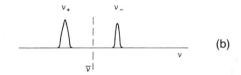

Figure 10.14.
Schematic diagram illustrating ENDOR spectroscopy for a centre with $S = \frac{1}{2}$ and $I = \frac{1}{2}$(see text): (a) levels; (b) ENDOR spectrum.

such as shown in Figure 10.14. For $A > 0$, $g_I > 0$, we obtain $v_{\mp} = (1/h)(A/2 \pm g_I\mu_nH)$ with the upper sign referring to $m_S = -\frac{1}{2}$ and the lower to $m_S = +\frac{1}{2}$. The middle frequency $v = \frac{1}{2}(v_+ + v_-) = \frac{1}{2}A/h$ in the spectrum yields the hyperfine coupling constant A. In contrast the line separation $v_+ - v_- = 2g_I\mu_nH$ is determined by g_I and thus is characteristic of the nucleus involved. The difference is used to identify chemically the nuclei involved in the ENDOR spectrum.

The argument can be extended to the case of several non-equivalent ligand nuclei. Two lines will appear for each of them. In the more general case of arbitrary electron spin S, the number of allowed lines in the ENDOR spectrum is $2m_S + 1$ for each ligand nucleus. The determination of the microscopic structure of the defect is accomplished by measuring and analysing the angular dependence of the ENDOR spectra. In general, the superhyperfine and quadrupole interactions are anisotropic. For the super-hyperfine interactions in equation (10.8) for the ENDOR frequency, A is replaced by an angular-dependent term, which contains the isotropic and anisotropic superhyperfine constants. From the symmetry pattern of these angular dependences the symmetry of the defect and the assignment of particular nuclei to particular ENDOR lines can be achieved.

The potential of the ENDOR technique is illustrated by the classical example of the F centre in alkali halides where the electron spin densities at some 13 shells around the anion vacancy have been measured. This provides a very detailed mapping of the electron wavefunction, therefore allowing for crucial tests of any theoretical model of the centre.

Atomic hydrogen in alkali halides is another classical example of ENDOR spectroscopy. A variety of structures are possible depending on whether hydrogen occupies a cation, an anion or an interstitial site. For interstitial hydrogen in KCl, the EPR spectrum only resolves the interaction with the four nearest-neighbour halogen nuclei, the interaction with the alkali nuclei contributing only to the linewidth. With ENDOR the superhyperfine and quadrupole splittings with two shells of potassium nuclei and three shells of chlorine nuclei surrounding the protons can be resolved. Figure 10.15 illustrates a section of the ENDOR spectrum[25] for iron in silicon after several data-handling operations.

The application of the ENDOR technique with conventional spectrometers is not easy if the paramagnetic defect has a low symmetry and if interaction with many lattice nuclei occur. It is then difficult, if not impossible, to assign unambiguously ENDOR lines to given nuclei. In this case the angular dependence must be measured in very small angular steps of a degree or less in order to be able to follow the angular pattern of each interacting ligand nucleus. It has proved necessary to use computer-controlled spectrometers to measure the angular dependence and to apply special digital filters and peak search techniques to determine the ENDOR line positions. A large amount of software to analyse the ENDOR spectra by solving the appropriate spin Hamiltonians has been developed. An example[26] of the application of these special algorithms is given in Figure 10.16.

As for EPR, optical detection of the resonance ENDOR lines can also be accomplished, obtaining the benefit of high sensitivity. The technique has been recently applied[27] to the As_{Ga} antisite in semi-insulating GaAs in order to probe the nature of the ligands surrounding the defect. It has been shown[28] that in the semi-insulating properties of the undoped liquid-encapsulated Czochralski grown material is an arsenic antisite−arsenic interstitial (As_{Ga}−As_{int}) pair. In this case the optically detected ENDOR spectrum was particularly complicated because of many additional line splittings caused by mutual coupling effects between the ligand nuclei via the unpaired electron (pseudodipolar couplings). The optically detected ENDOR effect was of the same order as the optically detected EPR effect, which in turn was about two orders of magnitude higher than the conventional EPR effect. Thus the total amplification was about four orders of magnitude compared with the conventional ENDOR technique.

10.6 MÖSSBAUER SPECTROSCOPY

One of the best-established techniques for investigating internal fields at atomic sites in solids is the effect discovered by Mössbauer in 1958[29]. A free

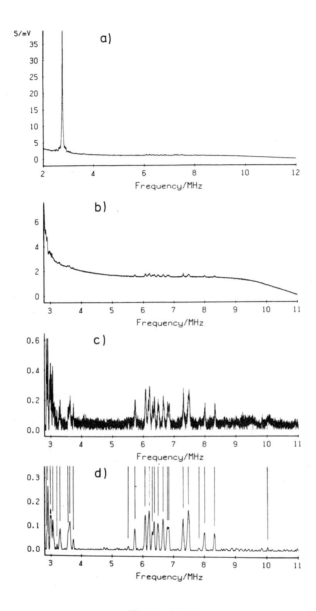

Figure 10.15.
Section of the ENDOR spectrum of interstitial Fe^0 in silicon: (a) as measured, X
band, 30 K; (b) after subtraction of ^{29}Si distant ENDOR line; (c) after subtraction
of the background line by a special algorithm; (d) after digital filtering (the vertical
lines mark the ENDOR line positions and are determined by a peak search
algorithm). (After Niklas)[24].

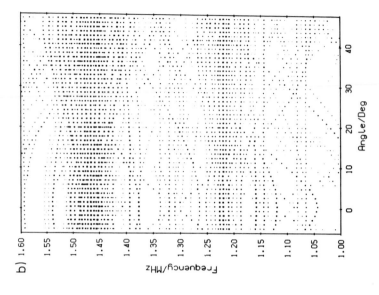

Figure 10.16.

Effect of deconvolution of ENDOR lines by special algorithms. The angular dependence of the ENDOR lines of $H_{S,A}^0$ centres in KCl (atomic hydrogen centres on anion sites in KCl). The crystal was rotated in a (001) plane (0^0 ΔB_0 [100]: $T = 40$ K: $B_0 = 3514$ G): (a) experimental results after digital filtering without deconvolution; (b) experimental results after digital filtering with deconvolution. (After Niklas[24].)

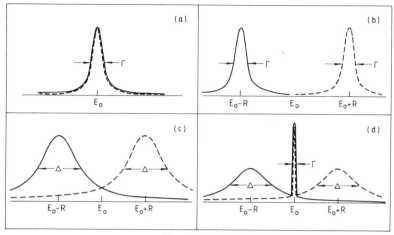

Figure 10.17.
Illustration of possible situations in resonant absorption of γ-rays, emitted in a transition of energy E_0: (a) static nuclei; (b) recoiling isolated nuclei; (c) recoiling nuclei inside a solid (widening of the γ lines from Γ to Δ); (d) occurrence of recoilless resonant absorption (the Mössbauer effect). Full line (emission band); dashed line (absorption band).

nucleus decaying from an excited level E_1 to the ground level E_0 emits a γ-ray with an energy $E_\gamma = E_1 - E_0 - R$, where

$$R = \frac{E_\gamma^2}{2Mc^2} \approx \frac{(E_1 - E_0)^2}{2Mc^2} \tag{10.9}$$

is the recoil energy of the nucleus having a mass M. For practical cases involving low γ transition energies, $R \approx 10^{-3} - 10^{-2}$ eV. The width of the γ emission line is determined by the uncertainty principle and is typically in the range $10^{-7} - 10^{-11}$ eV. Because the recoil energy is much larger than the linewidth the resonant absorption of the γ photon cannot be observed (Figure 10.17) unless the emitting nucleus is inside a solid sample, in which case the recoil energy spreads over many nearby atoms and leads to the creation of phonons or vibrational modes for a crystalline material. The width of the emission line is now much larger (about $10^{-3} - 10^{-2}$ eV) and allows for some resonant absorption which is enhanced by raising the temperature (Figure 10.17(c)). However, under certain circumstances, a fraction f of the decaying nuclei shows a recoiless γ emission with energy $E_\gamma = E_1 - E_0$ and natural linewidth because the recoil momentum of these zero-phonon γ-rays is transferred to the lattice as a whole (i.e. the Mössbauer effect). Specifically, $f = \exp(-k^2 < x^2 >_T)$, x being the nuclear displacement, which for a cubic Bravais lattice in the Debye approximation reduces to

$$f = \exp\left\{-\frac{3R}{2k\theta_D}\left[1 + \frac{2\pi^2}{3}\left(\frac{T}{\theta_D}\right)^2\right]\right\} \qquad (10.10)$$

where θ_D is the Debye temperature of the material. Equation (10.10) clearly illustrates the enhancement of f on reducing temperature.

Similarly to emission, recoiless absorption of γ-radiation ($E_\gamma = E_i - E_j$) can also be induced, making possible the so-called resonant recoiless absorption. Nuclei which provide decays with suitable energies, lifetime and linewidth are listed in Table 10.2. ^{57}Fe, derived from ^{57}Co, is preferred as it has a Mössbauer fraction approaching 100% at low temperatures. The usefulness of the Mössbauer technique lies in its extreme energy resolution (about 10^{-12}), which allows the study of very small perturbations on the energy levels of the emitting nuclei. In particular, the hyperfine interactions associated to the electric and magnetic fields induced at a nucleus can be measured by the particular arrangement of atoms and electron clouds around it. Its application is not restricted to any particular type of solid, but of course only defects in the vicinity of the iron impurity can be studied.

Experimentally the resonance between source and absorber is achieved by a Doppler shift generated by moving either the source or the detector[29].

Some of the hyperfine interactions that can be studied by Mössbauer spectroscopy are now briefly surveyed.

(a) Isomer shift. The isomer shift (IS) of the γ transition is due to the Coulomb interaction of the nuclear charge distribution with the electronic charge. In fact, the shift value in the nuclear energy level is proportional to the electronic charge density at the nucleus. The detailed expression reads

$$\Delta E_I = E_A - E_S = \tfrac{2}{3}\pi Z e^2 (R^{\star 2} - R^2)\{|\psi_A(0)|^2 - |\psi_S(0)|^2\} \quad (10.11)$$

Here, E_S and E_A refer to the source and absorber energy level, respectively, Z is the atomic number of the emitting nucleus, $R^\star$ and R are the radii of

Table 10.2.
Common Mössbauer nuclei: parameters

Parent Nucleus	Half-Life	Mössbauer nucleus	E_γ (kev)	Γ	$E_R \times 10^2$ (ev)
^{57}Co	270 days	^{57}Fe	14.4	4.6×10^{-9}	0.19
^{119}Sn	250 days	^{119}Sn	24	2.4×10^{-8}	0.26
^{191}Os	16 days	^{191}Ir	82.6	1.2×10^{-7}	1.9
^{67}Ga	78 h	^{67}Zn	93	4.8×10^{-11}	6.9

E_γ, energy of the γ transition; Γ, natural width of the γ emission line; E_R, recoil energy of the Mössbauer nucleus.

excited and ground states, and $|\psi_A(0)|^2$ and $|\psi_S(0)|^2$ are the s electron densities at the emitting and absorbing nuclei.

According to equation (10.11) the IS of the transition is only observed when the emitter and absorber electronic structures are not identical. Therefore the measurement of the IS provides information on the electronic wavefunction at the emitting or absorbing nuclear site and consequently on the nature of its bonding. In particular, the IS can very conveniently distinguish between different valence states of the same element in the material as well as enabling changes from one valence to another valence induced by appropriate treatments to be studied. For example the change $Eu^{2+} \rightarrow Eu^{3+}$ in magnesium caused by cold work can be detected.[30]

(b) Nuclear Zeeman interaction. The interaction of the magnetic moment of the nucleus with an external magnetic field induces a splitting of the nuclear levels, so that the resonance spectrum is made up of the transitions between the various excited- and ground-state sublevels.

The magnetic field causing the hyperfine splitting can be provided by the internal field in magnetically ordered materials or by the local field from the electron cloud in paramagnetic systems. For the latter case, the lifetime of the electronic spin levels has to be sufficiently long in comparison with the Larmor nuclear frequency in order for the hyperfine lines to be experimentally observed. A good example of this behaviour has been found for Fe^{3+} in α-Al_2O_3 or $LiNbO_3$[31]. In this case, the electronic spin relaxation line of the $^6S_{5/2}$ state at low temperatures is large in comparison with the Larmor precession frequency so that the hyperfine magnetic sextets are well observed for the crystal field states $|\pm\frac{5}{2}\rangle$ and $|\pm\frac{3}{2}\rangle$. Hence it can be confirmed that Fe^{3+} exists in these crystal lattices and changes caused by irradiation or thermal treatments can be monitored. (Fe^{2+} can also be recorded as it has a different spectrum.)

(c) Electric quadrupole interactions. Nuclei with spin $I > \frac{1}{2}$ present an electric quadrupole moment which can interact with the electric field gradient tensor at the nucleus. This tensor is made up of an electronic contribution from the electron cloud around the nucleus and a lattice contribution induced by the effective charges located at lattice sites. For axial symmetry (z axis) the electric quadrupole Hamiltonian is written as

$$\mathcal{H}_Q = \frac{e^2 Qq}{4I(2I-1)} [3I_z^2 - I(I+1)] \qquad (10.12)$$

For example this axial symmetry predicts a doublet for the ^{57}Fe case ($I = \frac{3}{2}$) whose separation gives the electric field gradient.

10.6.1   Summary of applications

In summary the parameters of the Mössbauer spectra can be used[29,32] to obtain information on a number of defect properties as described below.

From the IS the oxidation state of a given atom can be ascertained. Moreover, even for a given state, details about the bonding to neighbour atoms can be learnt.

The electric quadrupole interaction may be appropriate for the location of a given impurity in the lattice, e.g. substitutional or interstitial. Association or clustering of impurities or impurities and defects, such as vacancies, can also be conveniently investigated.

Magnetic hyperfine interactions give information on magnetic phases in alloys and internal fields.

Atomic vibrational motion can be studied by measuring the f parameter. The identification of local impurity modes and anisotropic vibrations is then possible. A particularly interesting application refers to the study of the anisotropies associated with the proximity of the Mössbauer atom to the solid surface[15].

Finally, long-range motion of atoms can be investigated through the width of the Mössbauer lines. The situation is similar to that found in NMR spectroscopy. The atomic diffusion causes an apparent reduction of the lifetime of nuclear levels related to the jump frequency from site to site.

Not only crystalline but also amorphous solids can be studied by Mössbauer spectroscopy. For amorphous systems the information provided by diffraction techniques is not very detailed and the analysis of the data has to rely on proposed structural models. The Mössbauer effect, as for the other local techniques, gives specific data on the atomic configuration around the active nucleus. A typical example is the study of the SiO_2-Fe system. The structure of the spectrum clearly confirms that a local order exists. The Fe^{2+} and Fe^{3+} ions lie in a rather well-defined environment although strongly distorted by comparison with a crystal.

10.7   SPIN PRECESSION METHODS

Alternative methods to those of Mössbauer and NMR (also EPR) spectroscopies have been developed to study the nuclear hyperfine interactions inside a material. They are classified under the general heading of spin precession techniques[15,33]. The objective is to measure directly the nuclear spin precession induced by the internal magnetic field or electric field gradient existing at the site of the active nucleus.

Let us first assume that a given population of nuclei having spin I is prepared with a preferential alignment along the magnetic field axis z. Then the nuclei will precess around such an axis with the Larmor frequency ω_L and consequently periodic variations (period, $2\pi/\omega_L$) in some appropriate signal (sensitive to spin direction) will be measured when observing along a perpendicular direction. In other words, the measured value $R(t)$ of such a signal will depend on time according to

$$R(t) = A_1[1 + A_2 \cos(\omega_L t)] \qquad (10.13)$$

where A_1 is the average amplitude and A_2 an oscillatory coefficient related to the degree of nuclear alignment or polarization. The Fourier transform $G(\omega)$ of (10.13) yields a resonant line at ω_L in the frequency spectrum. In a quantum version, this frequency is given by $\Delta E/\hbar = \omega_L$, where ΔE is the separation between consecutive Zeeman levels split by the magnetic field.

In practice, the situation is somewhat more complicated than indicated by equation (10.13), because of the relaxation processes leading to an exponential decay in $R(t)$, i.e.

$$R(t) = A_1[1 + A_2 \exp(-\sigma_d t) \cos(\omega_L t)] \qquad (10.14)$$

As a consequence of this relaxation, the resonance line at ω_L has a finite width related to the decay constant σ_d. The expected $R(t)$ and its Fourier transform for the nuclear precession ($I = \frac{5}{2}$) in a magnetic field is schematically illustrated in Figure 10.18.

For the electric quadrupole interactions, a similar analysis applies. However, the nuclear level splitting is not uniform and several resonant frequencies appear. For a nucleus with $I = \frac{5}{2}$ placed in an axial electric field, three frequencies ω_1, ω_2 and ω_3 appear; their values are in the ratio 1:2:3. This can be predicted from the formula giving the energy of the split quadrupole levels:

$$E_{\pm m} = \frac{3m^2 - I(I + 1)}{4I(2I - 1)} eQV_{zz} \qquad (10.15)$$

Q being the nuclear quadrupole moment, m the projection of I along the z axis and V_{zz} the corresponding field gradient component. The time precession function $R(t)$, in the absence of relaxation, will then be written

$$R(t) = A_0 + A_1 \cos(\omega_1 t) + A_2 \cos(\omega_2 t) + A_3 \cos(\omega_3 t) \qquad (10.16)$$

where the amplitudes A_1, A_2 and A_3 are related to the transition probabilities between the involved hyperfine levels. The Fourier transform $G(\omega)$ is constituted by three resonant lines at frequencies ω_1, ω_2 and ω_3.

In summary, by measuring the precession function $R(t)$ and then performing the Fourier transform, we obtain the hyperfine frequency spectrum, i.e. resonant frequencies and line intensities. From these data, the internal hyperfine field (either electric or magnetic) can be determined. In order to

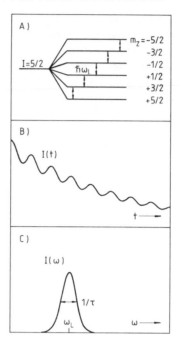

Figure 10.18.
Illustration of (A) level splitting of a nucleus with $I = \frac{5}{2}$ in an internal magnetic field, (B) associated precession in any physical property related to spin orientation and (C) Fourier transform of precession.

take advantage of these methods, the active nuclei must be preferentially oriented at the start of the experiments. In principle, the application of high magnetic fields and low temperatures is not practical. However, two efficient techniques have been developed and are discussed next.

(1) Time differential perturbed angular correlation (TDPAC).
(2) Muon spin rotation (μ^+SR).

10.7.1 Time differential perturbed angular correlation

In this technique[34] the initial orientation of the nuclei is achieved by the emission of the first γ-ray in a $\gamma-\gamma$ cascade. The measurement of the time variation in the intensity of the second γ-ray serves to monitor the nuclear precession. The time dependence of the γ-ray intensity is determined with the detector at a fixed position. A precession function $R(t)$ as given by equation (10.13) or (10.16) (for $I = \frac{1}{2}$) should be observed for each nonequivalent lattice site. The most commonly used nucleus is ^{111}In, which decays into ^{111}Cd ($\frac{5}{2}+$, 247 keV), with a half-life $T_{1/2} = 2.8$ days. The ^{111}Cd

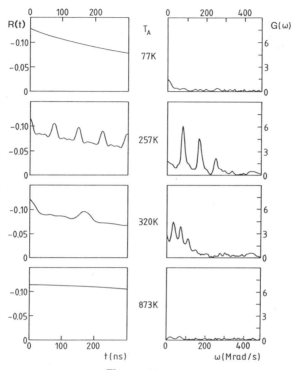

Figure 10.19.
TDPAC spectra $R(t)$ and corresponding Fourier transforms $G(\omega)$ for ^{111}In in gold after irradiation at 4.2 K and annealing at the temperatures indicated in the figure (after Th. Wichert and E. Recknagel[15]).

is the true probe nucleus, having a half-life $T_{1/2} = 84$ ns. Isotopes of ^{100}Pd and ^{181}Hf are sometimes used as alternatives.

An example of the usefulness of the technique is provided by experiments[15] on electron-irradiated gold. Figure 10.19 shows the TDPAC $R(t)$ spectra and the corresponding Fourier transforms $G(\omega)$ for ^{111}In in a gold sample irradiated at 4.2 K and annealed up to several temperatures. At 77 K, no structure is observed whereas, after annealing at 257 and 320 K, two different spectra appear which can be related to the quadrupole interaction with an axial symmetric defect. At about 900 K the isotropy is re-established. The two intermediate spectra have been attributed to some defects trapped in the vicinity of the ^{111}Cd. In particular, the spectrum observed after 2151 K annealing should correspond to vacancies since the same spectrum is observed for freshly quenched samples, where the occurrence of vacancies is well ascertained. At sufficiently high temperatures, the vacancies become untrapped and the cubic symmetry at the ^{111}Cd nucleus is restored.

10.7.2 Muon spin rotation

In this technique[15,33], positive muons (μ^+) resulting from the radioactive decay of positive pions (π^+) are introduced in the material. Since the μ^+ particles in a parallel beam are oriented along the momentum direction (i.e. polarized), they are well suited to precession studies. In the decay of the μ^+, one positron and two neutrinos are emitted: $\mu^+ \rightarrow e^+ + \nu_e + \nu_\mu$. The positron, which is preferentially emitted along the μ^+ spin direction, is used to monitor the spin orientation of the emitting μ^+. Oscillations occur in the intensity of the detected positron flux, under a magnetic field, which are superimposed on an exponential decay. In accordance with the previous discussion, the spin precession of the μ^+ provides information on the internal fields, either magnetic or electric.

In metals the μ^+ can be trapped at suitable interstitial positions of the material, although the nature of the traps and the trapping mechanisms are still objects of investigation[35,36]. The measurement of precession frequencies and relaxation rates yields data which allow the identification of the active trapping site. An illustrative example of the capability of this technique is provided by the experiments on iron irradiated with electrons (3 MeV) at 10 K. The Fourier transform $G(\omega)$ of the $R(t)$ spectra obtained at T_M, after an annealing treatment up to a temperature T_A, are shown in Figure 10.20. In the first spectrum ($T_A = 181$ K; $T_M = 92$ K), a single peak is observed which is associated with an interstitial position of the μ^+ in the iron lattice. When the measuring temperature is increased to $T_M = 160$ K, a new peak appears which should be attributed to the μ^+ trapped at some radiation-induced defect. This defect, which disappears after annealing at 249 K, has been identified as a vacancy.

The relaxation function $R(t)$, and therefore the width of the resonance line in the $G(\omega)$ spectrum, can be markedly influenced by muon motion. Motional narrowing effects similar to those considered in NMR spectroscopy have been used to investigate muon diffusion. Muons can be considered as a light isotope of hydrogen, and therefore the information gathered with the μ^+SR technique may contribute to a better understanding of hydrogen diffusion problems in solids. In fact, there is substantial evidence that μ^+ becomes localized at the same sites as the normal hydrogen isotopes ^{1}H, ^{2}D and ^{3}T. For example, in copper, muon migration occurs above 80 K with an activation energy of about 48 meV.

For semiconductors and insulating materials in which no efficient charge screening exists, the μ^+ may capture an electron and form a muonium atom, similar to a hydrogen atom. The hyperfine structure of the muonium can be probed by μ^+SR, so that interesting information can be obtained on the surrounding electronic structure. Normal muonium atoms, with isotropic hyperfine structure, have been mostly observed.

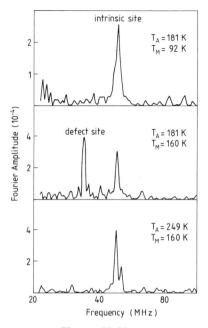

Figure 10.20.
Fourier transforms of the μ^+SR spectra for iron after electron irradiation at 10 K and subsequent annealing at the temperatures T_A indicated in the figure. The measuring temperature T_M is also indicated. (After Th. Wichert and E. Recknagel[15]).

More recently, π^+ particles have been used as probe particles for defect studies in solids. This illustrates the new horizons that are being opened up by the use of nuclear particles and techniques in solid-state physics. However, the limitations of μ^+SR are access to suitable sources and the need for relatively large defect concentrations (e.g. about 1% for vacancies).

10.8   POSITRON ANNIHILATION

The positron or positive electron (β^+) which was discovered in 1932 is being used to study defects in a variety of materials, mostly metals and alloys. When an energetic positron enters a solid, it rapidly loses energy and becomes thermalized in a time of 20 ps. At thermal velocities, it continues its motion through the lattice and finally annihilates with one electron of the material. In the vast majority of cases, the annihilation gives rise[37,38] to the emission of the two collinear γ-rays, with opposite momenta, and having an energy of 0.511 MeV each. Under certain simplifying assumptions the annihilation probability λ_a per unit time or the reciprocal of the lifetime τ can be expressed as

$$\lambda_a = \tau^{-1} = \pi r_o^2 c n_e \qquad (10.17)$$

where r_0 is the classical radius of the electron, c the velocity of light and n_e the average electron density seen by the positron. For a perfect solid, the lifetime τ has a well-defined value, between 100 and 500 ps, which is characteristic of the material. It should be mentioned that in certain materials (e.g. molecular crystals) the positron can capture one electron and form a positronium atom, in either a singlet (parapositronium) or a triplet (ortho-positronium) state. The lifetime of the parapositronium is 125 ps, whereas that for orthopositronium is 142 ns, although it may decrease to a few nanoseconds after capturing another electron with reversed spin.

If the material contains defects such as vacancies, microcavities or even dislocations which have a negative effective charge, then the positron may be trapped and form bound states. In such a case, its lifetime will become longer owing to localization in a region of low electronic density. Therefore the measurement of the positron lifetime constitutes an appropriate technique for detecting and studying those defects. In contrast, defects repelling the positron do not appreciably modify its lifetime. The method is sensitive to defect concentrations as low as 10^{-7} atomic sites.

In addition to an influence on the positron lifetime, the negatively charged defects induce a Doppler broadening of the energy spectrum of the emitted γ-rays, as well as a loss of collinearity between them. This can be easily understood with reference to Figure 10.21. As a consequence of the finite velocity of the centre of mass of the electron–positron system (total momentum p), the two emitted γ-rays will form a small angle $\theta = p_T/m_0c$, p_T being the transverse component of p, with regard to the emission direction. The energy of the γ photons differs from the ideal value of 0.511 MeV by $\Delta E = c p_L/2$, p_L being the longitudinal component of p. Since the positron is thermalized before annihilation, the major contribution to p comes from the electron, so that the experimental curves $N(\theta)$ showing the angular correlation between the γ-ray partners, and the $N(E)$ describing their energy spectra, will mostly reflect the momentum distribution of the electrons inside the material. In principle, the valence or conduction electrons, with small momentum, will contribute to narrow $N(E)$ and $N(\theta)$ distributions, whereas the deep electrons will yield wider curves. Practical values of the widths $\Delta\theta$ and ΔE of the angular and energy distributions are, respectively in millirads and kiloelectron volts.

When the positron annihilates after having been trapped at a defect, such as a vacancy, the relative density of deep ("core") electrons with regard to that of conduction or valence electrons will be smaller than in the perfect crystal. Consequently, a narrowing of the $N(\theta)$ and $N(E)$ distributions is measured.

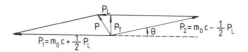

Figure 10.21.
Diagram showing the momenta p_1 and p_2 of the two γ-rays emitted during annihilation of a positron. Because of the finite momentum p of the electron–positron system, p_1 and p_2 are not opposite.

As a summary of the above discussion, three methods are available for studying defects by means of positron annihilation. They respectively refer to measurements of τ, $N(\theta)$ or $N(E)$. Positron lifetimes are mostly used. The positron source is, usually, made of ^{22}Na embedded in manganese. The source is sandwiched between the sample and a reference. The sample thickness (100 μm or more) should assure the total annihilation of the incident positrons. The 1.28 MeV γ-ray emitted in the decay of ^{22}Na sets the time origin and then the time elapsed until the detection of the annihilation γ-rays is determined by means of a time–amplitude converter. The required time resolution is a few picoseconds. As a detector, a crystal or plastic scintillator coupled to a photomultiplier tube is used.

Some selected examples of the application of this technique are as follows.

(1) The study of vacancies in metals was briefly mentioned in Chapter 7. The formation enthalpies of vacancies have been determined for a number of metals and they show good agreement with the values obtained by other methods. The analysis of the lifetime data is usually made by a simple model involving two positron states: the free or Bloch positron and the trapped positron. Consequently, two decay components are predicted. The concentration of vacancies can be determined from the ratio between the intensities of the fast and slow components.

(2) A study of the formation of vacancy clusters and microcavities can be made during annealing of irradiated solids. The clustering of vacancies induces an increase in the positron lifetime that is correlated with the size of the cluster. However, a saturation value is reached when the size becomes comparable with the de Broglie wavelength of the thermalized positron.

10.9   ION BEAM TECHNIQUES

The use of nuclear particles and ions as analytical probes has provided an extremely useful tool in materials science for investigating the structure and composition of surfaces (down to the order of a micrometre) and thin films[39-43]. Therefore, these techniques are particularly appropriate in connection with ion-implanted and diffused materials, thin-film protective coatings, etc., as required in microelectronics as well as modern metallurgy and ceramics. Some of the techniques are already well established and are routinely utilized, whereas others which offer new possibilities are gaining widespread acceptance. Here, we briefly focus on the best-established techniques and try to point out their interest for defect physics applications. Although some of them are restricted to one or a few atomic layers, they can be combined with techniques of successive layer removal by ion sputtering to explore deeper layers.

10.9.1 Rutherford backscattering

Rutherford backscattering spectrometry (RBS) is now a standard technique for impurity analysis and composition of near-surface layers. The principle of the method is very simple and the experimental arrangement is shown in Figure 10.22. Typically a light ion of H^+, He^+ or N^+ at $1-2$ MeV is fired at the target. Although most ions lose energy by electronic excitation and are only slightly deviated, a few ions suffer direct nuclear collisions and elastically recoil. These reflected ions are detected with an energy-sensitive detector such as a silicon barrier detector. The resultant charge pulse is proportional to the ion energy. The various pulses are sorted via a multi-channel analyser and hence an energy spectrum of the reflected particles is derived. The geometry used is for a deflection angle of 150 or 170° and about 10^{-6} of the incident ions are reflected. With various ion sources, geometries or detector size, we can probe to depths of about 1 μm, obtain a depth resolution of about 5 nm and separate masses to about 1%. In ideal situations of heavy impurity ions on a light target, we can achieve detection sensitivities approaching 10^9 impurities cm^{-2} of target surface[44].

The basic theory of these elastic recoils gives the reflected energy E_1 in terms of the incident energy as

$$E_1 = KE_0 \tag{10.18}$$

where the kinematic factor K in the laboratory frame is

$$K = \left(\frac{(M_2^2 - M_1^2 \sin^2 \theta)^{1/2} + M_1 \cos \theta}{M_2 + M_1} \right)^2 \tag{10.19}$$

for an incident ion mass M_1, a target ion mass M_2 and a deflection of θ.

If the collision does not occur immediately at the surface then electronic stopping dE/dx reduces E_0 to $E_0 - (dE/dx) \Delta x$ at a depth Δx. The reflected ion is similarly reduced in energy. Thus, even for each element, the observed energy spectrum has a sharply defined upper limit set by $E =$

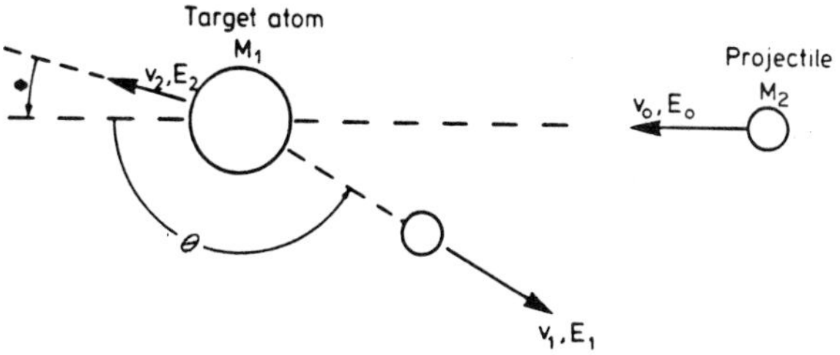

Figure 10.22.
Schematic diagram of an elastic collision event between a light projectile and a heavy target atom.

KE_0 for surface ions, and this broadens into an increasing lower-energy signal caused by deeper collisions.

To relate the signal intensity to the atomic concentration, we must consider the collision cross-section. For a simple Coulomb potential the form of the differential cross-section into a solid angle $d\Omega$ is

$$\frac{d\sigma}{d\Omega} = \left(\frac{Z_1 Z_2 e^2}{4E_0}\right)^2 \frac{4}{\sin^4 \theta} \times \frac{[\{1 - [(M_1/M_2) \sin \theta]^2\}^{1/2} + \cos \theta]^2}{[1 - \{(M_1/M_2) \sin \theta\}^2]^{1/2}} \qquad (10.20)$$

For the selected detector setting of θ, this implies that the scattering cross-section is proportional to Z^2; that is, RBS is most sensitive for heavy elements. If these elements are to be detected as impurities, then they are most obvious in the case of light element hosts as the impurity signals will occur at the higher-energy channels of the energy spectra. Figure 10.23 shows an idealized spectrum for this case.

In practice there are many refinements to both the analysis and the experimental techniques which are well documented, e.g. by Evans and Blattner[40], Reuter and Baglin[41] and Feldman and Poate[42], and the book by Chu et al.[39] is a standard reference.

A simple analysis of impurities limited to the surface layer is rarely required and in normal problems the depth distribution of the constituent ions must be determined. In so doing, there is need to allow for changes in the stopping power; hence the problem is relatively complex. A typical approach is to use a computer simulation of the structure until the simulation and data are in accord. An example of the approach is shown in Figure 10.24 for the complex problem of ion exchange between $AgNO_3$ and a sodium silicate glass[45]. The edges corresponding to the kinematic limit set by silver, calcium, aluminium and silicon can immediately be recognized and an edge for sodium atoms in the surface layers can just be noted. The samples were coated with a 10 nm layer of aluminium to avoid charging during the ion bombardment and this is apparent from the surface peak in

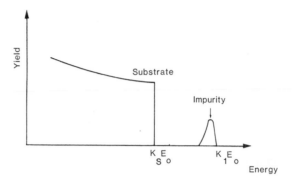

Figure 10.23.
Idealized spectrum obtained by RBS of a target with a heavy surface impurity. The reflected energy from the light substrate shows an edge at K_sE_0. The concentration and depth distribution of the heavy impurities is implied by the form of the higher-energy peak.

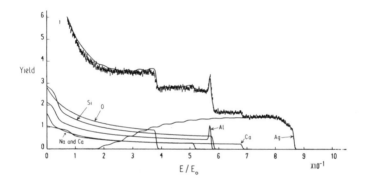

Figure 10.24.
A comparison of spectrum obtained by RBS with that derived by a computer simulation of the components near the surface of a silver ion exchanged sodium silicate glass.

the aluminium component. The computer simulation achieves a good fit by summation of the various components. More interestingly, Figure 10.25 shows that during the ion exchange silver ions entered the glass to a depth of about 1.5 μm and there is a depletion of sodium ions. The shape of the concentration profile resembles the expected error function profile for a diffusion-controlled process. Comparison with samples that have silver exchanged to depths of about 10 μm gives a similar diffusion profile but in this case it was ascertained by analysis of optical waveguide properties.

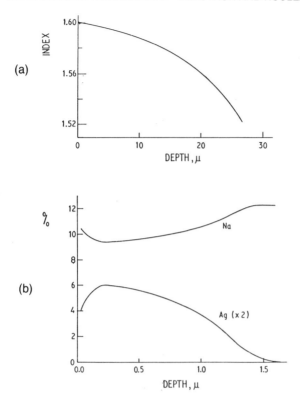

Figure 10.25.
(a) Refractive index enhancement caused by prolonged (40 h) silver exchange in sodium silicate and (b) silver and sodium depth distributions derived from RBS data (Figure 10.24) for a shorter reaction time (about 12 min).

10.9.2   Ion beam channelling

For single-crystal material the RBS yield will be modified if the beam is aligned with an open channel axis or plane of the crystal. Aligned beams undergo small angle deflections; hence there is a dip in the yield and the angular width ψ of the channel and the depth x_{min} of the minimum below the random-direction amorphous yield depends on the ion beam and the crystal perfection (Figure 10.26). Channelling has been detected for more than 20 years and the theory is well described in many places including the references given here[42,46-8].

The RBS channelling spectra (Figure 10.27) differ from the amorphous example in Figure 10.23 in that they show a sharp surface peak with a minimum in the lower-energy, i.e. random-yield, part of the spectrum.

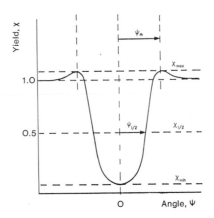

Figure 10.26.
Normalized RBS yield at a fixed interaction depth as a function of the angle between the incident ion beam and a channel axis.

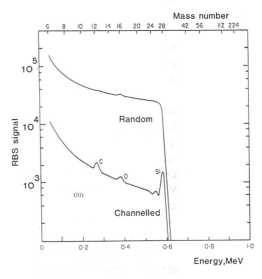

Figure 10.27.
A comparison of random and channelled RBS spectra for silicon with traces of oxygen and carbon impurities using an incident He$^+$ beam at 1.0 MeV. The upper scale shows the position of edges for different target masses.

Analysis of signals detected in the channelling direction allows interstitials or other defects which perturb the quality of the channels to be sensed. For example an interstitial impurity will be strongly detected compared with the host if viewed down such a channel. Figure 10.28 shows that simple

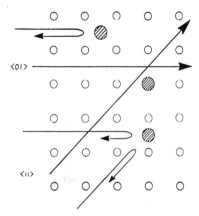

Figure 10.28.
An example of channelling to locate the lattice site of impurity atoms. In this two-dimensional view the ⟨01⟩ and ⟨11⟩ axis will both sense the interstitial impurity sites but the incoming ion will only be weakly scattered by the substitutional impurity.

geometry can be used to determine the site of such an interstitial as there will be strong midchannel signals for the impurity when viewing down the direction ⟨100⟩ or ⟨010⟩ but the impurity is shielded if viewed along ⟨110⟩.

As an example of the method the analysis of the irradiation-induced lattice sites of oxygen impurities in niobium can be considered. The impurity is lighter than the host element; so in simple random-direction RBS spectra the oxygen signal would be difficult to detect. However, in well-aligned channel measurements it is a strong signal. Nuclear reaction methods such as $^{16}O(d, \alpha)^{14}N$ discussed below may also be used. Palmer and co-workers[49,50] detected a variety of oxygen sites which depend on the irradiation temperature and mobility of niobium vacancies.

10.9.3   Nuclear reactions

Instead of scattering processes, nuclear reactions induced by the incoming beam particles can be used to investigate the presence of certain atomic species in the material. The reaction products are usually charged particles, which can be detected as in the RBS technique, or by γ-rays. The method is limited by the choice of targets and readily available ion beams but is particularly appropriate for low-Z materials and provides a complementary tool to RBS, which has a low sensitivity for these elements.

The nuclear reactions have a resonant character and therefore present a good yield for a small energy range of the incoming particles. By varying that energy, atomic layers at different depths can be explored, allowing for depth profiling of the active nuclei.

A number of reactions have been used; for example for the detection of oxygen, $^{16}O(d, \alpha)^{14}N$ or $^{16}O(d, p)^{17}O$, for detection of nitrogen, $^{14}N(d, \alpha)^{12}C$ and, for the detection of boron $^{10}B(\alpha, p)^{13}C$. Hydrogen is a particularly interesting case since it is an important impurity in many materials. Here, nuclear reactions such as $^{1}H(^{15}N, \alpha, \gamma)^{12}C$, with a sharp resonance at 6.38 MeV, provide an adequate detection and profiling method.

Particular cases of the nuclear reaction techniques are those involving the production of radioactive isotopes whose decay is measured (activation analysis). The best-established case is that induced by neutrons (neutron activation analysis), which constitutes a high-sensitive technique for chemical analysis. Charged-particle activation analysis has now become an alternative interesting tool. For example, when carbon, nitrogen and oxygen are bombarded with ^{3}He, they are converted into positron-emitting isotopes, with characteristic lifetimes as follows.

$$^{12}C(^{3}He, \alpha)^{11}C \qquad T_{1/2} = 20.4 \text{ min}$$

$$^{14}N(^{3}He, \alpha)^{13}N \qquad T_{1/2} = 9.96 \text{ min}$$

$$^{16}O(^{3}He, p)^{18}F \qquad T_{1/2} = 109.7 \text{ min}$$

Therefore, the study of the positron decay of a ^{3}He-bombarded sample provides a very sensitive method for chemical analysis. Indeed, detection limits of 8 ppb for carbon and 2 ppb for oxygen in silicon have been achieved, which are superior to those reached with neutron activation analysis.

10.9.4 Secondary-ion Mass Spectrometry

Bombardment of the material with $1-30$ keV ion beams induces sputtering of the atoms in the material. The small fraction of them, which are ejected as ions, can be detected by mass spectrometry techniques. The ions of the primary beam are produced with an ion gun and focused onto the sample to a typical spot size of $5-50$ μm. The sputtered ions are mass separated in a quadrupole mass filter or in a magnetic-type mass spectrometer and detected. For favourable conditions, detection limits below 1 ppm have been reported. One advantage of secondary-ion mass spectrometry (SIMS) over other electron or ion beam techniques (RBS) is that it can detect hydrogen. The reported mass resolution $m/\Delta m$ is about 1000 and even higher. The practical accessible depth depends on the emission rate and therefore on the material and size and intensity of the probing beam. Depths of the order of or greater than about 1 μm can be reached, with a depth resolution of about 100 Å. By using a low primary current, i.e. under the so-called static SIMS conditions, with negligible erosion rates, the accessible depth is one monolayer and very high sensitivities in the parts per billion range are attainable. However, it should be noted that absolute impurity analysis is very sensitive to surface treatments.

10.10 ELECTRON SPECTROSCOPY FOR CHEMICAL ANALYSIS

A beam of monochromatic X-rays with energy E_0, impinging on a material, ionizes the constituent atoms and electrons are emitted. The analysis of the energy of the emitted electron constitutes[51] one of the best established techniques for chemical analysis of surfaces and thin films, the so-called electron spectroscopy for chemical analysis (ESCA). It is, in fact, an X-ray-induced photoelectron spectroscopy, conceptually identical with ultraviolet and visible photoemission techniques.

The energy of ϵ_{ph} of the emitted photoelectrons is given by

$$\epsilon_{ph} = E_0 - E_b - \phi \qquad (10.21)$$

where E_b is the binding energy of the photoelectron (referred to the Fermi level) and ϕ the work function of the material or more precisely the work function of the material—detector system. Therefore, by measuring the energy spectrum of the emitted photoelectrons, the binding energy E_b and consequently the position of the corresponding energy level can be determined. Since E_b is influenced by the oxidation state of the ionized atoms as well as by the bonding structure of the material, these features can be adequately investigated by ESCA. The probing depth of the method is $5-20$ Å and is determined by the range of the photoelectrons before they suffer an inelastic collision. The sensitivity of the technique is around 0.5%. It is applicable to all elements except hydrogen and helium. In principle, the profile of atomic distributions inside the material can be determined by using ion sputtering methods in parallel, assuming that the sputtering does not cause impurity segregation, etc.

10.11 AUGER ELECTRON SPECTROSCOPY

Electrons can be used as incident particles to probe the electronic structure of materials[48,51] and to detect the presence of impurities or defects in composition. The Auger process used in Auger electron spectroscopy (AES) is schematically illustrated in Figure 10.29. The initial excitation event creates a hole in the deep level W. An electron from the level X jumps into level W and fills the hole. The energy liberated by this transition is communicated to one electron of level Y, which is ejected from the material. The

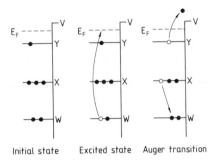

Initial state   Excited state   Auger transition

Figure 10.29.
Schematic diagram of the transitions involved in AES: E_F, Fermi level; V, vacuum level.

Auger process is designated in this case as WXY, explicitly indicating the three electronic levels involved. It represents an alternative channel to X-ray fluorescence emission, its cross-section being dominant for light elements. Levels W, X and Y may be atomic levels (K, L and M) or may belong to the higher electronic bands of the material.

The basic experimental arrangement consists of an electron gun providing a monochromatic beam which impinges on a sample placed in a chamber in which an ultrahigh vacuum has been attained. The gun usually works at energies of $2-3$ kV and yields currents in the microamperes range. The energy distribution of the electrons emitted is then analysed, the energy resolution being typically less than 1 eV. The Auger spectrum electrons are only a small fraction of the total secondary electron yield.

The energy ϵ_e of the Auger electrons corresponding to the WXY process is given by

$$\epsilon_e = E_W(Z) - E_X(Z) - E_Y(Z + \Delta) - \phi \qquad (10.22)$$

where the energies of the W, X, and Y levels are referred to the Fermi level and ϕ is the work function of the material. The energies E_W and E_X of the W and X levels correspond to the involved atom of atomic number Z, whereas E_Y refers to an equivalent atomic number $Z + \Delta$ ($\Delta \approx 1$), because of the presence of one extra hole in that shell. Different Auger peaks will appear in the spectrum depending on the levels involved. Moreover, the method takes advantage of the sensitivity of the level energies to oxidation state and chemical bonding of the Auger atom. As an example, the 1s energy level of sulphur varies by more than 6 eV between valence states 2^- and 6^+.

The information contained in the Auger peaks comes from a layer which is within a few ångströms of the surface; this is determined by the inelastic scattering cross-section of the Auger electrons. Therefore, AES provides a very selective probe of the surface of the material. In order to enhance the small Auger peaks over the continuum background, modulation techniques have to be used. The basic data obtained from AES refer to the chemical composition and stoichiometry of the surface, the presence and concentration of adsorbate atoms, and the bonding characteristics of both surface and adsorbate atoms. The sensitivity of the technique is better than 1% of a monolayer, i.e. about 10^{13} atoms cm^{-2}.

REFERENCES

1 W. Low, *Paramagnetic Resonance in Solids*, Academic Press, New York (1960).
2 A. Abragam, *The Principles of Nuclear Magnetism*, Oxford University Press, Oxford (1961).
3 J. W. Orton, *Electron Paramagnetic Resonance. An Introduction to Transition Group Ions in Crystals*, Iliffe, London (1968).
4 G. E. Pake and T. L. Estle, *The Physical Principles of Electron Paramagnetic Resonance*, W. A. Benjamin, London (1973).
5 C. P. Slichter, *Principles of Magnetic Resonance*, Springer, Berlin (1978).
6 D. L. Griscom, in Henderson and Hughes (eds), *Defects and Their Structure in Non-Metallic Solids*, Plenum, New York (1976).
7 R. J. Wagner, J. J. Krebs, G. H. Stauss and A. M. White, *Solid State Commun.* **36**, 15 (1980).
8 T. G. Castner and W. Kanzig, *Phys. Chem. Solids* **3**, 178 (1957).
9 R. A. Weeks, *J. Appl. Phys.* **27**, 1376 (1956).
10 R. A. Weeks and C. M. Nelson, *J. Appl. Phys.* **31**, 1555 (1960).
11 G. D. Watkins, in J. W. Corbett and G. D. Watkins (eds), *Radiation Effects in Semiconductors*, Gordon and Breach, New York (1971).
12 J. J. Davies, *Contemp. Phys.* **17**, 275 (1976).
13 B. C. Cavenett, *Adv. Phys.* **30**, 475 (1981).
14 K. A. McLauchlan, *Magnetic Resonance*, Oxford Chemistry Series, Clarendon, Oxford (1972).
15 E. Kaufmann and G. K. Shenoy (eds), *Nuclear and Electron Resonance Spectroscopies Applied to Materials Science, Materials Research Society Symposium Proceedings*, Vol. 3, Elsevier, New York (1981).
16 J. H. Strange, Principles of NMR: its uses in defect studies, in A. V. Chadwick and M. Terenzi (eds), *Proceedings of NATO Advanced Study Institute on Defects in Solids—Modern Techniques*, Cetraro, NATO ASI 147, Plenum, New York (1986).
17 J. J. Spokas and C. P. Slichter, *Phys. Rev.* **113**, 6 (1959).
18 D. C. Ailion, *Adv. Magn. Reson.* **5** (1971).
19 M. Mehring, *High Resolution NMR in Solids*, Springer, Berlin (1983).
20 O. Kanert, R. Kuchler and M. Mali, *J. Phys. (Paris), Colloq.* C **6 41**, 404 (1980).
21 R. W. Vaughan, *Annu. Rev. Phys. Chem.* **29**, 397 (1978).
22 C. P. Herrero, J. Sanz and J. M. Serratora, *Solid State Commun.* **53**, 151 (1985); *J. Phys. C: Solid State Phys.* **18**, 13 (1985).
23 J. M. Spaeth, in *Proceedings of the European Materials Research Society Conference on Advanced Materials for Telecommunications*, Strasburg, p. 117 (1986).

24 J. R. Niklas, *Habilitationsschrift*, Paderborn (1983).
25 S. Gruenlich-Seber, J. R. Niklas, E. R. Weber and J. M. Spaeth, *Phys. Rev. B* **30**, 6292 (1984).
26 G. Heder, J. R. Niklas and J. M. Spaeth, *Phys. Status Solidi (b)* **100**, 567 (1980).
27 D. M. Hoffmann, B. K. Meyer, F. Lohse and J. M. Spaeth, *Phys. Rev. Lett.* **53**, 1187 (1984).
28 J. M. Spaeth, *Proceedings of the Fourth Conference on Semi-insulating III−V Materials*, Hakone, 299 (1986).
29 U. Gonser, *Mössbauer Spectroscopy*, Springer, Berlin (1976).
30 R. G. Pirich, G. R. Burr, G. K. Stenog, B. D. Dunlop, B. Suit and J. D. Phillips, *Phys. Rev. Lett.* **38**, 1142 (1977).
31 S. K. Date, W. Keune, H. Engelmann, U. Gonser and I. Dezsi, *J. Phys. (Paris), Colloq. C6* **37**, 117 (1976).
32 R. L. Cohen, *Applications of Mössbauer Spectroscopy*, Vols 1 and 2, Academic Press, New York (Vol. 1, 1976; Vol. 2, 1981).
33 J. Chappert and R. I. Grynszpam (eds), *Muons and Pions in Materials Research*, North-Holland, Amsterdam (1984).
34 R. D. Gill, *Gamma Ray Angular Correlations*, Academic Press, New York (1975).
35 A. M. Stoneham, *Comments Solid State Phys.* **9**, 77 (1979).
36 E. Karlsson, *Hyperfine Interact.* **8**, 647 (1981).
37 P. Hautojavvi, *Positrons in Solids*, Springer, Berlin (1979).
38 R. W. Siegel, Positron annihilation spectroscopy, *Annu. Rev. Mater. Sci.* **10**, 393 (1980).
39 W. K. Chu, J. W. Mayer and M. A. Nicolet, *Backscattering Spectrometry*, Academic Press, New York (1978).
40 Ch. A. Evans, Jr, and R. J. Blattner, Modern experimental methods for surface and thin film chemical analysis, *Annu. Rev. Mater. Sci.* **8**, 181 (1978).
41 W. Reuter and J. E. E. Baglin, Secondary ion mass spectrometry and Rutherford backscattering spectroscopy for the analysis of thin films, *J. Vac. Sci. Technol.* **18**, 282 (1981).
42 L. C. Feldman and J. M. Poate, Rutherford backscattering and channelling analysis of interfaces and epitaxial structures, *Annu. Rev. Mater. Sci.* **12**, 149 (1982).
43 T. W. Conlon, *Contemp. Phys.* **26**, 521 (1985).
44 M. H. Mendenhall, R. P. Livi and T. A. Tombrello, *Nucl. Instrum. Methods B* **10−11**, 596 (1985).
45 D. J. O'Connor, D. W. Palmer, P. D. Townsend, J. Morris, C. W. Pitt, Z. Neuman and L. M. Walpita, *Nucl. Instrum. Methods* **182−183**, 797 (1981).
46 D. V. Morgan (ed.), *Channelling*, Wiley, New York (1973).
47 L. C. Feldman, J. W. Mayer and S. T. Picraux, *Materials Analysis by Ion Channelling*, Academic Press, New York (1982).
48 L. C. Feldman and J. W. Mayer, *Fundamentals of Surface Thin Film Analysis*, North-Holland, Amsterdam (1987).
49 R. E. Kaim and D. W. Palmer, *Philos. Mag. A* **40**, 279 (1979).
50 C. D. Meekison, R. E. Kaim and D.W. Palmer, *Phys. Rev. B* **33**, 3770 (1986).
51 D. A. Shirley, *Electron Spectroscopy*, Elsevier, New York (1972).

11
Computer Modelling Techniques

11.1 INTRODUCTION

The aim of computer simulation methods is to predict the properties of a system from an interatomic potential model. The methods are now applied widely throughout physical and biological sciences to topics ranging from liquid structure to protein conformation. Modelling methods have enjoyed notable success when applied to defect properties. In this chapter, we concentrate on methodologies and recent applications, mainly with respect to bulk defects, but we note that surface defect calculations are now possible. Our emphasis will be on ionic crystals the study of which has enjoyed great success in recent years. The study of defects in semiconductors generally requires the use of the quantum mechanical methods discussed in Chapter 12.

Simulation methods may be used in both quantitative and qualitative studies of defect processes. In the former category, we are concerned with calculations of the enthalpies and entropies of defect formation and migration and of the binding energies between defects and impurities. Moreover, dynamic simulation methods allow the direct study of diffusion in solids. More qualitative applications concern, for example, the evaluation of the critical parameters (e.g. ion size and polarizability) controlling defect energies, the prediction of the effects of temperature and pressure on defect properties, the elucidation of the modes of defect aggregation in non-stoichiometric oxides, and the determination of defect migration mechanisms in superionic conductors.

The simulation techniques used can be divided into three main categories. First, we consider static lattic defect simulations which involve calculation of the relaxation of the lattice around the defect and the determination of the energies and entropies of the defect configuration without, however, any explicit inclusion of the effects of thermal motions. In contrast, in the second technique, molecular dynamics simulations include thermal effects explicitly, by solving the equations of motion of a dynamic ensemble of particles representing the system and to which periodic boundary conditions are applied. Monte Carlo methods provide the third simulation technique; the method, which is based on statistical sampling procedures, is particularly effective for heavily disordered systems.

11.2   STATIC DEFECT SIMULATIONS

The central problem in static lattice defect calculation is the treatment of lattice relaxation around the defect. This is particularly important for ionic and semi-ionic crystals as defects are generally charged species in these solids, which owing to the long range of the Coulomb forces leads to a long-range relaxation field. However, it has proved possible to handle lattice relaxation around defects effectively by a procedure which is derived ultimately from the work of Mott and Littleton[1]. This approach, known as the two-region method is illustrated diagrammatically in Figure 11.1. It consists of the division of the lattice surrounding the defect into an inner region (I) which is treated atomistically and in which the coordinates of all ions are adjusted until they are at "zero force", i.e. there are no net forces acting upon them. This explicit atomistic method is essential for the region immediately surrounding the defect where the defect forces are strong.

In contrast, for the more distant weak-field regions, the defect forces are relatively weak, and lattice relaxation may be treated by more approximate methods. The Mott–Littleton approach, appropriate to ionic materials, calculates the polarization $P(r)$ at a point r with respect to the defect of charge q according to

$$P(\mathrm{r}) = \frac{q\boldsymbol{r}}{r^3}(1 - \epsilon_0^{-1}) \qquad (11.1)$$

where ϵ_0 is the static dielectric constant of the crystal. Equation (11.1) is strictly only applicable to cubic materials; for non-cubic crystals, more complex expressions are used as discussed by Catlow et al.[2] and Catlow and Mackrodt[3].

Within the two-region approximation, the energy E_D of defect formation may be written as

$$E_{\mathrm{D}} = E_{\mathrm{I}}(x) + E_{\mathrm{I-II}}(x, y) + E_{\mathrm{II}}(y) \qquad (11.2)$$

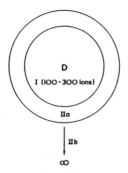

Figure 11.1.
Two-region strategy used in defect (D) simulations. Note the interface region IIa.

where E_I is the energy arising solely from interaction of atoms within region I, whose coordinates make up the vector x, E_{II} is the self-energy of region II for which y is the vector of coordinate displacements and E_{I-II} is an interaction energy term.

If the displacements y are sufficiently small, then we may assume the validity of the harmonic approximation and write

$$E_{II} = \tfrac{1}{2} y \cdot A \cdot y \qquad (11.3)$$

where A is the force constant matrix. Applying the equilibrium condition to region II, we have

$$\left(\frac{\delta E_{I-II}(x, y)}{\delta y} \right)_{y=\bar{y}} = -Ay \qquad (11.4)$$

Thus, substituting equation (11.4) into equations (11.3) and (11.1), we have

$$E_D = E_I(x) + E_{I-II}(x, y) - \tfrac{1}{2} \left(\frac{\delta E_{I-II}(x, y)}{\delta y} \right)_{y=\bar{y}} \cdot y. \qquad (11.5)$$

We have therefore removed the explicit dependence of E_D on E_{II}, which is convenient.

Defect calculations therefore consist of the relaxation of region I, followed by the evaluation first of $E_I(x)$ by direct summation and secondly of $E_{I-II}(x, y)$ and its derivative. For the latter, a subdivision is made of region II: for the inner part (region IIa) immediately surrounding region I, the displacements are calculated by the Mott–Littleton procedure as the sum of those due to all component defects in region I, and E_{I-II} and its derivative are calculated by direct summation of the interactions between the two regions. For the remainder of region II, the interaction is treated as arising purely from the net effective charge of the defect in region I, and the appropriate summations may be evaluated analytically. The use of the interface, region IIa, is found to be necessary in obtaining accurate results with inner regions of a modest size.

Further details of the mathematical development of the procedure outlined above are given in the report by Norgett[4] and the articles by Catlow and Mackrodt[3] and Catlow et al.[2] It is clear, however, that within the inherent limitation of the static lattice approximation the procedures summarized above yield reliable defect formation energies, given a sufficiently large size of the inner region (generally 100 atoms or more) and, of course, given reliable interatomic potentials. Moreover, automatic computer codes are now available for performing defect calculations: principally the HADES II code[5] which is confined to crystals of cubic symmetry, HADES III[2] which may be used for crystals of any symmetry and CASCADE[6] which has been optimized for use on CRAY computers. We illustrate later the success that has been achieved by calculations using these codes.

11.2.1 Calculation of defect entropies

We are concerned here with the vibrational contribution S_{vib} to the defect entropy. Configurational terms S_c, which arise from the possibility of different defect orientations, are easily estimated as $S_c = k \ln N_0$, where N_0 is the number of distinct orientations. Those configurational terms that arise from the distribution of the defects over a variety of sites are automatically included in mass action treatments of defect equilibria, discussed in Chapters 2 and 13.

S_{vib} arises from the perturbation of the phonon density of states by the defect. In the high-temperature limit, where $kT \gg \hbar\omega_i$, its value is given by

$$S_{\text{vib}} = -k \ln\left(\prod_{i=1}^{3N'} \omega_i' \Big/ \prod_{i=1}^{3N} \omega_i\right) + 3k(N' - N)\left[1 - \ln\left(\frac{\hbar}{kT}\right)\right] \quad (11.6)$$

where the primes represent the vibrational frequencies perturbed by the presence of the defect. Gillan and Jacobs[7] have discussed the calculation of the perturbed vibrational frequencies. They considered Green function methods but found that a "crystallite" method, which is closely related to the two-region approach in defect energy calculations, was more computationally viable. In essence, this method considers the perturbation of the frequencies by the defect only for an inner region, or crystallite, surrounding the defect.

Recently, a general code, SHEOL, has been written by Harding[8,9] for calculating vibrational entropy in ionic materials. This code promises to play an increasingly important role in defect physics, although the results of calculations using this technique reveal a far greater sensitivity to inter-atomic potential than has been found in calculations of defect energies.

Defect energy and entropy calculations can, of course, be applied to ground states and saddle points of defects, as well as to defect clusters, thus allowing us to obtain the values of the free energies of formation activation and clustering of defects. The successes and problems with this type of calculation will be considered in section 11.6.1.

11.3 MOLECULAR DYNAMICS SIMULATIONS

Unlike the techniques described in section 11.2, molecular dynamics methods explicitly include kinetic energy terms. The basic principles of the method are simple: a simulation "box" is specified, which in solid-state studies consists of a "supercell", i.e. an integral number of unit cells. The box, which usually contains several hundred particles, is repeated period-ically throughout space, thus generating an infinite system. The simulation therefore includes no effects arising from the surface. In running the simulation, the following procedure is then applied.

(1) A "start-up" configuration is generated, i.e. all particles in the simulation box are assigned positions x_i and velocities v_i. For solids, the positions are normally those observed in crystallographic studies; the velocities of the particles are normally taken as random in direction and of a magnitude corresponding to the kinetic energy appropriate to the temperature used in the simulation.

(2) The force f_i acting on each particle is calculated using specified interatomic potentials. A "time step" δt is then specified, and the coordinates and velocities of each particle are updated using a simple procedure based on the Newtonian equations of motion. The sim-plest zero-order version of these equations are as follows

$$x_i' = x_i + v_i \, \delta t \qquad (11.7)$$

$$v_i' = v_i + \frac{f_i}{m_i} \, \delta t \qquad (11.8)$$

where the primes indicate the values of x_i and v_i after the lapse of time δt. More sophisticated updating algorithms are available (and indeed are invariably used) to minimize the accumulation of errors on repeated updating owing to the finite size of δt. One of the more popular, that of Beeman[10], has been incorporated into solid-state simulation codes by Walker[11]. Obviously, δt must be shorter than the characteristic time of any important process in the solid; thus, δt should be at least an order of magnitude smaller than the periods of atomic vibrations. For this reason, values of $10^{-15}-10^{-14}$ s are commonly used.

(3) The updating procedure described above is repeated several thousand times, which allows study of the time evolution of the system. In the initial stages of the simulation, the system equilibrates, i.e. achieves equipartition of potential and kinetic energies, and achieves an equilibrium distribution of velocities. During this period, the temperature drifts and thus requires periodic scaling. Indeed, the observation of a constant temperature may, in general, be safely taken as the criterion for equilibrium. Once equilibrium is attained, the coordinates and velocities of successive time steps are stored and subsequently analysed to yield properties of interest.

In studies of transport properties, we are obviously most interested in calculating the diffusion coefficient D_α. The particular value of molecular dynamics methods is that they allow direct calculation of this important quantity via the relationship

$$6D_\alpha t + B_\alpha = \langle r_\alpha^2(t) \rangle \qquad (11.9)$$

where B_α is the thermal factor arising from vibrations of atoms and $\langle r_\alpha^2(t) \rangle$ is the mean-square displacement of particles of type α at time t, relative to their positions at $t=0$. Indeed, the occurrence of significant diffusion is shown by the observation of a linear plot of $\langle r_\alpha^2(t) \rangle$ against t. Typical results are shown in Figure 11.2; they were obtained for Li^+ conducting superionic Li_3N by Wolf et al.[12]

Conductivity can be calculated via the Nernst–Einstein equation provided that the values of the correlation factors f are known. More direct methods for calculating conductivity have been developed by Cicotti et al.[13] Perturbed molecular dynamics procedures are used in which small forces are applied to the ions (i.e. forces in addition to those arising from the interactions with other particles) to perturb the system. The calculated response is the induced current density

$$\mathcal{J}_x^Q(t) = \sum_i Q_i v_{ix}(t) \qquad (11.10)$$

where Q_i is the charge on the ith particle. The conductivity is then related to the statistical average $\langle \mathcal{J}_x^Q(\infty) \rangle$ of $\mathcal{J}_x^Q(t)$ in the limit of $t \to \infty$ by

$$\sigma = \frac{e\langle \mathcal{J}_x^Q(\infty) \rangle}{VF_x} \qquad (11.11)$$

where V is the volume of the molecular dynamic box and F_x is the magnitude of the perturbing forces.

The latter procedure, despite its elegance, has not commonly been used

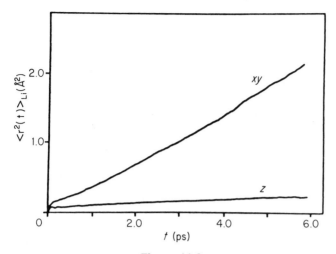

Figure 11.2.
Plot of mean-square displacement $\langle r^2 \rangle$ against time for Li^+ ions in Li_3N, for simultation at 400 K. (*xy* indicates transport in the *xy* plane of the layer structured crystal).

in simulation studies of solids, possibly because its application is less straightforward than the deduction of σ from D via the Nernst–Einstein relationship.

Structural information can be extracted from molecular dynamic studies via the radial distribution functions (RDFs) $g(r)$. The RDFs of the mobile sublattice in superionics are commonly (but not universally) more liquid like than those of normal solids. Thus, in Figure 11.3, we show the $Na^+ - Na^+$ RDF calculated for the layered superionic $\beta''-Al_2O_3$[14]. Note the diffuseness of the peaks corresponding to separations larger than nearest neighbour.

More sophisticated correlation functions can be obtained from the results of molecular dynamic simulations. The function $G_s^\alpha(r, t)$ is the probability of finding a particle of type α at a point r at time t, when that particle was at position 0 at time 0. As seen below, the behaviour of $G_s^\alpha(r, t)$ can yield information on mechanistic aspects of ion transport. In addition, Fourier transformation of $G_s^\alpha(r, t)$ yields the incoherent inelastic neutron-scattering cross-section of the system; this is an important relationship as neutron-scattering techniques have yielded a great deal of valuable information on particle dynamics in condensed matter; a good review by Lechner[15] is available.

A further function of importance is the intermediate scattering function $f^\alpha(k, t)$ defined as

$$F^\alpha(k, t) = \left\langle \sum_i \sum_j \exp[i\mathbf{k} \cdot \{\mathbf{r}_i^\alpha(t) - \mathbf{r}_j^\alpha(0)\}] \right\rangle \qquad (11.12)$$

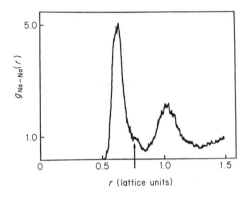

Figure 11.3.
$Na^+ - Na^+$ RDF in $\beta'' - Al_2O_3$. The results refer to simulation of non-stoichiometric material at 300 K.

where r_i is the position of the ith particle at time t, and $\mathbf{k}$ is the scattering vector defined as $\mathbf{k} = (2\pi/\lambda)\hat{\mathbf{k}}$, in which λ is the scattering length and $\hat{\mathbf{k}}$ is the unit scattering vector. The self part of $F^\alpha(k, t)$ is written $F_s^\alpha(k, t)$ and is simply defined by setting $i = j$ in equation (11.10). G_s^α and F_s^α are in fact related by the simple transformation

$$F_s^\alpha(k, t) = \int_0^\infty \exp(iQr) G_s^\alpha(k, t) \, dr \qquad (11.13)$$

so that the incoherent neutron scattering cross-section $S^{inc}(k, \omega)$ for scattering from nuclei of type α may be written as

$$S^{inc}(k, \omega) = \int_0^{+\infty} F_s^\alpha(k, t) \exp(-i\omega t) \, dt \qquad (11.14)$$

$F_s^\alpha(k, t)$ decays to an asymptotic value, and the nature of its decay may yield useful mechanistic information. It will attain the asymptotic value when a typical atom of type α has had sufficient time to sample the time-averaged distribution along the k direction and over distance λ. Thus, if λ is the vibrational amplitude of an atom in a solid, then vibrational motion is mainly responsible for the decay of F_s^α which will be followed by a slower diffusional decay. Important dynamic correlation functions include $Z_\alpha(t)$, the velocity autocorrelation function (VAF) and the van Hove self-correlation function $G_s^\alpha(r, t)$ referred to above.

The former is given by

$$Z_\alpha(t) = \frac{1}{N_\alpha} \sum_{i=1}^{N_\alpha} \frac{\langle v_i(t) \cdot v_i(0) \rangle}{\langle v_i(0) \cdot v_i(0) \rangle} \qquad (11.15)$$

where N_α is the number of particles of type α, and $v_i(t)$ is the velocity of particle i at time t; $v_i(0)$ is the velocity at $t = 0$. The VAF is the ensemble average of the projection of the velocity vector of a particle along its velocity vector at a time t earlier. If a particle is vibrating about a lattice site, $Z_\alpha(t)$ will show oscillatory behaviour, which will decay, however, to zero if the oscillations are damped. For a particle exercising random diffusive motion the VAF rapidly decays to zero.

The molecular dynamics technique has very considerable power, yielding a large amount of detailed information on the simulated system. However, there are a number of serious restrictions of the method in simulating transport in solids. The principal restrictions are as follows.

(1) The technique allows us to study only rapid diffusion (i.e. diffusion coefficients greater than 10^{-7} cm^2 s^{-1}). For slower diffusion, the number of atomic displacements during the period sampled by the diffusion is insufficient (in most practical simulations the sampling time is rarely more than 100 ps). A number of possible extensions of the technique are being examined to tackle this problem but, to date, molecular dynamics simulations of transport in solids have generally been applied to superionics only.

(2) The technique does not readily permit the extraction of Arrhenius energies. They can only be extracted by running the simulation at several temperatures, which is a costly process. Furthermore, once calculated, the Arrhenius energies cannot easily be decomposed into defect formation and migration terms, which is of course achieved naturally and simply by the static simulation methods.

(3) It is not possible, except with the expense of very large amounts of computer time, to include any treatment of ionic polarization, essentially because the electron distribution responds to changes in nuclear positions instantaneously. Thus for each time step the equilibrium ionic dipole moments (i.e. shell coordinates in shell model treatments, see section 11.5) must be calculated. This can be achieved only by an iterative process, the use of which greatly increases the time required. This factor is in addition to that arising from the increased number of species in the simulation (i.e. cores and shells for each ion in shell model simulations). The exclusion of polarizability is potentially a serious problem, as the inclusion of these effects is known to be important in defect calculations.

It is probable, however, that advances both in computational technique and in computer hardware will reduce or remove some of these difficulties in the near future and that the scope of the technique will greatly expand.

11.4 MONTE CARLO SIMULATIONS

Monte Carlo techniques are essentially procedures for statistical sampling via random number generation. The field has been extensively reviewed recently by Murch[16-20]. When applied to the study of atomic diffusion processes, the procedure is normally as follows.

(1) A simulation box is defined and periodic boundary conditions are applied, as with molecular dynamic techniques. However, it is generally possible to include larger numbers of particles.

(2) Defects are created in the lattice. Thus, vacancies may be introduced by abstracting atoms from a certain number of sites.

(3) A defect is selected at random, as is a corresponding jump direction for the defect. The environment of the defect is examined and the jump frequency ω_i obtained. The jump frequency appropriate to all possible defect environments must be supplied as input to the program or they must be calculated from Arrhenius energies, for example, that are appropriate to the given environment. To ensure maximum efficiency of the simulation, all jump frequencies are scaled so that the maximum value is unity.

(4) A random number R in the range $0-1$ is generated, and the value of the number is compared with ω_i. If $R < \omega_i$, the jump is deemed to be successful; if $R > \omega_i$, it is considered unsuccessful. This procedure effectively weights the probability of a jump according to its frequency.

(5) The procedures in steps (3) and (4) above are repeated several thousand times. In the initial stages of the simulation the system equilibrates, as in molecular dynamic simulations. After equilibration the successive configurations generated by the simulation can be used to follow the process of defect migration.

Monte Carlo simulations of transport are valuable in complex systems containing several types of jump frequency, e.g. alloys[18,19] and non-stoichiometric compounds[20]. The technique is particularly effective for calculating correlation coefficients f_α which are obtained from the simulation using the Einstein expression

$$f_\alpha = \frac{\langle R_\alpha^2 \rangle}{na^2} \tag{11.16}$$

where $\langle R_\alpha^2 \rangle$ is the mean-square displacement of atoms of type α after an average of n jumps, each of length a. Monte Carlo techniques may become a very powerful technique for studying transport in heavily defective compounds. In these applications the frequencies ω_i are obtained from Arrhenius expressions using activation energies calculated by the static simulation techniques. An example of such an application is given later in this chapter. The Monte Carlo technique, in common with other simulation techniques, rests ultimately on the quality of the description of the interatomic forces; the models in current use are discussed in section 11.5.

11.5 INTERATOMIC POTENTIALS

An interatomic potential model is a mathematical representation of the potential energy of the system as a function of particle coordinates. This total potential $V(r_1, \ldots, r_N)$ is normally decomposed into a sum of functions of two, three or more particle coordinates:

$$V(r_1, \ldots, r_N) = \sum_{i>j} \varphi_{ij}(r_i, r_j) + \sum_{i>j>k} \varphi_{ijk}(r_i, r_j, r_k) + \ldots \quad (11.17)$$

where the φ_{ij} and φ_{ijk} are analytical or numerical functions. The vast majority of simulations of both solids and liquids take only the first term on the right-hand side of equation (11.17), i.e. the potentials are purely "two body" in nature. A further simplification assumes "central-force" models in which the φ_{ij} depend only on the distance r_{ij} between particles i and j, i.e. $\varphi_{ij}(r_i, r_j) = \varphi(r_{ij})$. In ionic materials it is then convenient to decompose $\varphi(r_{ij})$ into Coulombic and non-Coulombic components, i.e.

$$\varphi_{ij}(r_{ij}) = \frac{q_i q_j}{r_{ij}} + V_{ij}(r_{ij}) \quad (11.18)$$

where q_i and q_j are the charges on ions i and j, and $V_{ij}(r_{ij})$ is usually referred to as the short-range potential. The simulation codes then handle the Coulomb term separately, normally via the Ewald summation, which involves a transformation into reciprocal space (see the paper by Tosi[21] for a good discussion); the short-range terms are, in contrast, handled in real space and are normally cut off (i.e. set to zero) beyond a specified distance. A number of analytical functions have been used for $V_{ij}(r_{ij})$; these include Lennard-Jones and Morse functions, but the most popular in ionic crystal simulations has been the Buckingham potential:

$$V(r) = A \exp\left(-\frac{r}{\rho}\right) - Cr^{-6} \quad (11.19)$$

in which it is tempting to associate the attractive r^{-6} term with genuine dispersion (i.e. induced dipole–induced dipole forces). However, in practice, this term will normally include contributions from other attractive forces including small covalent terms. There is no ambiguity, however, about the interpretation of the exponential repulsive term which describes the Pauli repulsion which comes into play when atomic charge clouds overlap. We should note that there is no need other than convenience to use analytical short-range potentials and that numerical potentials have been used extensively and successfully by Mackrodt and co-workers[22].

In simulating defect properties, it has been found essential to include a description of ionic polarizability, as charged defects extensively polarize the surrounding lattice. Earlier studies using point dipole models were found to be unsatisfactory, and the pioneering work of Faux and Lidiard[23] showed the value of the shell model in defect calculations. The model, originally developed by Dick and Overhauser[24] describes a polarizable atom or ion in terms of a core into which the mass of the ion is concentrated, which is connected by a harmonic spring to a massless shell, the latter representing the polarizable valence shell electrons. When an electric field is applied to the atom, the shell will be displaced from the core and hence a dipole moment will develop. Short-range forces are normally taken as acting between the shells, thus giving rise to a coupling between short-range energies and polarization—an important factor whose omission in simpler point dipole treatments was largely responsible for the inadequacies of the potentials.

Ionic two-body shell model potentials thus form the basis of almost all static and dynamic simulation work on polar crystals in recent years. Such models must, of course, be parameterized, i.e. the variable parameters adjusted so as to correspond to the particular crystal under investigation. These parameters include ionic charge, short-range parameters (e.g. A, ρ and C in equation (11.19)), shell charges and spring constants.

As regards ionic charges, in most studies of ionic and semi-ionic systems, these have been fixed at the fully ionic, i.e. integral, values. We should stress that, as argued by Catlow and Stoneham[25], the use of integral ionic charges does *not* imply an electron distribution corresponding to that of a fully ionic system, and' that the validity of a potential model is assessed purely by its ability to reproduce known properties of the crystal. Nevertheless, as the field extends to the study of semicovalent materials such as silicates, there may be increasing use of potential models based on partial charges, and indeed a successful model for Mg_2SiO_4 has already been developed by Parker and Price[26].

The short-range repulsive terms may be parameterized by two procedures. First, this may be done by empirical methods in which variable parameters are adjusted via a least-squares fitting routine to achieve the best possible match of calculated and experimental crystal properties, the latter including structural properties, elastic and dielectric constants and, where available, phonon dispersion curves. By "fitting" to the structural properties, we mean the adjustment of potential parameters until the potential gives the observed structure (including cell dimensions and atomic coordinates) as close as possible to equilibrium, and we should note that complex low-symmetry structures may contain several variables, each of which is a datum to be used in the fitting procedure. An empirical potential has been developed for several halide and oxide crystals (see, for example, the papers of Catlow *et al.*[27] and Sangster and Atwood[28] for studies of the alkali halides, and the papers of Sangster and Stoneham[29], Lewis and Catlow[30] and Lewis[31] for work on transition-metal oxides), and these models have been successfully used in defect calculations. However, the empirical parameterization procedure is clearly limited in that it can only be applied to crystals for which empirical data are available (although in some cases extrapolation procedures may be used, as in the study of Lewis and Catlow[30]). A more fundamental problem is that empirical methods yield information only on potentials at internuclear spacings close to those observed in the perfect lattice, but in defect configurations, especially those involving interstitials, the internuclear separations will differ considerably from perfect lattice values. In such cases, empirically derived potentials will be reliable only if the form of the analytical potential is accurate over a wide range of internuclear distances—a point on which there may be considerable uncertainty.

For this reason there have been considerable efforts in recent years in developing theoretical methods for deriving interatomic potentials in ionic materials. The major contribution has been made here by Mackrodt and co-workers who have studied both "electron gas" and *ab initio* methods. The former approach derives from the work of Gordon and Kim[32] and Wedepohl[33]. First, electron densities are obtained for the isolated interacting atoms. This is normally achieved by solving the Hartree−Fock equations for each atom, and Mackrodt and co-workers have stressed the importance of "crystal-adapted" wavefunctions, i.e. of solving the equations in the Madelung potential appropriate to the crystal; this is of particular importance for anions for which the wavefunctions are more diffuse. When these electron densities have been obtained, the Coulomb interactions are calculated and approximate expressions based simply on the total electron density are used to calculate the kinetic energy, exchange and correlation contributions to the interaction energy. For details, we refer to the paper of Gordon and Kim[32]. In *ab initio* studies of interatomic potentials, calculations are performed on a "supermolecule", i.e. a molecule consisting of those atoms, ions or molecules whose interaction we require; the calculations are performed as a function of the internuclear spacings from which the interatomic potential is extracted. The method has been used recently by, for example, Mackrodt *et al.*[34] and Saul *et al.*[35] in solid-state studies; the latter work, which concerned the derivation of a potential suitable for modelling the OH$^-$ ion in ionic solids also stressed the importance of performing the calculations in the Madelung field appropriate to the solid.

In performing *ab initio* studies of interatomic potentials, considerable care must be exercised concerning two further aspects. The first concerns the question of basis sets of the component species in the supermolecule. These must be large and flexible to allow for the electron density redistribution which occurs when the species interact. Moreover, if the wavefunctions yield energies for the component species which are close to the Hartree−Fock limit, then this will minimize the "basis-set superposition error", which arises from the lowering of the intra-atomic energies of the interacting atoms due to the greater flexibility of the composite wavefunctions in the supermolecule compared with the single-atom wavefunction. The second technical aspect concerns the "level" of the calculation; the Hartree−Fock approximation does not include a representation of the effects of electron correlation. If it is desired to include such effects, then recourse must be to multiconfigurational wavefunctions, obtained from either configurational interaction or multiconfigurational self-consistent field procedures (see, for example, the book by Szabo and Ostlund[36] for a discussion of these methods).

On going beyond the Hartree−Fock approximation, the computer time required for the calculation will increase very considerably and, for heavier atoms, the Hartree−Fock method itself may become prohibitively expensive in computer resources. Nevertheless, with the continuing expansion in computer power, we may expect increased use of *ab initio* Hartree−Fock methods in deriving reliable interatomic potentials for solids.

At present, only empirical methods may be used to derive shell model parameters (although recent work of Fowler and Pyper[37] suggests that polarizabilities may be accurately calculated). Measured static and high-frequency dielectric constants must be available if these are to be fitted with any reliability; otherwise, extrapolation methods must be used. Recent work of Cormack[38] has demonstrated the sensitivity of calculated defect energies in a complex oxide to shell model parametrization. The derivation of reliable and general procedures for deriving these parameters is therefore one of the most urgent requirements of the field.

As noted earlier, the great majority of interatomic potential models for ionic solids are of the two-body form. Our recent work has shown that this is clearly inadequate for one class of materials, namely framework structures such as silicates. Thus, for $\alpha-SiO_2$, for example, two-body central-force models cannot reproduce the observed crystal properties. However, Sanders *et al.*[39] found that a simple extension of the model, i.e. the inclusion of bond-bending forces around the $O-Si-O$ bond angles dramatically improved the potential which was then capable of reproducing the observed elastic, dielectric and lattice dynamic properties with reasonable accuracy. The form of the additional terms used was particularly simple: the bond-bending energy $E(\theta)$ is written as

$$E(\theta) = \tfrac{1}{2}K_B(\theta - \theta_0)^2 \qquad (11.20)$$

where θ_0 is the tetrahedral angle and K_B is the bond-bending force constant. This type of function may be relatively simply incorporated into the simulation codes and seems unquestionably to improve the performance of potentials in several silicate crystals. It is at present an open question as to how useful such bond-bending terms will be in the study of other types of compound.

11.6 APPLICATIONS

Our account will focus on recent examples of defect simulations which illustrate both the 'state of the art' of the field and the problems that may be encountered in performing reliable studies.

11.6.1 Static defect simulations

An extensive range of defect energy calculations are now available in the literature for several classes of ionic and semi-ionic material, including both oxides and halides. A review of recent results is available from Mackrodt[40]. Table 11.1 gives a small selection of results on simpler materials, for which we compare calculated and experimental defect energies. The comparison between theory and experiment is seen to be very satisfactory and, in general, it is found that, for those compounds for which good interatomic potentials are available, accurate defect energies may be calculated.

The current direction of the field is increasingly towards complex and semi-ionic materials. Examples are provided first by recent work of Lewis and Catlow[41] on $BaTiO_3$ (an important electronic ceramic material), some of the results of which are summarized in Table 11.2. The calculations shows that the basic disorder in the material is of the Schottky type; the defect energies are high, however: $2-3$ eV per defect, suggesting that intrinsic disorder will play a very minor role except possibly at the highest temperatures. Lewis[42] and Lewis and Catlow[43] also undertook a detailed study of the behaviour of impurities in the material which play a vital role in determining its electronic properties.

A second example is provided by Doherty's recent work on Mg_2SiO_4 (a material of great important as it forms the major component of the upper part of the earth's mantle), a summary of which is given in the recent article of Catlow et al.[44]. Doherty examined several models for the intrinsic defect structure of the material. His work suggests that the lowest-energy defects are Mg^{2+} Frenkel pairs (with a formation energy of about 5 eV), although Schottky pairs consisting of Mg^{2+} and oxygen vacancies are close in energy.

Table 11.1
Defect energy calculations

Crystal	Process	Calculated energy (eV)
$NaCl^{45}$	Schottky pair formation	2.4−2.7 (2.3−2.7)
$NaCl^{45}$	Cation vacancy migration	0.66 (0.7−0.8)
CaF_2^{46}	Anion Frenkel pair formation	2.6−2.7 (2.6−2.7)
CaF_2^{46}	Anion vacancy activation	0.35 (0.38−0.47)
CaF_2^{46}	Anion interstitial activation	0.91 (0.77−0.79)
MgO^{47}	Schottky pair formation	7.5−7.7 (5.7)
MgO^{47}	Cation vacancy activation	1.8−2.2 (2.0−2.3)
NiO^{48}	Schottky pair formation	6−7
NiO^{48}	Cation vacancy activation	1.86 (1.5)

The experimental values are given in parentheses and discussed in the literature cited; see also Table 5.1.

Table 11.2.
Defect energies for undoped $BaTiO_3$

Compound	Schottky energy (eV per defect)	Element	Frenkel energy (eV per defect)
$BaTiO_3$	2.29	Oxygen	4.49
TiO_2	2.90	Barium	5.94
BaO	2.58	Titanium	7.56

A low value of about 0.3 eV was calculated for the Mg^{2+} interstitial migration activation energy; a larger value of 2 eV was found for the Mg^{2+} vacancy migration energy. There are at present very few definitive data with which to compare the results of the calculations which have therefore a clear predictive value. However, there does seem to be some evidence for Arrhenius energies for Mg^{2+} diffusion in the range of 3−4 eV (see the article of

Catlow et al.[44] for details), which would be compatible with the results of the calculations.

As noted earlier in this chapter, the calculations have been applied in a more "qualitative" sense to the study of complex modes of defect aggregation. Earlier applications to non-stoichiometric oxides, e.g. $Fe_{1-x}O$, UO_{2+x} and TiO_{2-x} are reviewed by Catlow[49]. A particularly relevant recent study concerned the widely investigated rare-earth-doped alkaline-earth fluorides. Controversy has surrounded the nature of the aggregates formed by the substitutional rare-earth dopants and their charge-compensating interstitials in heavily doped crystals (i.e. crystals containing more than 5 mol% rare earth). The calculations of Bendall et al.[50] were of value in suggesting that there is a change in cluster structure on going from larger rare earths (e.g. La^{3+} and Nd^{3+}) to smaller cations (e.g. Er^{3+}). For the former, relatively small clusters consisting of two dopant ions and two or three interstitials (Figure 11.4) were calculated to have the greatest stability; we note that these clusters are stabilized by a coupled lattice–interstitial relaxation mode that was identified in an early theoretical study of Catlow[51]. In contrast, for the smaller dopant ions, the beautiful symmetrical cubo-octahedral cluster shown in Figure 11.4(b) has the greatest stability; the cluster consists of six rare-earth ions grouped around a central interstitial site. The eight F^- lattice sites of the cube are vacant; the 12 F^- ions are situated each above one of the cube edges. Greatest stability is achieved when the central cube contains a pair of F^- interstitials oriented along the $\langle 111 \rangle$ direction.

The predictions of the calculations have recently been strongly supported by an extended X-ray absorption fine-structure spectroscopy (EXAFS) study of the local environment of the rare-earth cation in CaF_2. The EXAFS technique (see Chapter 8 and the paper by Hayes and Boyce[52] for a good review) allows us to probe the local structure of particular atomic species. Spectra obtained by EXAFS were collected for CaF_2 doped with 10 mol.% of a range of dopants, the data being collected using the synchrotron radiation source at the Daresbury Laboratory, Science and Engineering Research Council, UK. The spectra for the larger rare-earth ions (e.g. Nd^{3+}) could be fitted accurately assuming the formation of the type of cluster shown in Figure 11.4(a), whereas the spectra of the smaller ions (e.g. Er^{3+}) indicated the presence of the cubo-octahedral clusters. The work (details of which are available in the work of Catlow et al.[53]) is a nice illustration of the way in which simulations and experiments may be used in a concerted way to investigate complex problems in defect physics and chemistry.

Simulation techniques have been less extensively applied to point defects in metals, which is largely attributable to difficulties in deriving suitable interatomic potentials. A selection of results are collected in Table 11.3 and

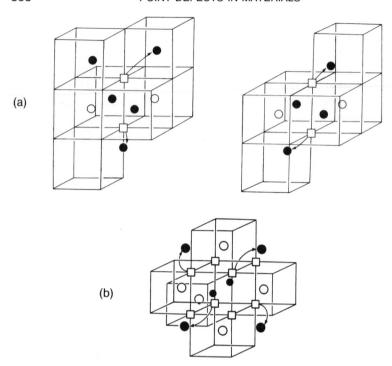

Figure 11.4.
Dopant interstitial clusters in rare-earth-doped CaF_2 for (a) clusters containing two dopants ions with two and three interstitials and (b) a cubo-octahedral dopant hexamer: $\bigcirc$, dopant; $\bullet$, interstitial; $\square$, vacancy.

Table 11.3.
Calculated and experimental vacancy formation energies in metals

Metal	Lithium	Sodium	Potassium	Aluminium
Calculated energy (eV)	0.48	0.29	0.31	0.19
Experimental energy (eV)	0.34	0.36	0.39	0.66

For details of calculations and references to theoretical and experimental studies, we refer to the work of Taylor[54,55]; see also Table 7.3.

are compared with experiment where data are available. For further discussion, we refer again to the excellent review of Taylor[54].

As noted earlier, reliable methods are now available for studying defect entropies as well as energies. A number of recent applications to ionic materials have been reported by Gillan and Jacobs[7] and Harding[8,9]. Their work opens up an exciting future for this area of defect physics, as it will enable us to calculate both absolute concentrations of defects and absolute rates of defect transport.

11.6.2 Molecular dynamics simulations

As argued in section 11.3, the main applications of molecular dynamics simulations to solids have concerned superionics. Simulations have been reported on several superionic materials including high-temperature CaF_2[56], $SrCl_2$[57,58], AgI[59], Ag_2S[60], Li_3N[12,61,62] and $\beta''-Al_2O_3$[14,61]. In this section, we concentrate on Li_3N and $\beta''-Al_2O_3$; a number of reviews e.g. by Gillan[63] and Vashishta[60], are available in which work on the fluorite-structured materials and on the superionic silver compounds is discussed.

The study of $\beta''-Al_2O_3$[14,61] was particularly illuminating as it showed how molecular dynamics techniques can be used to obtain subtle mechanistic information.

$\beta''-Al_2O_3$ is an important layer-structured superionic. As illustrated in Figure 11.5, conduction planes containing the mobile Na^+ ions are sandwiched between spinel-structure alumina layers which also contain Mg^{2+} (the formula of the stoichiometric compound being $Na_2O.MgO.5Al_2O_3$). The compound is closely related to the celebrated $\beta-Al_2O_3$, shown in

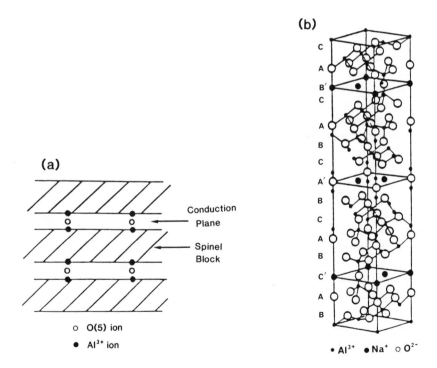

Figure 11.5.
Structure of $\beta''-Al_2O_3$: (a) schematic illustration of layer structure; (b) actual structures.

Figure 6.7 in which, however, the spinel blocks are thicker and in which there is no Mg^{2+}.

As normally prepared, both β- and $\beta''-Al_2O_3$ are non-stoichiometric. In the latter case the non-stoichiometry arises from the replacement of Mg^{2+} by Al^{3+} with the consequent creation of Na^+ ion vacancies in the conduction plane. Amongst the many unanswered questions concerning this material, one of great importance concerns the role of non-stoichiometry in controlling the conductivity of the material. Other questions concern the variation in the conductivity with temperature, as there is good evidence from experimental studies[64] that there is a change in the ion migration mechanism at about 500 K. In general, the nature of the dynamics of the mobile Na^+ ions is poorly characterized.

To investigate these problems, Wolf[61] and Wolf et al.[14] performed an illuminating dynamic simulation study. They used a simulation box in which the triply primitive hexagonal unit cell of the crystal (containing 90 ions) had been quadrupled normal to the c axis. The interatomic potentials were taken from those of the appropriate binary oxides; the time steps used in the simulation were 5×10^{-14} or 4×10^{-14} s depending on the temperature which was varied from 300 to 700 K.

The first problem to be investigated concerned the effect of non-stoichiometry on the conductivity. In non-stoichiometric $\beta-Al_2O_3$, one in six of the Na^+ sites are vacant; the closest approach to this composition which could be achieved with the box size which could be used in the calculations was that in which one in eight of the Na^+ ions is missing. Simulations were therefore performed on the latter composition and on the stoichiometric material.

The contrast between the behaviour of the two systems is very marked. Plots of the mean-square displacement of Na^+ ions against time (the slopes of which are proportional to the diffusion coefficient) are illustrated for the two compositions at 550 K in Figure 11.6. They show that, in the stoichiometric material, the Na^+ diffusion coefficient is small whereas, in the non-stoichiometric solid, it is high. Table 11.4 gives the calculated conductivities (obtained from the diffusion coefficient via the Nernst–Einstein relationship assuming an Na^+ vacancy migration mechanism) for the non-stoichiometric material at three temperatures. Agreement between theory and experiment is certainly acceptable at the two higher temperatures. The less satisfactory result for 300 K is explicable, however, as diffraction data[65] indicates the presence of a vacancy superlattice at this temperature. The simulation box is too small to permit the formation of such a supercell. The calculated conductivity will therefore be too high as the supercell would be expected

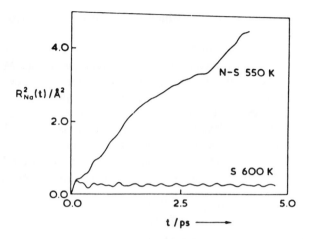

Figure 11.6.
Plots of mean-square displacement against time for Na^+ ions in $Na^+-\beta''-Al_2O_3$: curve S, stoichiometric system at 600 K; curve N–S, non-stoichiometric system at 550 K.

to reduce the Na^+ conductivity by locking the Na^+ vacancies into an ordered array.

Table 11.4.
Calculated and experimental conductivities of non-stoichiometric $Na^+-\beta''-Al_2O_3$

T (K)	Calculated D ($cm^2\ s^{-1}$)	Calculated σ ($\Omega^{-1}\ cm^{-1}$)	Observed σ ($\Omega^{-1}\ cm^{-1}$)
300	1.41×10^{-5}	1.53	0.014–0.160
550	2.24×10^{-5}	1.23	0.80
700	5.78×10^{-5}	2.49	1.24

Information on the structural consequences of the non-stoichiometry is provided by the Na^+-Na^+ radial distribution functions, which are illustrated in Figure 11.7 for both stoichiometric and non-stoichiometric solids at 300 K. The greater diffuseness of the RDF for the latter is indicative of the greater disorder in the conduction plane of the non-stoichiometric compound. We also note that the first peak in the RDF of the non-stoichiometric material has a shoulder at high r, which is indicated by the arrow in the diagram. This feature can be explained in terms of relaxation of the nearest-neighbour Na^+ ions towards an Na^+ vacancy, and its occurrence indicates that there are well-defined vacancies in the conduction plane. Migration would therefore be expected to take place via a conventional vacancy hopping mechanism. In contrast, at 700 K, the shoulder has disappeared, suggesting that vacancies are present at lattice sites for too short a period for the occurrence of appreciable relaxation of the surrounding lattice, indicating a much more "continuous" type of migration mechanism.

Further evidence for this fascinating change in the nature of the Na^+ ion dynamics is provided by study of the moment ratio $P_\alpha(t)$ defined as

$$P_\alpha(t) = \frac{3\langle r_\alpha(t)^4\rangle}{5[\langle r_\alpha(t)^2\rangle]^2}$$
(11.21)

It has been shown by Hansen and MacDonald[66] and Rahman[67] that this ratio will tend to unity at large t when a continuous diffusion mechanism is operative. In contrast, strong deviations from unity indicate a hopping mechanism. Figure 11.8 shows plots of $P(t)$ against t for three temperatures. At higher temperatures, $P(t)$ approaches unity from which it deviates strongly, however, at the lower temperature.

The work of Wolf and co-workers referred to above provides far greater details of the nature of the Na^+ ion migration revealed by the simulations. It is clear that subtle effects may be predicted which would be of great interest

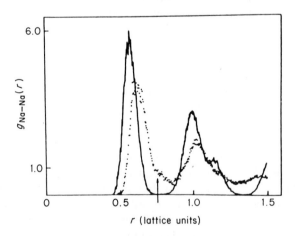

Figure 11.7.
RDFs for stoichiometric (——) and non-stoichiometric (●) $\beta''-Al_2O_3$ at 300 K.

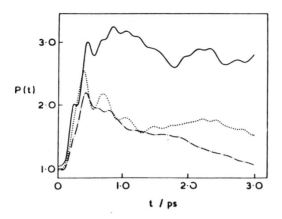

Figure 11.8.
$P(t)$ against t for three temperatures in $Na^+-\beta''-Al_2O_3$: ——, 300 K; ....., 550 K; —·—, 700 K.

to investigate experimentally.

A second superionic where molecular dynamics methods have provided valuable insight into migration mechanisms is Li_3N. This is an intriguing material with a unique structure (illustrated in Figure 11.9) in which Li_2N layers containing hexagonal networks of Li^+ ions are linked together by bridging Li^+ ions. The material is an exceptionally good Li^+ conductor at relatively low temperatures, with a conductivity of $10^{-3}\ \Omega^{-1}\ cm^{-1}$ at 400 K. Moreover, the conductivity is anisotropic, as expected for a layered material, with a much higher conductivity parallel to the plane than perpendicular to it. Thus, at 400 K, $\sigma_{\parallel}/\sigma_{\perp} \approx 13$ (see the work by Rabenau[68] for a discussion of these experimental data).

However, recent crystallographic studies by Shulz[69] found no evidence for extensive disorder in the material, which raises the obvious question of the factors responsible for the exceptionally high conductivity. In this context the molecular dynamics study of Wolf et al.[12] and Wolf and Catlow[62] has been most revealing. The simulations were performed at two temperatures (300 and 400 K) using the box size of 192 ions and employing interatomic potentials derived by fitting to bulk crystal properties of Li_3N.

The simulations clearly demonstrated the occurrence of rapid and anisotropic diffusion in the material. This is clearly shown by the plots of $\langle r_{Li}^2 \rangle$ against t in Figure 11.2, which become linear at larger t, allowing transport coefficients to be calculated. Indeed, the calculated conductivity is in excess of the experimental value. We attribute this to inaccuracies arising from the use of rigid-ion potentials.

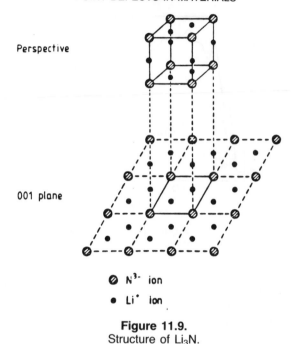

Perspective

001 plane

⊘ N³⁻ ion

• Li⁺ ion

Figure 11.9.
Structure of Li_3N.

The simulations also qualitatively reproduce experiment by revealing a highly ordered structure for the material, despite the high conductivity. This is clear from the RDFs for both $Li^+ - Li^+$ and $N^{3-} - N^{3-}$ correlations, which are shown in Figure 11.10. They consist of a series of well-separated Gaussians, which is normal behaviour for an ordered solid.

However, detailed examination of the individual configurations in the simulation demonstrates the presence of a low level of "interstitial" Li^+ ions, i.e. ions in the regions between the Li_2N layers. Their concentration (less than 1%) is too low to be detected by crystallographic techniques, but the simulations show that they play a vital role in effecting ion transport. Moreover, the process of interstitial creation, which involves a complex and concerted Li^+ migration process, itself effects ion transport. The mechanism involves the concerted motion of six Li^+ ions, which ends with the ejection of an Li^+ ion into the intergap region and the creation of a vacancy in the Li_2N layer. The reverse process also occurs in which Li^+ interstitials re-enter the layer via a concerted mechanism and a vacancy is annihilated, as shown in Figure 11.11. Vacancy migration within the layers may also occur via the type of concerted process illustrated in Figure 11.11, as well as by conventional vacancy hops.

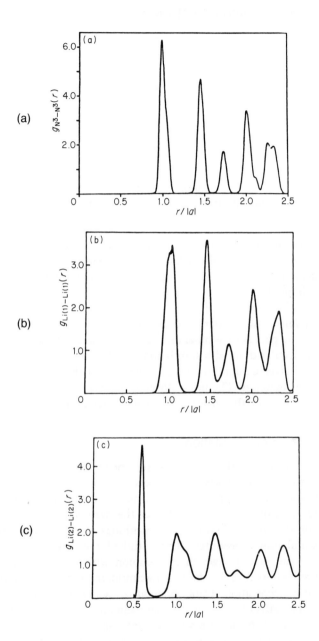

Figure 11.10.
RDFs in Li₃N: (a) N−N; (b) Li(1)−Li(1); (c) Li(2)−Li(2).

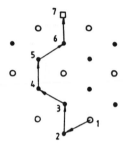

Figure 11.11.
Correlated migration mechanism in Li_3N: $\bigcirc$, N^{3-} ion; $\bullet$, Li^+; $\square$, Li^+ vacancy.

Equally fascinating are the mechanisms that effect transport perpendicular to the layers. Interstitials are also vital in this process, which occurs via the exchange mechanisms illustrated in Figure 11.12. Interstitial Li^+ ions displace the bridging Li^+, which then re-enter the Li_2N layers. Two such mechanisms are illustrated in Figure 11.12.

Other mechanisms were identified by the simulations, and the reader should consult the papers of Wolf et al.[12] and Wolf and Catlow[62] for details. Clearly, the molecular dynamics simulations provide unique information on the complex and diverse migration mechanisms responsible for the superionic behaviour of this material. The unravelling of this type of mechanistic detail is the greatest power of the molecular dynamics technique.

11.6.3   Monte Carlo simulation

As noted, the Monte Carlo technique has been extensively applied to complex systems, and the reviews and papers of Murch[16-19] and co-workers and de Bruin and Murch[20] give a good indication of the scope and range of the technique. To illustrate the potential of the method, we have chosen a recent study of Murch et al.[70] which also combines Monte Carlo techniques with static simulation procedures. They studied the intriguing problems posed by Y^{3+}−doped CeO_2 which in common with other doped fluorite oxides (e.g. Ca^{2+}−doped ZrO_2) shows a maximum in the plot of conductivity against dopant concentration. Data obtained by Wang et al.[71] are shown in Figure 11.13. The effect is difficult to understand as the conductivity would normally be expected to increase with dopant concentration until the concentration of the mobile oxygen vacancies which are created as charge compensators for the low-valence impurity ions is 50 mol.%. However, as noted in Figure 11.13, the maximum occurs at a relatively low impurity concentration (∼5 to 7 mol% vacancies) and the decrease in the conductivity

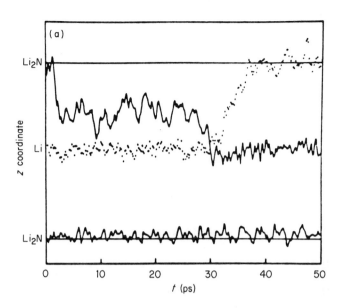

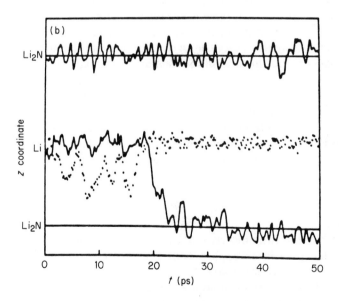

Figure 11.12.
Mechanism effecting transport parallel to c axis in Li_3N. Two coordinates of selected ions are plotted as a function of time.

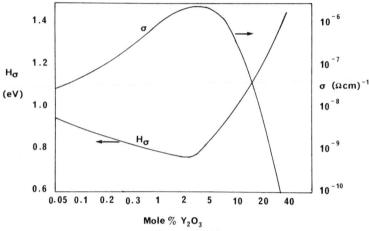

Figure 11.13.
Conductivity data in Y_2O_3-doped CeO_2. (After Wang et al.[71].)

after the maximum is dramatic. We also note the maximum in the conductivity corresponds to a minimum enthalpy of activation.

To investigate this problem, Murch et al.[70] undertook Monte Carlo simulation study of Y^{3+}-doped CeO_2 for a wide range of dopant concentrations, assuming a random distribution of dopant ions. The procedure used was essentially that described in section 11.4, but the jump frequencies were taken as proportional to the Arrhenius factor, $\exp(\Delta E_i/kT)$, with ΔE_i being calculated using the static simulation techniques, for all possible dopant environments for oxygen ions jumping by a vacancy mechanism. In defining the dopant environment, only first-neighbour cation sites with respect to the two oxygen ions involved in the jump were considered. A weak direct-current electric field was applied, enabling us to calculate an important and interesting quantity, i.e. the conductivity correlation factor f_I defined as

$$f_I = \frac{2kT\langle X \rangle}{nqa^2E} \tag{11.22}$$

where $\langle X \rangle$ is the drift distance of ions of charge q after an average of n jumps of length a in an applied field E. f_I represents the efficacy of ion jumps in effecting conductivity.

The results of the calculation are summarized in Figures 11.14–11.16. We note that the maximum in the conductivity is reproduced qualitatively. As regards the enthalpy, the increase at higher dopant concentrations is well reproduced but the calculated behaviour at low concentrations is less satisfactory. Of particular interest is the variation in f_I with dopant concentration

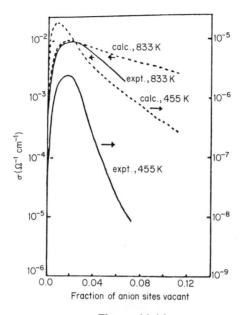

Figure 11.14.
Calculated and experimental conductivities in Y_2O_3–doped CeO_2 for two temperatures.

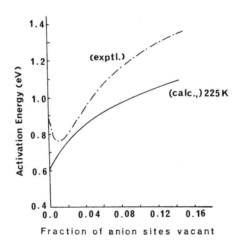

Figure 11.15.
Experimental and calculated Arrhenius energies in Y_2O_3–doped CeO_2.

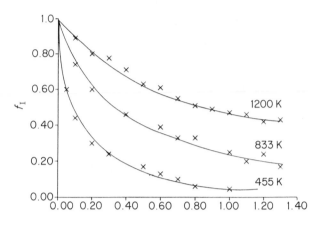

Fraction of anion sites vacant(10^{-1})

Figure 11.16.

Variation on the conductivity correlation factor f_I with dopant concentration in Y_2O_3–doped CeO_2.

where a pronounced decrease is observed. Thus, at higher concentrations, vacancy jumps are becoming decreasingly effective in leading to bulk ionic conductivity—an observation that yields qualitative insight into the origins of the observed maximum.

Clearly, further work remains to be done on these fascinating systems, but the results obtained to date indicate the power of the Monte Carlo technique in examining diffusion in complex solids, and the way in which that power is enhanced by combination of Monte Carlo with static simulation techniques.

11.7 SUMMARY AND CONCLUSIONS

Computer simulation of defect properties of solids is now a well-established technique. The field is currently moving in the direction of materials of greater complexity and materials for which simple Born model potentials are inadequate. A particularly important development is the combination of classical simulation with quantum mechanical calculations, where considerable progress can be expected in the near future and which will be considered in the next chapter.

REFERENCES

1 N. F. Mott and M. J. Littleton, *Trans. Faraday Soc.* **34**, 485 (1938).
2 C. R. A Catlow, R. James, W. C. Mackrodt and R. F. Stewart, *Phys. Rev. B* **25**, 1006 (1982).

3 C. R. A. Catlow and W. C. Mackrodt (eds), *Computer Simulation of Solids*, *Lecture Notes in Physics*, Vol. 166, Springer, Berlin (1982).
4 M. J. Norgett, *Report* No. AERE−R 7650, Atomic Energy Research Establishment (1974).
5 M. J. Norgett, *J. Phys. C: Solid State Phys.* 4, 298 (1971).
6 M. Leslie, *Report* No. DL−SCI−TM31T, Daresbury Laboratory, Science and Engineering Research Council (1981).
7 M. J. Gillan and P. W. M. Jacobs, *Phys. Rev. B* 28, 759 (1983).
8 J. H. Harding, *Physica B* 131, 13 (1985).
9 J. H. Harding, *Phys. Rev. B* 32, 6861 (1985).
10 D. Beeman, *J. Comput. Phys* 20, 130 (1976).
11 J. R. Walker in C. R. A. Catlow and W. C. Mackrodt (eds), *Computer Simulation of Solids*, *Lecture Notes in Physics*, Vol. 166, Springer, Berlin (1982).
12 M. L. Wolf, J. R. Walker and C. R. A. Catlow, *J. Phys. C: Solid State Phys.* 17, 6623 (1984).
13 G. Cicotti, G. Jacucci and I. R. MacDonald, *Phys. Rev. A* 13, 428 (1976).
14 M. L. Wolf, J. R. Walker and C. R. A. Catlow, *Solid State Ionics* 13, 33 (1984).
15 R. Lechner, in F. Bénière and C. R. A. Catlow (eds), *Mass Transport in Solids*, Plenum, New York (1983).
16 G. E. Murch, in G. E. Murch and A. S. Nowick (eds), *Diffusion in Crystalline Solids*, Academic Press, New York (1984).
17 G. E. Murch, in G. E. Murch, H. K. Birnbaum and J. R. Cost (eds), *Non Traditional Methods in Diffusion* AIME, Warrendale, PA (1984).
18 G. E. Murch, *Philos. Mag. A* 46, 151 (1982).
19 G. E. Murch, *Philos. Mag. A* 46, 575 (1982).
20 H. J. de Bruin and G. E. Murch, *Philos. Mag.* 27, 1475 (1973).
21 M. Tosi, *Solid State Physics* 16, 1 (1964).
22 W. C. Mackrodt, E. A. Colbourn and J. Kendrick, *Report* No. CL−R/81/1637/A ICI Corporate Laboratory, (1981).
23 I. D. Faux and A. B. Lidiard, *Z. Naturforcsh. (a)* 26, 62 (1971).
24 B. G. Dick and A. W. Overhauser, *Phys. Rev.* 112, 90 (1958).
25 C. R. A. Catlow and A. M. Stoneham, *J. Phys. C: Solid State Phys.* 16, 4321 (1983).
26 S. C. Parker and G. D. Price, *Phys. Chem. Miner.* 10, 209 (1984).
27 C. R. A. Catlow, K. M. Diller and M. J. Norgett, *J. Phys. C: Solid State Phys.* 10, 1395 (1977).
28 M. J. L. Sangster and R. M. Atwood, *J. Phys. C: Solid State Phys.* 11, 1541 (1978).
29 M. J. Sangster and A. M. Stoneham, *Philos. Mag.* 43, 597 (1980).
30 G. V. Lewis and C. R. A. Catlow, *J. Phys. C: Solid State Phys.* 18, 1149 (1985).
31 G. V. Lewis, *Ph.D. Thesis*, University of London (1984).
32 R. G. Gordon and Y. S. Kim, *J. Chem. Phys. C* 12, 431 (1972).
33 P. T. Wedepohl, *Proc. Phys. Soc., London* 92, 79 (1967).
34 W. C. Mackrodt, R. F. Stewart, J. C. Campbell and I. M. Hillier, *J. Phys. (Paris), Colloq. C7*, 64 (1980).
35 P. Saul, C. R. A. Catlow and J. Kendrick, *Philos. Mag. B* 51, 107 (1985).
36 A. Szabo and N. S. Ostlund, *Modern Quantum Chemistry*, Macmillan, London (1984).
37 P. W. Fowler and N. C. Pyper, *Proc. Roy. Soc.* A398, 377 (1985).
38 A. N. Cormack, to be published.

39 M. J. Sanders, M. Leslie and C. R. A. Catlow, *J. Chem. Soc., Chem. Commun.* 1271 (1984).
40 W. C. Mackrodt, in G. Petot-Ervas, H. J. Matzke and C. Monty (eds), *Transport in Non-Stoichiometric Compounds*, North-Holland, Amsterdam (1984).
41 G. V. Lewis and C. R. A. Catlow, *J. Phys. Chem. Solids* **47**, 89 (1986).
42 G. V. Lewis *Ph.D. Thesis*, University of London (1984).
43 G. V. Lewis, C. R. A. Catlow and R. W. Casselton, *J. Am. Ceram. Soc.*, **68**, 555 (1985).
44 C. R. A. Catlow, M. Doherty, G. D. Price, M. J. Sanders and S. C. Parker, *Mater. Sci. Forum* **7**, 163 (1986).
45 C. R. A. Catlow, J. Corish, K. M. Diller, P. W. M. Jacobs and M. J. Norgett, *J. Phys. C: Solid State Phys* **12**, 451 (1979).
46 C. R. A. Catlow, M. J. Norgett and T. A. Ross, *J. Phys. C: Solid State Phys* **10**, 1627 (1977).
47 W. C. Mackrodt and R. F. Stewart, *J. Phys. C: Solid State Phys* **12**, 431 (1979).
48 C. R. A. Catlow, W. C. Mackrodt, M. J. Norgett and A. M. Stoneham, *Philos. Mag.* **35**, 177 (1977).
49 C. R. A. Catlow, in O. T. Sorenson (ed.), *Non-Stoichiometric Oxides*, Academic Press, New York (1984).
50 P. J. Bendall, C. R. A. Catlow, J. Corish and P. W. M. Jacobs, *J. Solid State Chem.* **51**, 159 (1984).
51 C. R. A. Catlow, *J. Phys. C: Solid State Phys.* **6**, L64 (1973).
52 T. L. Hayes and T. B. Boyce, *Solid State Phys.* **37**, 173 (1983).
53 C. R. A. Catlow, A. V. Chadwick, G. N. Greaves and L. M. Moroney, *Nature (London)* **312**, 601 (1984).
54 R. Taylor, in C. R. A. Catlow and W. C. Mackrodt (eds), *Computer Simulation of Solids, Lecture Notes in Physics*, Vol. 166, Springer, Berlin (1982).
55 R. Taylor, *Physica B* **151**, 103 (1985).
56 G. Jacucci and A. Rahman, *J. Chem. Phys.* **69**, 4117 (1978).
57 M. Dixon and M. J. Gillan, *J. Phys. C: Solid State Phys.* **13**, 1901 (1980).
58 M. J. Gillan and M. Dixon, *J. Phys. C: Solid State Phys.* **13**, 1919 (1980).
59 P. Vashishta and A. Rahman, in P. Vashishta *et al.* (eds), *Fast Ion Transport in Solids*, North-Holland, Amsterdam (1979).
60 P. Vashishta, *Solid State Ionics* **18-19**, 3 (1986).
61 M. L. Wolf, *Ph.D. Thesis*, University of London (1984).
62 M. L. Wolf and C. R. A. Catlow, *J. Phys. C: Solid State Phys.* **17**, 6635 (1984).
63 M. J. Gillan, *Physica B* **131**, 157 (1985).
64 G. C. Farrington and J. Briant, in P. Vashista *et al.* (eds), *Fast Ion Transport in Solids*, North-Holland, Amsterdam (1979).
65 G. Collin, R. Comes, J. P. Boilot and P. Colomban, in M. Kleitz, B. Sapoval and D. Ravaire (eds), *Solid State Ionics 83*, North-Holland, Amsterdam, p. 311 (1983).
66 J. P. Hansen and I. R. MacDonald, *The Theory of Simple Liquids*, Academic Press, New York (1976).
67 A. Rahman, *J. Chem. Phys.* **65**, 4845 (1976).
68 A. Rabenau, *Solid State Ionics* **6**, 277 (1982).
69 H. Schulz, *Ann. Rev. Mater. Sci.* **12**, 351 (1982).
70 G. E. Murch, C. R. A. Catlow and A. D. Murray, *Solid State Ionics* **18-19**, 196 (1986).
71 D. Y. Wang, D. S. Park, J. Griffiths and A. S. Nowick, *Solid State Ionics* **2**, 95 (1981).

12
Quantum Mechanical Methods

12.1 INTRODUCTION

The simulation techniques discussed in Chapter 11, although clearly very powerful, are restricted in their range of applications. They cannot be used to investigate colour centres or any defect in a covalent material in which bonds are broken. Thus their range of applications to defects in semiconductors is very limited. In contrast, the class of methods discussed in this chapter is generally applicable to *all* types of defect. These methods are based on attempts at varying levels of approximation to solve the Schrödinger equation for the defect and a surrounding region of crystal. As in simulations, defect calculations can in principle be performed in two ways, i.e. supercell methods may be employed, or an isolated defect or defect cluster may be embedded in a surrounding region of lattice. Related, but distinct from the latter, are the Green's function methods which attempt to calculate the responses of the lattice to the perturbation provided by the defect. With expansion of computer power, the supercell technique is likely to grow in importance. To date, however, the majority of calculations have been based on some type of embedded cluster or use of the Green's function methods. In the next few sections, we therefore discuss the type of technique that has been employed in the different classes of calculation concentrating on the former approach which tended to be more useful for the type of defect problem treated in this book; supercell methods will be discussed subsequently. First, however, we describe the simplest method based on effective mass theory.

12.2 EFFECTIVE-MASS THEORY

This simplest of approaches is appropriate when we have a defect electron (or hole) weakly trapped by an impurity or defect and lying close in energy to the conduction (or valence) band (Figure 12.1). The system may then be treated as a pseudo-hydrogen-atom-like state with the electron moving in the effective field of the dopant, requiring us to solve the Schrödinger equation

$$\left(\frac{-\hbar^2}{2m^\star}\nabla^2 + V_{\text{eff}}\right)\chi(r) = E\chi(r) \tag{12.1}$$

where $m^\star$ is th effective mass of the electron or hole in the nearby band. E is the energy of the state *relative* to the band edge and $\chi(r)$ is not the total wavefunction $\psi(r)$; this is given by

$$\psi(r) = U_0(r)\chi(r) \tag{12.2}$$

where $U_0(r)$ is the Bloch function of the hole or electron state at the neighbouring band edge; $\chi(r)$ is essentially an envelope function. It is assumed that the interaction between the impurity and electron or hole is Coulombic in origin in which case V_{eff} is given by

$$V_{\text{eff}}(r) = \frac{-e^2}{\epsilon r} \tag{12.3}$$

where ϵ is the static dielectric constant of the solid.

Equation (12.1) is now perfectly hydrogen-atom like in its structure, and its solution will yield the usual Rydberg states which are scaled, however, by the effective mass and the dielectric constant. Thus,

$$E_n = -\frac{R}{n^2}\frac{m^\star}{m\epsilon^2} \tag{12.4}$$

where R is the Rydberg constant and n is the integral quantum number. The Rydberg series converge on the band edge which plays the role of the continuum in conventional atomic structure calculations.

To illustrate the application of effective-mass theory, we give a comparison in Table 12.1 of calculated and experimental energies of donor states in phosphorus-doped germanium. The agreement is generally good and illustrates the value of the approach.

For detailed discussions of the applications of effective mass methods, the reader should refer to the work of Stoneham.[3,4] Probably the most critical is that the envelope function $\chi(r)$ varies slowly in space—a requirement which essentially confines the method to states with diffuse wavefunctions. This generally limits the method to shallow defects for which, however, it has considerable utility.

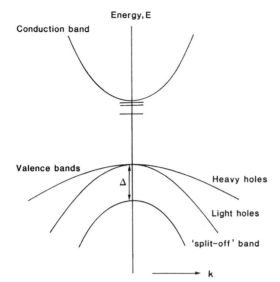

Figure 12.1.
A simple schematic illustration of the energy levels of shallow donor states near a conduction band edge. The ordinate represents one-electron energy levels and the abscissa, the wavevector of non-localized states. (After Lidiard[1].)

Table 12.1.
Comparison of the experimental and the theoretical (effective-mass) energy levels of shallow (group V) donors in silicon and germanium

System	Energy level (10^{-3} eV) for the following states							
	1S	$2P_0$	2S	$2P_1^{\pm}$	$3P_0$	3S	$3P_1^{\pm}$	$4P_0$
Silicon (theory)	31.27	11.51	8.83	6.40	5.48	4.75	3.12	3.22
Si:P	45.5, 33.9, 32.6	11.45	—	6.39	5.46	—	3.12	3.38
Si:As	53.7, 32.6, 31.2	11.49	—	6.37	5.51	—	3.12	—
Si:Sb	42.7, 32.9, 30.6	11.52	—	6.46	5.51	—	3.12	—
Germanium (theory)	9.81	4.74	3.52	1.73	2.56	2.01	1.03	1.67
Ge:P	12.89, 9.88	4.75	—	1.73	2.56	—	1.05	—
Ge:As	14.17, 9.96	4.75	—	1.73	2.56	—	1.04	—
Ge:Sb	10.32, 10.01	4.74	—	1.73	2.57	—	1.04	—

The units (10^{-3} eV) for the energy levels are below the conduction band minima. The experimental ground levels are split as a result of the mixing of states from different but equivalent conduction band minima. This effect lies outside the effective-mass approximation and is unobservably small for the excited states
(After Lidiard[1] and Bassani[2].)

12.3 EMBEDDED-CLUSTER CALCULATIONS

There are two important aspects to such calculations. The first concerns the way in which the Schrödinger equation is solved for the defect cluster; the second is the effect of the treatment of the surrounding region of the crystal. We review the different approaches in order of their level of sophistication.

12.3.1 Simple one-electron calculation

These techniques are applicable to colour centres such as the F centres in the alkali halides and the F^+ centres in MgO. The simplest form involves solving the Schrödinger equation for the electron in the Madelung field of the surrounding lattice, i.e. we write

$$\mathcal{H}\psi(r) = V\psi(r) \tag{12.5}$$

where

$$V = \sum_i \frac{q_i e}{r_i}$$

in which q_i is the charge of the surrounding ions and r_i the distance between the electron and the ith charge. As in other quantum mechanical calculations, $\psi(r)$ may be written as an expansion of Slater or Gaussian orbitals, i.e.

$$\psi(r) = \sum_j C_j r^{n-1} \exp(-\zeta_j r) \tag{12.6a}$$

for the Slater expansion, and

$$\psi(r) = \sum_j C_j r^{n-1} \exp(-\zeta_j r^2) \tag{12.6b}$$

for the Gaussian expansion. The angular dependence of the wavefunction is described using spherical harmonics. The variational principle is used to obtain the optimum values of the coefficients C_j, and in more sophisticated calculations the exponents ζ_j may be optimized. The approach was pioneered by Gourary and Adrian[5], who obtained useful results on F centres in alkali halides, which are summarized in Table 12.2.

The method is clearly extremely crude. Two obvious omissions concern first the effects of lattice relaxation, which as emphasized in Chapter 11 is of major importance in defect calculations, and secondly the oversimplification of using a point charge representation of the ions surrounding the colour centre electrons; this neglects the fact that the Pauli principle excludes the F-centre electrons from the region occupied by electron charge clouds of the neighbouring ions. These effects can be modelled by the use of pseudo-potentials on the ions, i.e. functions to which the F-centre wavefunction

Table 12.2.
Results of simple one-electron point ion calculations of the F Centre in KCl

Type of function for F-centre electron	1s energy (eV)	2s energy (eV)	Energy difference (eV)
One Slater	−5.39	−3.51	1.88
Two Slater	−5.96	−3.97	1.99

(After Gourary and Adrian[5].)

must be orthogonal. A variety of pseudopotential approaches have been adopted (see, for example, the work of Stoneham[3] for a detailed discussion). A related method adopted by Bartram et al.[6] induces "ion-size" corrections by using effecting potentials of the form

$$V_{eff} = V_{PI} + \sum_i C_i \delta(r - R_i) \qquad (12.7)$$

where the C_i are derived from a pseudopotential treatment developed by Cohen and Heine[7]. The C_i coefficients are found to be small, however, although the use of this more sophisticated potential does effect significant improvements[8,9].

Even with such improvements the range of this class of calculation is very limited. They are clearly confined to the simplest types of colour centre. Moreover, the inability to handle lattice relaxation is a major weakness; it should be noted that the methods fail badly when applied to the relaxed excited states of colour centres. More modern colour centre calculations include the nearest neighbour and even more distant neighbours in the quantum mechanical cluster. These are discussed later in the chapter after simpler cluster calculations have been described.

12.3.2 Defect molecule calculations

The classic and pioneering example of the use of this method was the work of Coulson and Kearsley[10] on the vacancy in diamond. The vacancy is, of course, surrounded tetrahedrally by four carbon atoms. The creation of the vacancy results in unsatisfied valencies on these four atoms. The resulting "dangling bonds" will be sp^3 hybrids pointing towards the vacancy as shown in Figure 12.2. The wavefunction is written in terms of linear symmetry-adapted combinations of the four sp^3 hybrids. The resulting wavefunctions are of A_1 (non-degenerate) and T_2 (triply degenerate) character. We note that the vacancy may have three charge states: V^+ (configuration

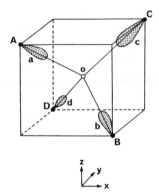

Figure 12.2.
The vacancy in diamond with sp^3 hybrid orbitals on the four atoms (A, B, C and D surrounding the central vacancy site O.

$a_1{}^2t_2{}^1$, i.e. two electrons in A_1 and one in T_2), V (configuration $a_1{}^2t_2{}^2$) and V$^-$ (configuration $a_1{}^2t_2{}^3$). The equivalent analysis for the silicon vacancy was considered in section 7.4.

Having written down the one-electron wavefunctions, the many-electron functions may be written using Slater determinants (further discussions of which follow below). Energies may then be obtained by diagonalization of the Hamiltonian matrix. The original paper by Coulson and Kearsley gave the results in terms of several independent parameters. Moreover, their work concluded that the multiplet splitting (arising from electron–electron interactions) is a dominant feature in contrast with the experimental evidence of Watkins[11]. It appears that the Coulson–Kearsley approach exaggerates the effects of electron–electron repulsions which are reduced by delocalization of electron density from the dangling bonds into the surrounding C–C bonds. This problem was treated by Lannoo[12] and Lannoo and Bourgoin[13] who show that the Coulson–Kearsley method may be approximated by a two-parameter model, the critical parameters being Δ (the splitting of A_1 and T_1 levels) and U (the electron–electron interaction parameter). The problem devolves into calculating these two parameters. Lannoo[14] discusses a variety of approaches including larger-cluster calculations (of the type that will be discussed in section 12.3.3) and the self-consistent X_α method used by Messmer and Watkins[15].

The situation is complicated by the occurrence of Jahn–Teller distortions in all three charge states of the vacancy; these arise from the coupling of the degenerate wavefunctions with the surrounding lattice; the distortion has a tetragonal symmetry (with an additional trigonal component for the V$^-$ centre). The theory has been considered in detail by Lannoo (see the review by Lannoo[14]). It turns out that the one-electron models provide a reasonable approximation to experiment and that detailed calculations of this for the vacancy in both diamond and silicon provide results in reasonable agreement with those obtained using the one-electron method. Figure 12.3 illustrates the one-electron states for the different charge states of the vacancy in silicon and shows the effect of Jahn–Teller distortions.

The work summarized above shows the usefulness of the defect molecule approach. The method is limited, however. It takes no proper account of the interaction of the defect with the surrounding lattice and relies on more sophisticated theories to provide estimates of the critical parameters Δ and U. The methods described in section 12.3.3 aim to overcome these difficulties.

12.3.3 Extended cluster calculations

By an extended cluster, we mean a cluster which includes the defect site itself and at least the surrounding nearest-neighbour atoms which are fully described by the quantum mechanical method. A wide range of such calculations have been performed; they vary both in the nature of the quantum mechanical method used for the cluster and in the way in which the cluster is embedded in the surrounding lattice. Four classes of quantum

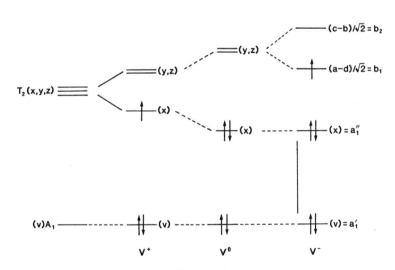

Figure 12.3.
One-electron energy levels for distorted configurations of the three charge states of the vacancy in diamond. In the V^+ and V^0 centres the T_2 levels are split by a tetragonal distortion and there is an additional trigonal distortion in V^-. See Lannoo[14] for further discussion.

mechanical procedures can be used in embedded calculations. These are as follows.

(1) Non-self-consistent methods.
(2) Self-consistent Hartree–Fock methods with empirical parametrization.
(3) *Ab initio* Hartree–Fock (SCF) methods.
(4) Configurational interaction (CI) methods.

We concentrate on the first three methods as CI techniques have not been extensively used in defect studies. Since these methods rely on molecular orbital approaches, it is first necessary to review the basic theory of these methods.

12.3.4 Approximate molecular orbital techniques

The orbital approximation on which these methods rest assumes that many-electron wavefunctions can be written as a product of one-electron functions or orbitals. Since electrons are fermions, the many-electron wavefunctions must be antisymmetric with respect to particle exchange; this property can be elegantly and neatly guaranteed by using the determinantal form of the wavefunction given below in equation (12.8) as:

$$\Psi(1\dots n) = N^{-1/2} \begin{vmatrix} \psi_1(1)\alpha(1)\psi_1(1)\beta(1) & \dots & \psi_n(1)\alpha(1)\psi_n(1)\beta(1) \\ \psi_1(2)\alpha(2)\psi_1(2)\beta(2) & \dots & \psi_n(2)\alpha(2)\psi_n(2)\beta(2) \\ \cdot & & \cdot \\ \cdot & & \cdot \\ \cdot & & \cdot \\ \psi_1(2n)\alpha(2n)\psi_1(2n)\beta(2n) & \dots & \psi_n(2n)\alpha(2n)\psi_n(2n)\beta(2n) \end{vmatrix}$$

(12.8)

where N is a normalization constant. The use of the determinantal expression ensures that the $2n$ electrons are permuted over the $2n$ spin orbitals. Note that each orbital ψ_i consists of two spin functions $\psi_i\alpha$ and $\psi_i\beta$. These determinantal wavefunctions, known as Slater determinants can be written in a short-hand notation as

$$\psi(1 \dots n) = |\psi_1\overline{\psi}_1 \dots \psi_n\overline{\psi}_n| \tag{12.9}$$

where the bars in for example, $\overline{\psi}_1$, indicate the β spin function.

For a closed-shell system, a single determinant may be used as an approximation to the many-electron wavefunction. However, it is an approximation with important limitations; in particular, it omits the effects of electron correlation. Multideterminantal wavefunctions must therefore be used if correlation is to be included; this is achieved in the CI method.

Let us pursue the use of the single-determinantal function, however. The energy of the many-electron system will be written as

$$E = \langle \psi^\star(1 \ldots n) | \mathcal{H} | \psi(1 \ldots n) \rangle \tag{12.10}$$

We can solve the Schrödinger equation within the single-determinantal approximation by application of the variational principle, i.e. adjustment of the parameters within a general expression for the wavefunction until the minimum energy is obtained.

When the above energy expression is multiplied out, three types of integral are obtained; first, we obtain purely one-electron terms of the type

$$\mathcal{H}_{ii} = \int \psi_i^\star(n) \left(\frac{-\hbar^2 \nabla^2}{2m} - \sum_A \frac{Z_A e}{r_{An}} \right) \psi_i(n) \, d\tau_n \tag{12.11}$$

where Z_A is the charge on the nucleus A and r_{An} is the distance of the nth electron to nucleus A. These terms are readily evaluated. Far more difficulty arises from the two-electron integrals which are a consequence of the electron repulsion term $1/r_{12}$ in the Hamiltonian, and from which two types of integral follow: first, Coulomb integrals of the form

$$\mathcal{J}_{ij} = \iint \psi_i^\star(1) \psi_j^\star(2) \frac{1}{r_{12}} \psi_i(1) \psi_j(2) \, d\tau_1 \, d\tau_2 \tag{12.12}$$

which can be envisaged as arising from the Coulomb repulsion between the two components of charge density represented by ψ_i^2 and ψ_j^2; secondly, *exchange* integrals which take the form

$$K_{ij} = \iint \psi_i^\star(1) \psi_j^\star(2) \frac{1}{r_{12}} \psi_i(2) \psi_j(1) \, d\tau_1 \, d\tau_2 \tag{12.13}$$

and which has no simple classical interpretation. Expansion of the total electron energy expression then gives

$$E = 2\sum_i^n \mathcal{H}_{ii} + \sum_i^n \mathcal{J}_{ii} + \sum_i^n \sum_{j(\neq i)}^n (2\mathcal{J}_{ij} - K_{ij}) \tag{12.14}$$

detailed derivations of which may be found in several standard texts, e.g. those by Pople and Beveridge[16], Atkins[17] and Szabo and Ostlund[18]. Application of the variational principle to the above expression, subject to the orthonormality conditions

$$S_{ij} = \int \psi_i(1) \psi_j(1) \, d\tau_i = \delta_{ij} \tag{12.15}$$

leads to the well known Hartree–Fock equations discussions of which are given in these texts and which are discussed further below. The normal procedure is to expand the orbitals, using the linear combination of atomic orbitals (LCAO) approximation, i.e.

$$\psi_i = \sum_\mu C_{\mu i} \Phi_\mu \tag{12.16}$$

where the functions Φ_μ are atom-centred functions written in terms of Slater or Gaussian orbitals. Solution of the Hartree–Fock equations will yield a set of coefficients and hence orbitals. One-electron orbital energies ϵ_i can be defined as

$$\epsilon_i = \mathcal{H}_{ii} + \sum_j^n (2\mathcal{J}_{ij} - K_{ij}) \tag{12.17}$$

where the total electronic energy ϵ_{tot} is written as

$$\epsilon_{tot} = 2\sum_i^n \epsilon_i - \sum_i^n \sum_j^n (2\mathcal{J}_{ij} - K_{ij}) \tag{12.18}$$

i.e. *not* as a sum of one-electron terms.

When the LCAO approximation is used, the integrals in these equations may be written as

$$\mathcal{J}_{ij} = \sum_{\mu\lambda\nu\sigma} C_{\mu i}{}^\star C_{\lambda j}{}^\star C_{\nu i} C_{\sigma j} \iint \Phi_\mu(1)\Phi_\nu(1)\frac{1}{r_{12}}\Phi_\lambda(2)\Phi_\sigma(2)\,d\tau_1\,d\tau_2 \tag{12.19}$$

and

$$K_{ij} = \sum_{\mu\lambda\nu\sigma} C_{\mu i}{}^\star C_{\lambda j}{}^\star C_{\nu i} C_{\sigma j} \iint \Phi_\mu(1)\Phi_\lambda(1)\frac{1}{r_{12}}\Phi_\nu(2)\Phi_\sigma(2)\,d\tau_1\,d\tau_2 \tag{12.20}$$

The Hartree–Fock equations can be written in short-hand as

$$\mathbf{F}C = \epsilon\mathbf{S}C \tag{12.21}$$

where **F** is a matrix of the Fock operator which is a sum of the kinetic energy, nuclear, Coulomb and exchange terms which are defined above. **S** is the overlap matrix whose elements $S_{\mu\nu}$ are simply $\int \varphi_\mu \rho_\nu d\tau$. Solution of the Hartree–Fock equations thus consists of evaluation of the terms in the Fock and overlap matrix and then solution of the secular equation

$$\det\|F - \epsilon S\| = 0 \tag{12.22}$$

the roots of which are the energies ϵ_i and the eigenvectors give the co-efficients C_μ from which the wavefunction can be constructed. The matrix **F** is second order in the coefficients C and can only be solved iteratively. This iterative procedure is known as the self-consistent field (SCF) component of the calculation. The most time-consuming task is commonly the evaluation of the many-electron integrals. If the atomic orbitals Φ_μ, Φ_λ, Φ_ν and Φ_σ

are centred on different atoms, these integrals may be very small. Many of the approximate SCF methods rely on this fact in neglecting or approximating various integrals. In this context we now consider the schemes referred to above, i.e. the non-SCF methods based on the Hückel theory and the approximate SCF procedures, notably the complete neglect of differential overlap (CNDO) and closely related intermediate neglect of differential overlap (INDO) methods.

12.3.5 The Extended Hückel method

In this simplest of approaches, no attempt is made to achieve truly self-consistent solution of the Hartree−Fock equations. The method, developed originally by Hoffmann, uses a simple parametrized representation of the Fock matrix, writing the diagonal elements $F_{\mu\mu}$ as

$$F_{\mu\mu} = -I_\mu \qquad (12.23)$$

where I is the approximate ionization potential. The off-diagonal terms $F_{\mu\nu}$ are written as

$$F_{\mu\nu} = \frac{K}{2}\,(F_{\mu\mu} + F_{\nu\nu})S_{\mu\nu} \qquad (12.24)$$

where K is a constant, commonly taken to be 1.75. The total energy of the system is then commonly treated as a sum of the one-electron energies ϵ_i.

The extended Hückel theory was applied in the 1960s and early 1970s to a number of defect cluster problems (see, for example, the work of Messmer and Watkins[19,20] and Larkins[21] and the review of Lidiard[22]. The economic nature of the method allowed some useful results to be obtained. For example, Watkins et al.[23] performed a set of calculations on the interstitial in diamond. However, it is clear from calculations on the *perfect* clusters that the extended Hückel method gives a poor representation of the band structure of the material, which suggests that the results for defective crystals will be similarly inadequate. We believe that the status of the extended Hückel method in defect physics is best summarized by Lidiard[22] who considered that it "would be naive to accept the Hückel energy function as providing more than an initial and rough guide to the possible static configurations" (of defects).

12.3.6 Complete neglect of differential overlap (CNDO)

In this method, integrals of the type

$$\int \Phi_\mu(1)\Phi_\nu(1)\frac{1}{r_{12}}\Phi_\lambda(2)\Phi_\sigma(2)\,\mathrm{d}\tau = \langle \mu\nu|\lambda\sigma\rangle \qquad (12.25)$$

(in which we have used the short-hand notation employed by Pople and Beveridge[16]), are set to zero, except where $\mu = \nu$ and $\lambda = \sigma$, i.e.

$$\langle \mu\nu | \lambda\sigma \rangle = \langle \mu\mu | \lambda\lambda \rangle \, \delta_{\mu\nu} \, \delta_{\lambda\sigma} \qquad (12.26)$$

Moreover the remaining integrals are approximated according to

$$\langle \mu\mu | \lambda\lambda \rangle = \gamma_{AB} \qquad (12.27)$$

where γ_{AB} depends only on the atom types A and B and not on the orbitals μ and λ; it is a mean electron–electron repulsion term. The method also neglects overlap integrals of the type $\int \Phi_\mu(1)\Phi_\nu(1) \, d\tau$ in normalizing the wavefunction.

CNDO makes a number of other approximations. First, only valence electrons are considered; these are assumed to move in a field of the type $V = \Sigma_A V_A$ where the V_A are the potentials due to nucleus and core electrons of the atoms of type A. When the LCAO approximation is implemented, two types of term appear. Consider first the diagonal terms:

$$\int \Phi_\mu | \mathcal{H} | \Phi_\mu \, d\tau = U_{\mu\mu} - \sum_{B \neq A} \int \Phi_\mu | V_B | \Phi_\mu \tau \, d\tau \qquad (12.28)$$

where Φ_μ is an orbital on atom A and $U_{\mu\mu}$ is an atomic property of atom A which is parametrized semiempirically; the second term on the right-hand side of the equation gives the electrostatic interaction of the electron in orbital μ with the cores of atoms B, for which further approximations are made.

Furthermore, off-diagonal terms of the type $\int \Phi_\mu | \mathcal{H} | \Phi_\nu \, d\tau$ are approximated as

$$\int \Phi_\mu | \mathcal{H} | \Phi_\nu \, d\tau = \beta_{AB} S_{\mu\nu} \qquad (12.29)$$

where the β_{AB} are parameters that are determined empirically and the $S_{\mu\nu}$ are overlap integrals $\int \Phi_\mu(1)\Phi_\nu(1) \, d\tau$.

The method therefore simplifies the Hartree–Fock equations by ignoring many of the integrals and parametrizing others, with the parametrization using empirical data on ionization energies (for the U terms) and bond strengths (for the β terms). A number of parametrizations are available, as discussed by Pople and Beveridge[16], who also present a detailed account of the theory of CNDO.

The method has been applied extensively especially in the 1960s and 1970s to the study of a wide range of molecules, particularly organic molecules. This work, much of which is again reviewed by Pople and

Beveridge[16], clearly established the utility of the technique especially when more qualitative information such as charge distribution is required.

The economy of the method has also encouraged its application to defect cluster calculations, as large clusters can be handled with relatively small amounts of time on modern computers. Indeed the method has been widely and successfully applied to a range of problems in defect physics notably by Stoneham and co-workers[24-28]. These applications are wide ranging, varying from studies of interstitial defects in silicon and diamond to calculations of the processes responsible for silicon oxidation; mechanisms of radiation damage have also been studied.

To illustrate the use of such calculation, we take first the work on interstitials in diamond. The question of the structure of the self-interstitial in diamond has been controversial. A number of possible models shown in Figure 12.4 were examined by Mainwood et al.[24]. They used a reparametrized CNDO code on clusters containing up to 38 atoms with the unsatisfied valences on the edge of the cluster being saturated by hydrogen atoms. The different proposed interstitial configurations have tetrahedral or hexagonal symmetry; they involve insertion of an additional atom at a bond centre site or a split interstitial configuration as shown in Figure 12.4(c) has been suggested. The CNDO calculations allowed relaxation of the lattice atoms around the interstitial. They found that the most stable state is the $\langle 100 \rangle$ split interstitial in a singlet rather than a triplet state. The hexagonal interstitial was found to be about 1.5 eV higher in energy than the singlet, whereas the least-favoured configuration is the bond-centred structure which earlier extended Hückel calculations had suggested was close in energy to the split configuration.

More recently a similar CNDO study by Masri et al.[26] on silicon has also found that the $\langle 100 \rangle$ split interstitial configuration has the greatest stability. This study also examined positive and negatively charged as well as neutral interstitials. Their results suggested that, for positively charged defects, the hexagonal configuration would be more stable. The results have important consequences for an understanding of interstitial mobility. The hexagonal configuration provides a saddle point for migration of the split interstitial. Since the positively charged interstitial is predicted to be more stable in the hexagonal form, a migration mechanism can be envisaged that is accelerated by hole and electron capture processes of the type

$$I_{(s)}^0 + h \rightarrow I_{(H)}^+$$

$$I_{(H)}^+ + e^- \rightarrow I_{(s)}^0$$

where $I_{(s)}^0$ indicates a neutral split interstitial and $I_{(H)}^+$ a positively charged hexagonal interstitial. The hole–electron capture mechanism described above

(a)

(b)

(c)

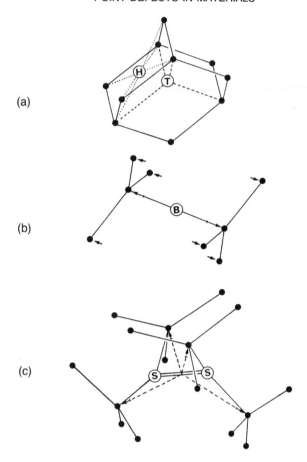

Figure 12.4.
(a) Interstitial configurations in diamond tetrahedral (T) and hexagonal (H) structures; (b) a bond centre configuration; (c) a (100) split configuration.

was proposed by Bourgoin and Corbett and is discussed by Masri *et al.*[26]. The CNDO calculations proved to be most valuable in confirming its plausibility on energetic grounds.

Our final example of the use of the CNDO method is the imaginative study by Stoneham and co-workers[27,28] of the mechanism of oxidation of silicon. Both "dry" and "wet" oxidation conditions are important. In the former case, the basic reaction is

$$Si(s) + O_2 \rightarrow SiO_2(s)$$

and, in the latter

$$Si(s) + 2H_2O \rightarrow SiO_2(s) + 4H$$

In both cases, it appears that neutral molecules diffuse through the growing SiO_2 layer and attack the silicon and the $Si-SiO_2$ interface. Hagon et al.[27] performed calculations on a 52-atom cluster of quartz and examined first the energetics of diffusion of the oxygen molecule down the c-axis channel in quartz. They found that the most stable configuration of the sorbed molecule had an energy of +0.34 eV compared with the energy of the vacuum; the migration energy down the channel was 0.84, giving a total Arrhenius energy for oxygen diffusion of 1.18 eV. This compares well with the experimental Arrhenius energy for silicon oxidation of 1.23 eV. In a companion study, Hagon et al.[28] examined water diffusion down the open c-axis pore of SiO_2. They showed that a spiral motion of this molecule provides a low-energy migration mechanism with an activation energy of about 1.75 eV, giving a resulting Arrhenius energy for water diffusion which is approximately twice the experimental value. The difficulty probably arises from the failure of the CNDO method to include any representation of ionic polarization, which might be expected to lower the incorporation energy of the water molecule by an appreciable amount.

These illustrative examples clearly show the extent to which the economical CNDO method can be used in defect studies. Growth in computer power will allow even larger clusters to be considered, and better parametrizations can be expected to improve the reliability of the technique.

12.3.7   Intermediate neglect of differential overlap (INDO)

This method is closely related to CNDO but differs in that certain of the two-electron integrals are retained. These are the intra-atomic exchange interaction of the type

$$\int\int \Phi_\mu(1)\Phi_\nu(2)\frac{1}{r_{12}} \Phi_\nu(1)\Phi_\nu(2) \, d\tau_1 \, d\tau_2 \qquad (12.30)$$

which are responsible for singlet−triplet splitting energies, for example and the neglect of which is one of the severer approximations of the CNDO method. In the INDO method, integrals of this type are retained.

INDO and the closely related modified MINDO method have been very extensively applied to the study of organic molecules. Applications to defects have been limited.

12.3.8   The *ab initio* Hartree—Fock methods

In these methods, all integrals are evaluated either numerically or analytically (although many codes will omit the evaluation of an integral if its value is expected to be below some lower limit). Both Gaussian and Slater orbitals may be used, and there are now several automated general computer programs for carrying out such calculations.

The accuracy of the calculations is limited first by the inherent approximations of the Hartree—Fock method, notably the omission of electron correlation effects. Secondly the choice of basis sets can be crucial in influencing the results of the calculations. It would be ideal to work with large basis sets, close to the Hartree—Fock limit (i.e. basis sets which yield the lowest possible total energy for the system obtainable from the Hartree—Fock method). In practice, it is generally necessary to work with more restricted basis sets, in view of the rapid increase in computer time with the number n of basis functions (the time needed for integral evaluation varies as approximately n^4). Considerable care must be exercised in using more restricted basis sets. If the sets are "unbalanced", i.e. better on some atoms than on others, then electron density will tend to flow to the atom with the superior set. A second and related problem concerns the basis set superposition error which was referred to in the previous chapter and which distorts the interaction energy between atoms. This subtle effect can be understood as follows. When a calculation is performed on a pair of interacting atoms, described by incomplete basis sets, the purely intra-atomic energy of each atom is lowered by the presence of the basis functions on the neighbouring atoms. The effect, which can be quite large when small basis sets are used, is particularly important when Hartree—Fock methods are being used to study interatomic potentials.

Given that adequate basis sets are available, there remains the crucial problem of embedding techniques when Hartree—Fock cluster methods are applied to defects. Several researchers have used the simple expedient of terminating clusters with hydrogen atoms. For covalent materials, the method may be suitable, particularly when large clusters are used. Recent successful applications have been reported by Mackrodt et al.[29] to defects of SiC and by Mombourquette and Weil[30] to defects in quartz. In more ionic materials it is almost certainly necessary to include the effects of the Madelung potential of the surrounding lattice on the quantum mechanical cluster. Such calculations were recently performed by Grimes et al.[31,32] on defects in MgO. Thus, in studying for example a magnesium vacancy, calculations were performed on an MgO_6^{10-} and O_6^{12-} clusters surrounded by about 100 point charges. The formation energy for this and other defects for *unrelaxed* nuclear coordinates surrounding the defect are given in Table 12.3, where they are compared with results of Mott—Littleton calculations. An encouraging measure of agreement is found.

Table 12.3.
Energies of unrelaxed defects in MgO

Defect	Energy (eV) obtained by the following techniques	
	Quantum mechanical cluster	Mott–Littleton
Magnesium vacancy	36.69	34.25
Oxygen vacancy	42.31	41.12
Li^+ Substitutional	17.67	18.97
F^- Substitutional	20.20	19.94

The Mott–Littleton calculations allows shell relaxations of (100) ions but no core displacements, as discussed in text.
(After Grimes et al.[32].)

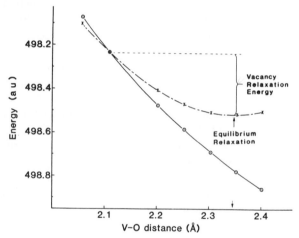

Figure 12.5.
Results of quantum mechanical calculations on MgO clusters: ⊙, results in which only nearest neighbours surrounding the vacancy are included in the quantum mechanical cluster; x, energies calculated when a shell of six Mg^{2+} ions was also included explicitly.

The work of Grimes et al.[32] also emphasized the crucial role of nuclear relaxation in defect calculations—a point that had long been apparent from Mott–Littleton calculations—and the difficulty of including these effects in quantum mechanical cluster calculations. Grimes et al. first attempted to model relaxation around the magnesium vacancy by relaxing outwards the surrounding oxygen ions in quantum mechanical embedded-cluster calculations. As shown in Figure 12.5, the calculation failed totally, with the calculated cluster energy simply decreasing monotonically with increasing oxygen ion displacements. The effect is easy to understand; relaxation of these oxygen ions takes them closer to neighbouring positive point charges

Table 12.4.
Energies of relaxed defects in MgO

Defect	Energy (eV) obtained by the following techniques	
	Mott—Littleton	Quantum mechanical cluster
Magnesium vacancy	24.52	30.90
Oxygen vacancy	25.203	30.01
Li^+ substitutional	16.55	15.56
F^- substitutional	15.31	17.00

(After Grimes et al.[32].)

which simulate the surrounding Mg^{2+} ions. As there is no short-range repulsion between the oxide ions and the point changes, there are no forces acting to counterbalance the attractive electrostatic interaction. However, it was found that by increasing the size of the quantum mechanical cluster to include six surrounding Mg^{2+} ions, this difficulty could be overcome, as shown in Figure 12.5. Indeed, the calculated magnitudes of the oxygen ion displacements now compare favourably with values obtained from Mott—Littleton calculations. Table 12.4 compares quantum mechanical cluster and Mott—Littleton calculations of this formation energy for relaxed cluster geometries of a variety of defects in MgO. Although there are significant discrepancies for vacancies, the results for the impurity states are encouraging and suggest that embedded-cluster calculation could be particularly effective in studying the properties of impurities in ionic solids.

The most detailed treatment of the embedding problem using point charges has been reported by Vail et al.[33], who examined carefully the problem of consistency between the cluster and point charge configuration. This was achieved by interfacing the quantum mechanical cluster calculations with the Mott—Littleton methodology. The procedure they adopted is as follows:

(1) The quantum mechanical cluster is "simulated" by a set of point charges and the lattice is relaxed to equilibrium around the point charge simulators.

(2) The Hartree—Fock equations are solved for the quantum mechanical cluster in the field of the relaxed surrounding lattice which is represented by point charges.

(3) From the electron density distribution obtained for the quantum mechanical cluster a new set of point charge simulators is defined. These are chosen so as to reproduce the electrostatic multipole moments of the cluster but, in practice, moments higher than the quadrupole moment have not been considered.

(4) The relaxation calculation is repeated with the new simulators.

(5) The quantum mechanical calculation is repeated with the new relaxed point charge distribution.

(6) The iterative process is continued until we have full consistency between the quantum mechanical cluster and the relaxed, embedding point charges. In practice, it is found that this is achieved after not more than three quantum mechanical calculations have been performed.

In their original study, Vail et al.[33] examined the problem of the F^+ centre in MgO. The results which are summarized in Table 12.5 were encouraging for the ground-state properties but disappointing for the relaxed state. The latter is indeed a very difficult problem. In the excited state the electron is only weakly bound to the vacancy, and as a consequence the wavefunction is diffuse. Electron density therefore tends to "spill out" of the quantum mechanical cluster into the point charge region, where the lack of Pauli repulsion leads to unrealistically low energies. The problem can be mitigated by using restricted basis sets which constrain the electron to remain within the cluster, but the procedure is an unsatisfactory one as the results of the calculations then become dependent on the choice of basis set. It seems that very-large-cluster calculations will be needed to overcome this problem.

In concluding this discussion, we draw attention to two important recent developments. The first concerns the use of pseudopotentials which were discussed earlier in this chapter and the role of which in many calculations is to provide a simple and economical description of the effects of core electrons. Pseudopotentials may be used both within the quantum mechanical cluster in place of an explicit description of core electrons, thereby reducing

Table 12.5.
Optical transition energies for the F^+ centre in MgO

Procedure	Absorption energy (eV)	Emission energy (eV)
Experiment	4.96	3.13
One-electron calculation (without point charges)	4.94	—
Cluster with nearest-neighbour Mg^{2+} surrounding vacancy	3.84	—
Cluster with nearest and next-nearest neighbour around vacancy	5.28	4.62

See Vail et al.[33] for details of the procedures.

the magnitude of the calculations. Alternatively, in the point charge region, their use in place of simple point charges should reduce the "spillage" problem referred to above, by providing Pauli repulsion between the electrons within the cluster and the surrounding point charges. Few calculations have been reported to date, but their use in the future is expected to increase.

Secondly, we note that expansion in computer power is making good calculations on larger clusters increasingly feasible and that the use of *ab initio* quantum mechanical embedded-cluster calculation is expected therefore to expand considerably in the near future.

In concluding this section, we note that the Hartree–Fock molecular orbital method is not the only way of solving the Schrödinger equation, and we will now give a short account of alternative procedures.

12.4 LOCAL-DENSITY METHODS

These methods differ from these discussed above in that the energy of the system is written directly as a function of the electron density; they aim to obtain directly the variation in the local electron density $\rho(r)$ as a function of the nuclear coordinates l. The fundamental basis of the approach is the theorem of Hohenberg and Kohn[34] which showed that the ground-state energy of a system is a unique function of the electron density.

The ground-state energy $E(\rho)$ of the system can be written as

$$E(\rho) = \int t_r(\rho) \, \mathrm{d}r + \int \rho V_N \, \mathrm{d}r + \tfrac{1}{2} \int \rho V_e \, \mathrm{d}r$$

$$+ \int \epsilon_{xc}(\rho) \, \mathrm{d}r + V_{N-N} \qquad (12.31)$$

where $t_r(\rho)$ is the single-particle kinetic energy density, $\epsilon_{xc}(\rho)$ is the energy density of exchange (x) and correlation (c) interactions, V_N is the potential energy of the nuclei, V_e is the potential energy of the electron cloud $\rho(r)$ and V_{N-N} is the nuclear–nuclear repulsion energy.

The variation principle requires that we minimize E with respect to ρ, which can be achieved by the usual Lagrangian methods (see, for example, the work by March[35]). The most difficult term to evaluate is, of course, ϵ_{xc} which includes the many–electron effects. Evaluation of this term generally involves the use of approximations. However, the fact that the exchange–correlation term is of a local nature has generally allowed the inclusion of correlation effects more readily than by Hartree–Fock methods.

The method has been widely applied to the study of perfect solids[35-38], and can give good values for the ground-state properties of crystals, including structural, cohesive and lattice dynamic properties; a good recent review is available in reference 37. When applied to defective solids, the local-density methods have been most commonly used in conjunction with Green's function techniques discussed in section 12.5.

12.5 THE GREEN'S FUNCTION METHODS

This is a large field. Many of the more recent applications have concentrated on metals, although there are several studies of semiconductor systems. Our account will be brief, and it should be supplemented by other references: 3, 12, 37, 38, 39, 40, 41, 42, 43, 44, 45, 46. The idea of the Green's function methods can best be introduced by reference to the classical defect calculations. In the present context, the Green's functions are response functions. Thus let us consider the defect simulation techniques described in section 11.2. We recall the division of the lattice into two regions, with the outer region being treated using the harmonic approximation so that, from equation (11.3), we write

$$E_{II} = \tfrac{1}{2} \mathbf{y} \cdot \mathbf{A} \cdot \mathbf{y} \tag{12.32}$$

where $\mathbf{A}$ is the force constant matrix describing how the energy of region II respond to displacements. We can equate $\mathbf{A}^{-1}$ to $\mathbf{G}^0$, the perfect lattice Green's function for the system. We also recall that the equilibrium condition required that

$$-\frac{\partial E_{I-II}}{\partial y} = \mathbf{A} \cdot \mathbf{y} \tag{12.33}$$

Now, in the treatment we gave in Chapter 11, E_{I-II} and its derivative are evaluated by explicit summation. We now wish to generalize the approach and to note that the inner region might include a quantum mechanically described species containing a number of variational parameters λ used in optimizing the wavefunction. We then expand the derivative to first order in y:

$$-\frac{\partial E_{I-II}}{\partial y} = \mathbf{F}^0(\lambda, x) + \mathbf{F}^1(\lambda, x)\mathbf{y} = \mathbf{A} \cdot \mathbf{y}$$

Therefore,

$$\mathbf{F}^0(\lambda, x) = [\mathbf{A} - \mathbf{F}'(\lambda, x)] \, \mathbf{y}$$

and

$$\mathbf{y} = \mathbf{G}\mathbf{F}^0(\lambda, x)$$

where

$$G = [A - F^1(\lambda, x)]^{-1} \qquad (12.34)$$

where **G** is now the perturbed Green's function, knowledge of which allows us to calculate the response of the crystal to the defect.

For more quantum mechanically based discussions of the Green's function we refer to the book by Stoneham[3]. Clearly the calculation of the function requires knowledge of the forces in the crystal. A detailed discussion is beyond the scope of the previous text; a review has been given by Pisani[40].

The method has been used extensively to study the properties of impurities in metals and semiconductors (see, for example, the work of Dederichs et al.[38], Lannoo[12] and Baraff et al.[43,44], Scheffler et al.[45] and the recent review of Zunger[45] who considers the special case of transition metal impurities). Related techniques have been applied to ionic systems by Scheffler et al.[40]. The method is powerful but may become unsuitable for defects where there is extensive relaxation or rearrangement of the surrounding lattice ions. For such systems, the Hartree–Fock cluster methods described earlier are probably more useful.

12.6 SUPERCELL METHODS

To date, there have been relatively few examples of the application of these methods, although the approach is attractive. Using any of the methods described in the previous sections, a supercell is set up containing a defect which is periodically repeated. If the geometry of the supercell is optimized, then lattice relaxation is included in an automated way. However, the computational demands are high as the unit cell used must be large.

An excellent discussion of the role and methodology of supercell calculations is given in a series of papers by Harker and Larkins[47,48,49], who show what modifications are needed to the Hartree–Fock equations in order to satisfy the translational periodicity of the supercell. Their work concentrated on the use of CNDO and INDO. Using modest-sized unit cells (whose sizes ranged from eight to 64 atoms), they were able to calculate cohesive energies, band gaps and band widths for carbon and silicon which were in reasonable agreement with experiment. They also extended their approach to ionic crystals, considering in detail the case of LiF. Using an Li_8F_8 supercell and a reparametrized CNDO set, they again obtained good agreement between calculated and experimental crystal properties. The results are presented in Table 12.6 where they are compared with experiment; a similar comparison is made for an $Li_{14}F_{13}$ cluster calculation. The results demonstrate in the words of Harker and Larkins "the great potential [of the technique] for the study of ionic solids with and without defects".

Table 12.6.
Properties of the LiF crystal using a CNDO method

	Parameter set		Experiment
	J1	J2	
LiF crystal			
Li_8F_8 supercell			
Internuclear distance (Å)	1.86	2.01	2.0
Cohesive energy (eV per atom pair)	12.3	10.8	8.7
Valence p bandwidth (eV)	4.4	4.9	6.1, 4.6±0.3
Direct band gap (eV)	21.6	21.8	14.2
Bulk modulus (10^{-11} dyn cm^{-1})	17.9	23.8	6.7
Ionicity	0.63	0.61	
$Li_{14}F_{13}^+$ cluster			
Internuclear distance (Å)	1.80	1.94	2.01
Cohesive energy (eV per atom pair)	10.7	9.5	8.7
Valence p bandwidth (eV)	2.9	3.0	6.1
Direct band gap (eV)	20.6	21.3	14.2
Ionicity	0.72	0.69	
LiF molecule			
Internuclear distance (eV)	1.51	1.65	1.51
Dissociated energy (eV)	6.5	5.7	6.6
Dipole moment (Debyes)	6.6	6.2	6.3
Ionicity	0.62	0.66	

Two different parameter sets were used. (After Harker and Larkins[48].)

Application of supercell methods to defects in metals and semiconductors is being made increasingly in conjunction with band structure techniques. Examples are given in references 47–49. To date, however, most calculations of this type have been confined to relatively small supercells (between eight and 16 atoms are typical), and further expansion in computer power will be needed for these methods to become routine and reliable. There is no doubt, however, that available computer power will continue to grow and that the use of supercell methods in defect studies will expand.

12.7 QUANTUM MECHANICAL METHODS—CONCLUSION

Quantum mechanical methods now have a major role to play in defect studies. The techniques are well established and, with modern supercomputers, many of the approximations that were used in earlier work have ceased to be necessary. Future work is likely to concentrate first on very large Hartree–Fock cluster calculations with a detailed treatment of lattice relaxation and on the supercell methods. There is no doubt that, with the type of computer power that is becoming available, the methods will become increasingly predictive.

REFERENCES

1 A. B. Lidiard, in N. H. March (ed.), *Orbital Theories in Molecules and Solids*, Oxford University Press, Oxford, p. 123 (1974).

2 F. Bassani, in *Theory of Imperfect Crystalline Solids*, International Atomic Energy Agency, Vienna, p. 265 (1971).

3 A. M. Stoneham, *Theory of Defects in Solids*, Oxford University Press, Oxford (1975).

4 A. M. Stoneham, *Philos. Mag. B.* **188**, 1611 (1985).

5 B. S. Gourary and F. J. Adrian, *Phys. Rev.* **105**, 1180 (1957).

6 R. H. Bartam, A. M. Stoneham and P. Gash, *Phys. Rev.* **176**, 1014 (1968).

7 M. H. Cohen and V. Heine, *Phys. Rev.* **122**, 1821 (1961).

8 A. H. Harker, *J. Phys. C: Solid State Phys.* **7**, 3224 (1974).

9 A. H. Harker, *D.Phil. Thesis*, Oxford University (1974).

10 C. A. Coulson and M. J. Kearsley, *Proc. R. Soc. London, Ser. A* **241**, 433 (1957).

11 G. D. Watkins, in P. Baruch (ed.), *Radiation Effects in Semiconductors*, Dunod, Paris, p. 97 (1965).

12 M. Lannoo, *Defects and Radiation Effects in Semiconductors; Inst. Phys. Conf. Ser.* **46**, 1 (1979).

13 M. Lannoo and J. C. Bourgoin, *Point Defects in Semiconductors: I. Theoretical Aspects, Springer Series in Solid State Science*, Vol. 22, Springer, Berlin (1981).

14 M. Lannoo, in A. M. Stoneham (ed.), *Current Issues in Semiconductor Physics*, Adam Hilger, Bristol, p. 27 (1986).

15 R. P. Messmer and G. D. Watkins, in J. W. Corbett and G. D. Watkins (eds), *Radiation Damage in Semiconductor*, Gordon and Breach, New York, p. 23 (1971).

16 J. A. Pople and D. L. Beveridge, *Approximate Molecular Orbital Theory*, McGraw-Hill, New York (1970).

17 P. W. Atkins, *Molecular Quantum Theory*, Oxford University Press, Oxford (1973).

18 A. Szabo and N. S. Ostlund, *Modern Quantum Chemistry*, Macmillan, London (1984).

19 R. P. Messner and G. D. Watkins, *Phys. Rev. Lett.* **25**, 656 (1970).

20 R. P. Messner and G. D. Watkins, *Phys. Rev. B* **7**, 2568 (1973).

21 F. P. Larkins, *J. Phys. C: Solid State Phys.* **4**, 3065 (1971).

22 A. B. Lidiard, in J. H. Whitehouse (ed.), *Defects in Semiconductors*, Institute of Physics, London, p. 228 (1973).

23 G. D. Watkins, R. P. Messner, C. Weigel, D. Peak and J. W. Corbett, *Phys. Rev. Lett.* **27**, 1573 (1971).

24 A. Mainwood, F. P. Larkins and A. M. Stoneham, *Solid-State Electron.* **21**, 1431 (1978).

25 V. J. B. Torres, P. H. Masri and A. M. Stoneham, *Mater. Sci. Forum* **10–12**, 73 (1986).

26 P. Masri, A. H. Harker and A. M. Stoneham, *J. Phys. C: Solid State Phys.* **16**, L613 (1983).

27 J. P. Hagon, A. M. Stoneham and M. Jaros, *Philos. Mag.* **55**, 211 (1987).

28 J. P. Hagon, A. M. Stoneham, and M. Jaros, *Philos. Mag. B* **55**, 225 (1987).

29 W. C. Mackrodt, D. Birnie and D. Kingry, in C. R. A. Catlow and W. C. Mackrodt (eds), *Advances in Ceramics* Vol. 23, in the press.

30 M. J. Mombourquette and J. A. Weil, *Can. J. Phys.* **63**, 1282 (1985).
31 R. W. Grimes, C. R. A. Catlow and A. M. Stoneham, *J. Phys. C: Solid State Phys.* in the press.
32 R. W. Grimes, C. R. A. Catlow and A. M. Stoneham, in C. R. A. Catlow and W. C. Mackrodt (eds), *Advances in Ceramics*, Vol. 23, in the press.
33 J. M. Vail, A. H. Harker, J. H. Harding and P. Saul, *J. Phys. C: Solid State Phys.* **17**, 3401 (1984).
34 P. Hohenberg and W. Kohn, *Phys. Rev.* **136**, B864 (1964).
35 N. H. March, in C. R. A. Catlow and W. C. Mackrodt (eds), *Computer Simulation of Solids, Lecture Notes in Physics*, Vol. 166, Springer, Berlin, p. 97 (1982).
36 C. S. Wang and W. E. Pickett *Phys. Rev. Lett.* **51**, 597 (1983).
37 M. Cohen, *Materials Research Society Symposium Proceedings*, Materials Research Society, Pittsburgh, PA (1986).
38 P. H. Dederichs, R. Zeller, H. Akai, S. Blügel and A. Oswald, *Philos. Mag. B* **51**, 137 (1985).
39 J. Callaway and N. H. March, in H. Ehrenreich and D. Turnbull (eds), *Solid State Physics* **38**, 136 (1984).
40 M. Scheffler, J. P. Vigneron and G. B. Bachelet, *Phys. Rev. B.*, **31**, 6541 (1985).
41 C. Pisani, *Philos. Mag. B* **51**, 89 (1985).
42 S. T. Pantelides, *Rev. Mod. Phys.* **50**, 797 (1978).
43 G. A. Baraff, E. O. Kane and M. Schlüter, *Phys. Rev.* **B21**, 5662 (1980).
44 G. A. Baraff and M. Schlüter, *Phys. Rev.* **B33**, 7346 (1986).
45 M. Scheffler, J. Bernholc, N. O. Lipari and S. T. Pantelides, *Phys. Rev.* **B29**, 3269 (1984).
46 A. Zunger, in H. Ehrenreich and D. Turnbull (eds), *Solid State Physics* **39**, 276 (1986).
47 A. H. Harker and F. P. Larkins, *J. Phys. C: Solid State Phys.* **12**, 2487 (1979).
48 A. H. Harker and F. P. Larkins. *J. Phys. C: Solid State Phys.* **12**, 2497 (1979).
49 A. H. Harker and F. P. Larkins. *J. Phys. C: Solid State Phys.* **12**, 2509 (1979).

13
Statistical Mechanical Models

13.1 INTRODUCTION

In this chapter, we introduce the reader to the more sophisticated type of statistical mechanical models that have been used in attempting to construct thermodynamic descriptions of defective crystals. The mass action approximation described in Chapter 2 and elsewhere is essentially based on an ideal solution theory and as such is limited to dilute defect solutions. The development of models for more heavily defective systems has proved difficult. The related field of alloy thermodynamics has drawn heavily on statistical mechanical models developed for liquids, e.g. regular solution theory. Such models are less appropriate for disordered ionic solids which have borrowed considerably from the theory of electrolyte solutions. In the present chapter, we therefore first recall the basic statistical mechanical principles which lead to the ideal solution formalism; the simplest elaboration, namely the description of activity coefficients using the Debye—Hückel theory, is then described. More elaborate theories based on cluster expansion and site exclusion techniques are then considered. We conclude with a fascinating case study, namely the properties of the heavily defective system $U_{1-y}Pu_yO_{2-x}$.

13.2 IDEAL SOLUTION THEORY

The ideal solution model assumes that we may write the variation in the free energy $G(x)$ of the system as a function of the molar concentration x of defects according to

$$G(x) = G_0 + xg_0 + RT[x \ln x + (1 - x) \ln(1 - x)] \quad (13.1)$$

where G_0 is the free energy of the hypothetical perfect crystal and g_0 is a "standard" free energy of defect formation, i.e. the defect free energy excluding the configurational entropy contribution which is expressed by the third and fourth terms on the right-hand side of equation (13.1) and may be derived by the following simple statistical mechanical arguments. Let us consider defects permuted over N sites. The total probability of this system is given by the simple expression

$$W = \frac{N!}{n!(N - n)!} \quad (13.2)$$

However,

$$S = k \ln W = k[N \ln N - n \ln n - (N - n) \ln(N - n)]$$

which on rearrangement and substitution of x for n/N and of R for kN gives

$$S = R[x \ln x + (1 - x) \ln(1 - x)] \quad (13.3)$$

which is, of course, the ideal entropy of mixing. From the free-energy expression the effective defect chemical potential μ_D can be derived by differentiation. This will have the form

$$\mu_D(x) = \mu_D{}^0 + RT \ln x \quad (13.4)$$

where $x \ll 1$. Use of such expressions is, of course, the basis of the mass action theory of defect equilibria which is the basis of most statistical mechanical theories of defect equilibria.

It is important to enumerate the conditions under which such expressions hold. The most important are as follows.

(1) Interactions between the defects are negligible.
(2) The defects are randomly distributed.

We note that restriction (1) may be modified in that defect clustering is allowed within the framework of the ideal solution theory, provided that the clusters themselves can be described in terms of random non-interacting defects.

Both these assumptions break down; indeed the more sophisticated treatments discussed below are based on various attempts to deal with the failure of the two approximations.

13.3   THE DEBYE–HÜCKEL THEORY

This theory, which has been imported directly from the theory of electrolyte solutions, treats the effects of the long-range Coulomb interactions on deviations from ideality. As discussed in many standard physical chemistry textbooks (see, for example, those by Atkins[1] and Moore[2], the Debye—Hückel theory is based on the concept of the ionic atmosphere in which ions (in this case defects) are surrounded by a "cloud" of species of opposite (effective) charge. The theory proceeds by combining the Poisson equation with the Boltzmann distribution function. The latter may then be approximated by a linear expression at a sufficiently high dilution; from solution of the linearized Poisson—Boltzmann equation, it is possible to write the defect activity coefficient γ_{D-H} as

$$\log(\gamma_{D-H}) = \frac{-q^2}{2\epsilon kT} \frac{\chi}{(1 + \chi r)} \qquad (13.5)$$

where

$$\chi^2 = \frac{8\pi q^2 x}{V\epsilon kT} \qquad (13.6)$$

in which q is the charge of the ion, x is the ion concentration, V is the molar volume of the crystal, ϵ is the dielectric constant and r is the distance of closest approach of the defects.

It is well known that the Debye—Hückel theory is inappropriate for all but very dilute solutions. Indeed for 1:1 electrolyte solutions the upper limit of applicability is normally taken as 10^{-3} M. Nevertheless the theory is still applied very often to more concentrated defect solutions (see, for example, the work of Corish and Jacobs[3,4]). It is certainly the simplest improvement on ideal solution theories. However, there is no doubt that a satisfactory theory of all but very dilute solutions must go beyond this approximation.

13.4 CLUSTER METHODS

The basis of these methods is to consider explicitly clustering involving the formation of ion pairs, triplets, quartets, etc. The method is more generalized than the simple mass action theory of defect clustering and can include effects arising from interactions between defect clusters, for example. The basis of the method is due to Mayer[5] who for electrolyte solutions expresses the activity coefficients using the Debye—Hückel limiting law plus a power series in the ionic concentrations. The coefficients are cluster integrals and in the treatment given in reference 6 the cluster is defined in terms of the distance of closest approach. If a pair of ions i and j are closer together than d, they are members of the same cluster. A third ion k may form a component of the cluster provided that it is closer to i or j (but not necessarily both) than d. The activity a_α of a given cluster α can be written as

$$a_\alpha = Z_\alpha K_\alpha \qquad (13.7)$$

To explain the two terms in equation (13.7), we assume that α consists of K_1 point defects of type 1, K_2 of type 2 and K_σ of type σ. We then denote the free energy of formation of the cluster within the perfect crystal as A_α. Z_α and K_α are then given by

$$K_\alpha = \exp(-\beta A_\alpha)\Big/ \prod_{s=1}^{\sigma} K_s! \qquad (13.8)$$

$$Z_\alpha = \prod_{s=1}^{\sigma} [\exp(\beta \Delta s)]^{K_s} \qquad (13.9)$$

where Δs is the change in chemical potential on forming.

Evaluation of the terms in the above equation is non-trivial. For a good introductory discussion, we refer the reader to the work of Allnatt and Loftus[6]. Simple treatments only include pair clusters and enable expressions to be derived for activity coefficients which perform better at a higher concentration than the simple Debye–Hückel theory does.

Allnatt and Loftus considered the effects of allowing the expansion to include terms due to three and four clusters. They showed that, for concentration of defects much greater than 10^{-3} in, for example, cadmium-doped AgCl, triplet or larger clusters made significant contributions to the activity coefficients.

Cluster models are unquestionably of value in describing heavily defective systems. However, the models described above are most appropriate when the attractive forces between the defects are our main concern. The models described in section 13.5 include explicitly the effects arising from defect repulsions.

13.5   SITE EXCLUSION METHODS

The aim of these methods is to include in an explicit manner the effects of defect–defect repulsions. The simplest way of proceeding is to define an envelope of sites surrounding the defect which exclude occupation by like defects; the excluded volume might, for example, consist of the nearest-neighbour sites. Such models are, of course, crude. They imply an infinite defect–defect repulsion energy for defects within the excluded volume, whereas outside the volume the repulsion is effectively zero.

A more realistic approach was suggested by Atlas[7] who developed a model in which there are a range of defect–defect repulsive energies depending on the defect spacing. A discrete range of repulsive energies U_0,

..., U_i are defined depending on the range of possible defect repulsions. Atlas considered the specific case of the reduced fluorite oxides, e.g. CeO_{2-x}, which contain both vacancies and reduced cations. For the system he wrote down the following partition function Q_D of the defective crystal:

$$Q_D = \sum_i \Omega_{ve(i)}(n_j, m_k) \frac{q'(N_a - N_v)}{q(N_a)}$$

$$\times \exp - \{(N_v \bar{e}_f + \sum_j n_j U_j + \sum M_k \epsilon_k + N_{ve}(N_v) \bar{e}_{ve})/kT\} \qquad (13.10)$$

In this equation, $\Omega_{ve}(n_j, m_k)$ is the configurational degeneracy of the defective crystal, and $q'(N_a - N_v)$ and $q(N_a)$ are the vibrational partition functions of the ions in the defective and perfect crystals, respectively. N_v is the total number of vacancies and $\bar{e}_f$ their mean energy. N_j and U_j are the number of vacancies in each subset each of repulsive energy U_j. M_k is the number of reduced cation in each subset of repulsive energy ϵ_k. N_{ve} (N_v) is the total number of reduced cation–vacancy pairs and $\bar{e}_{ve}$ is the mean reduced cation–vacancy attraction energy.

The development of the Atlas theory is mainly concerned with simple evaluations of the values of U_i and ϵ_i, and with parametrization or evaluation of the other terms in the partition function. The theory can achieve an impressive measure of agreement with experiment. Figure 13.1–13.3 show calculated and experimental data for CeO_{2-x}.

The method is therefore of proven success for heavily defective materials, but its applicability may be limited by the need for extensive parametrization.

13.6 STATISTICAL MECHANICAL MODELS FOR THE MIXED OXIDE $U_{1-y}Pu_yO_{2-x}$—A CASE STUDY

The non-stoichiometric mixed oxide formed by dissolving PuO_2 into the isostructural UO_2 is an important technological material as it is used as the fuel in breeder reactors. It is also of very considerable interest theoretically owing to the complexity of the defect structure and the availability of a wide range of data on the variation in oxygen partial pressure p_{O_2} with x and y. The aim of the theories discussed in this section are to reproduce this variation.

The mixed oxide tends to lose oxygen according to the equation

$$(UPu)O_2 \rightarrow (UPu)O_{2-x} + \frac{x}{2}O_2(g) \qquad (13.11)$$

the simplest defect model of which is

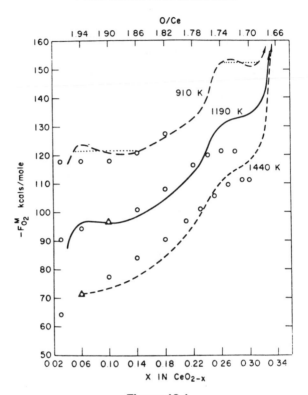

Figure 13.1

Calculated and experimental variation in the free energy $F_{O_2}{}^M$ of oxidation as a function of x in CeO_{2-x}, after Atlas[7], using experimental data from Bevan and Kordis[8], ($\bigcirc$): $\triangle$, points used to compute theoretical curves.

$$2Pu^{4+} + O_O \rightarrow 2Pu^{3+} + V_O^{\cdot\cdot} + \tfrac{1}{2}O_2(g) \qquad (13.12)$$

where we assume that the plutonium is preferentially reduced owing to the higher 4+ ionization potential of plutonium compared with uranium. Clusters might be expected between the oxygen vacancy and the Pu^{3+} ion. The simplest models of the system[9,10] proposed the formation of a simple substitution vacancy cluster, giving for the reduction reaction

$$\begin{array}{cccc} 2Pu^{4+} & + O_O{}^{2-} \rightarrow (Pu^{3+}-V_O) + & Pu^{3+} & + \tfrac{1}{2}O_2 \\ (y-2x) & x & x & p_{O_2} \end{array} \qquad (13.13)$$

where the concentrations of each species are given under the equation as a mass action treatment; we then obtain

$$\frac{x}{y-2x} \propto p_{O_2}{}^{-1/4} \qquad (13.14)$$

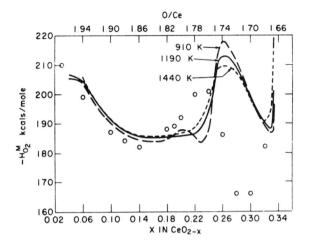

Figure 13.2.
Calculated and experimental enthalpy $H_{O_2}{}^M$ of oxidation in CeO_{2-x}: O, data at 1353 K from Bevan and Kordis[8].

Although this model has some limited success for a restricted range of data, it is not generally applicable; in particular the $p_{O_2}{}^{-1/4}$ behaviour does not fit a wide range of data. Models of the type discussed in section 13.5 were developed[11], with improved reproduction of the data, but with large numbers of variable parameters. Recent attempts have been made to derive a less-parametrized model. The first is the simple random-plutonium model proposed by Catlow and Tasker[12]. In this model the plutonium ions are considered to be distributed at random. The theory focuses on the anion vacancies and classifies anion sites according to the number of plutonium ions in the nearest- and next-nearest neighbour sites. This can be written using simple combinatorial expressions. Thus, if i is the number of ions in the nearest-neighbour site (for which the maximum is 4) and j is the number in the next-nearest-neighbour site (maximum, 12), we can write for the fraction P_{ij} of anion sites with this distribution

$$P_{ij} = (2x)^{i+j}(1 - 2x)^{16-i-j}\frac{4!}{i!(4 - i)!}\frac{12!}{j!(12 - j)!} \qquad (13.15)$$

The next step is to calculate the energies of the oxygen vacancy in each of the sites. This can be done using the simulation methods described in Chapter 11. Using these energies, it is possible to calculate the number of vacancies in each site readily. Since only one vacancy can occupy each site, Fermi−Dirac statistics hold, i.e. we can write

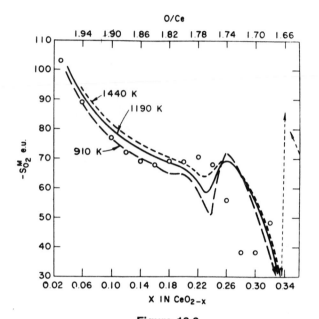

Figure 13.3.
Calculated and experimental entropy $S_{O_2}^M$ of oxidation in CeO_{2-x}: O, data at 1353 K from Bevan and Kordis[8].

$$n_{ij} = \frac{2NP_{ij}}{1 + \exp(\epsilon_{ij} - \epsilon_0)/kT} \qquad (13.16)$$

where ϵ_{ij} is the binding energy of the vacancy to site i, j and ϵ_0 is the "Fermi energy" which may be defined by the normalization condition

$$\sum_{ij} n_{ij} = Nx. \qquad (13.17)$$

Enthalpy and entropy terms are then evaluated in the former case by straightforward summation, and in the latter by combinational expressions using the relationship $S = k \ln W$.

This approach enjoyed some modest success, but the reproduction of experimental data over a wide range of y and x values is not impressive. The model has been developed, however, by Harding and Pandey[13], who introduced a site-blocking model such that each vacancy "controls" a number λ of surrounding sites which may not be occupied by another vacancy. Harding and Pandey carefully examined the question of the distribution of the electrons which are created by reduction over the plutonium sites; in contrast, Catlow and Tasker[12] assumed a random distribution of reduced cations. The Harding–Pandey model preferentially localizes the electrons

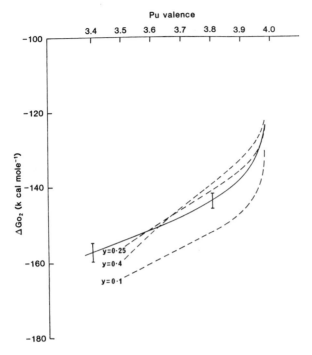

Figure 13.4.
Calculated (— — —) and experimental (———) variation in ΔG_{O_2} for oxidation of
$UPuO_{2-x}$ plotted against the degree of reduction expressed as mean plutonium
valence. The results taken from the bound-electron model are discussed by
Harding and Pandey[13].

that are created on reduction on sites that are either nearest or next-nearest
neighbour to the vacancies; electrons that cannot be so accommodated are
then distributed at random as in the simpler model.

The Harding–Pandey model therefore improves on the preceding theory
by including explicitly vacancy–vacancy interactions and by more accurately describing the interaction between reduced cations and oxygen
vacancies. A detailed survey of this theory is beyond the scope of this book.
Figure 13.4, however, gives a comparison between calculated and experimental oxygen potentials. The agreement is encouraging.

13.7 SUMMARY

In the study of heavily defective materials, it is clearly essential to extend
theories beyond the ideal solution model or the simplest correction to such

models given by the Debye–Hückel theory. Cluster and site exclusion modes contain important physical content. However, there is greater promise, we believe, in the type of model described in section 13.6 where calculated defect energies are used to input to statistical mechanical models. In addition the Monte Carlo simulation techniques of the type discussed in Chapter 11 are likely to play an increasing role in statistical mechanical studies of complex defects.

REFERENCES

1 P. W. Atkins, *Physical Chemistry*, Oxford University Press, Oxford (1976).
2 W. J. Moore, *Physical Chemistry* (5th edition), Longmans, Harlow (1974).
3 J. Corish and P. W. M. Jacobs, in M. W. Roberts and J. M. Thomas (eds), *Surface and Defect Properties of Solids*, Vol. 2, Chemical Society, London, p. 160 (1973).
4 J. Corish and P. W. M. Jacobs, in M. W. Roberts and J. M. Thomas (eds), *Surface and Defect Properties of Solids*, Vol. 11, Chemical Society, London, p. 218 (1977).
5 J. E. Mayer, *J. Chem. Phys.* **18**, 1426 (1950).
6 A. R. Allnatt and E. Loftus, *J. Phys. (Paris), Colloq.* C9 395 (1973).
7 L. M. Atlas, *J. Phys. Chem. Solids* **29**, 91 (1968).
8 D. J. M. Bevan and J. Kordis, *J. Inorg. Nucl. Chem.* **26**, 1509 (1964).
9 F. Schmitz, *J. Nucl. Mater.* **58**, 357 (1975).
10 C. R. A. Catlow, *J. Nucl. Mater.* **67**, 236 (1977).
11 L. Manes and B. Manes-Pozzi, in *Plutonium and Other Actinides*, North-Holland, Amsterdam, p. 145 (1975).
12 C. R. A. Catlow and P. W. Tasker, *Philos. Mag.* **48**, 649 (1983).
13 J. H. Harding and R. P. Pandey, *J. Nucl. Mater.* **125**, 125 (1984).

14
Applications

14.1 SELECTION OF EXAMPLES

Defect studies have a long and distinguished academic history which has developed in parallel with more pragmatic approaches for a wide variety of techniques. Progress has been made not only in the detailed understanding of defect structures but also in the associated methods of measurement and theoretical analysis. For example, such familiar experimental techniques as optical absorption, electron paramagnetic resonance, deep-level transient spectroscopy, and electron microscopy of electronic measurements were often developed directly to satisfy the needs of defect analysis. Academic curiosity has focused attention on the more tractable defect studies such as the colour centres in alkali halides whereas by contrast the technologically driven efforts have frequently been concerned with poorly characterized materials. Although success in generating improved materials may be empirical, subsequent understanding may allow improved intuition for the future. A vast list of defect-controlled applications could be given with examples from the Bronze Age to high-temperature superconductors, from electron emission in valve filaments to large-scale integrated circuits in silicon or GaAs, from colour centres in gem stones to optical fibre lasers or from surface corrosion to new selective catalysts. With such a formidable array of examples to discuss, it is mandatory that a book concerned with defect properties offers a few specific examples. The choice of examples is somewhat arbitrary and in this case the selection is unrelated to commercial importance but has been made purely on the basis of our own research interests.

Despite this narrow viewpoint, the range of examples includes items of current and future technological interest with major commercial possibilities as well as areas with a limited appeal (e.g. only to archaeologists). In general, it will be apparent that in many applications we have a broad appreciation of the defect processes but lack an in-depth knowledge of specific defects. This level of understanding may not be ideal for a text, but it is intellectually challenging as it requires a broad perspective of the possibilities and need not inhibit progress in marketing of the products. The classic example of development and usage being followed only later by the detailed defect study was the introduction of photography.

The following list of applied topics which are defect dominated and for which we have some degree of understanding are thermoluminescence dosimetry, ion beam implantation, colour centre lasers, photorefractive devices, chemistry of nuclear fuels and sensors.

14.2 THERMOLUMINESCENCE DOSIMETRY APPLICATIONS

The release of energy in the form of luminescence and the sensitivity of photon detection makes thermoluminescence an attractive method for measurement of small quantities of stored energy. In particular, this is used to measure radiation history of samples. The two major areas are the age determinations of archaeological or geological samples and personnel dosimetry.

14.2.1 Personnel dosimetry

For personnel dosimetry the most common material is based on LiF as it has a stopping cross-section for electrons and X-rays similar to body tissue, it is compact and of low cost, it is simple to read and, in principle, it can be optically re-excited if an initial measurement is uncertain. After high-temperature processing, the material is re-usable. These are desirable features and have caused it to supersede photographic film dosimetry.

Pure LiF is not particularly sensitive but, with hindsight, we can now justify the choice of impurities and thermal processing which leads to the material used. Three dopants are involved in the material, namely magnesium (at 100 ppm), titanium (about 15 ppm) and oxygen (in uncertain amounts). The metals are added during growth but the oxygen is not normally well controlled. Both the Mg^{2+} and the Ti^{4+} ions can substitute onto Li^+ sites as their relative ionic radii are 0.65 Å, 0.68 Å and 0.70 Å, respectively. To allow for charge compensation, the Mg^{2+} ions associate with a Li^+ vacancy to form a dipole whereas the Ti^{4+} forms a complex with three neighbouring O^{2-} ions substituted on F^- sites. Once again there is a good match between the ionic sizes for the substitution, with O^{2-} and F^- radii being 1.40 Å and 1.33 Å, respectively. In some cases OH^- substitutions are also considered. The defect complexes provide both charge compensation and minimal lattice strain, and their models are sketched in Figure 14.1. To achieve a uniform distribution of the magnesium, the material is heated to at least 400°C and rapidly cooled to freeze in the $Mg-Li_{vac}$ dipoles. After irradiation there is charge trapping at the magnesium sites. On subsequent heating this leads to a glow curve of the form shown in Figure 14.2 in which peaks 2 and 3 are probably associated with dipoles and peaks 4 and 5 with a higher cluster of trimers. It appears that the trimers develop during the thermoluminescence heating cycle, even for heating rates 2°C s^{-1}. Very high heating rates (greater than 100°C s^{-1}) inhibit the trimer formation because diffusion processes are too slow compared with the electronic charge transfer.

For dosimetry the low, i.e. 120°C, peaks are unsuitable as they will fade with room-temperature storage. Thus, it is preferable to convert all the dipoles into trimers. This is achieved by a heat treatment at about 100°C for 24 h before the dosimeter is irradiated. The thermoluminescence dosimetry is then based on the 200°C peak.

The model for the dosimeter is that the $Mg-Li_{vac}$ trimers trap charge which when released causes emission at the $Ti-O$ complexes. To optimize the sensitivity, the impurity content must be increased but by levels of, say, 400–1000 ppm Mg, the magnesium enters into a Suzuki phase (see section 5.4) rather than into the trimers. Thus, 100 ppm Mg probably generates less than 30 ppm of trimers as some magnesium may be in the form of dipoles and some locked in the Suzuki phase. Therefore the concentration of trimers may be matched by the 15 ppm Ti. The dosimeter has excellent linearity for doses from millirads up to some 10^4 rad (10^2 Gy) and, allowing for a supralinear response, could be used to 10^5 rad (10^3 Gy). The lower end

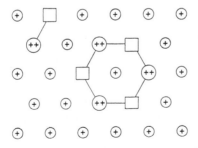

Figure 14.1.
Possible magnesium defect sites in LiF. The figure shows a (111) plane of lithium ions with an Mg^{2+} ion substituted next to an Li^+ vacancy and a cluster of three such defects.

POINT DEFECTS IN MATERIALS

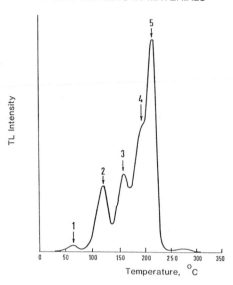

Figure 14.2.
A typical thermoluminescence (TL) curve from a LiF: Mg dosimeter.

of the scale is ideal for personnel dosimetry. The upper end of the scale is of use for laboratory irradiations. Overall, some 0.05% of the incident radiation is re-emitted as light energy.

From a pragmatic viewpoint this is an efficient dosimeter but nevertheless there are still many unresolved features of the defect behaviour. For example it is not directly obvious how the release of an electron from the magnesium site can stimulate luminescence by hole capture at the titanium site. Further, there is a strong suspicion that the two sites are not independent; for example the emission spectra of LiF:Ti is modified up to 100°C in the case of LiF:Ti:Mg samples. Less direct evidence comes from a study of the kinetics. Peak 5 exhibits a classical first-order thermoluminescence shape but the activation energy and pre-exponential factor are unreasonably high (e.g. 2.06 eV and 5×10^{20} s^{-1}). Various explanations have been proposed including a direct association of the two centres in which charge transfer between the magnesium and titanium sites operates in two stages. If each proceeds with values of E and S the model combines them in the observed kinetics as $2E$ and S^2, i.e. the curve shape is still first order but the parameters need a new interpretation.

Academic studies may refine the model but empirically the material is already adequate for personnel dosimetry.

14.2.2   Archaeological age determinations

In the ideal situation the dosimeter is initially set to zero and then ionization produces some trap filling at a constant rate so that a measurement of the number of filled traps (by thermoluminescence), the irradiation dose rate and the efficiency of trap filling directly gives the age from

$$\text{age (years)} = \frac{\text{measured thermoluminescence}}{\text{dose rate (rad/year)} \times \text{efficiency (i.e. thermoluminescence/rad)}}$$

For pottery the first stage of emptying the shallow traps occurs during firing of a pot at some $600-800°C$. Trap filling from radioactive components in the clay (uranium, thorium, potassium, etc) is relatively constant as they have long half-lives and the decay can be accounted for. Laboratory estimates of the background irradiation rate are possible and, in approximate terms, we expect the sum of the internal irradiation from the clay of the pottery plus background materials (e.g. for buried artefacts) and cosmic rays to provide about 1 rad year^{-1}. The efficiency term can also be estimated by noting the change in signal from a known laboratory irradiation. Figure 14.3 shows how the age determination might operate. Curve B is the accumulated thermoluminescence between the firing of the pot and the present. Curve A is for a known laboratory dose of say 1000 rad, and curve C is the black-body radiation. The laboratory irradiation also produces short-lived signals at low temperatures. At higher temperatures the signals should be parallel in curves A and B (the so-called plateau test). If this is so and the ratio of the signal on curve A to that on curve B is 2, then for a dose rate of 1 rad year^{-1} the pottery is 500 years old.

Although the principle is trivial, in practice considerable care must be taken and correction factors introduced. Nevertheless there are many examples where the thermoluminescence ages agree well within 10% with more classical archaeological dates. The major difficulty is in extracting a suitable sample from the pottery. The normal practice is to use a few milligrams of quartz separated out of the clay matrix. Quartz tends to be "pure" with few impurities and the key irradiation sources are in the clay. It responds linearly over a wide dose range and dose rates. The latter point is important as the natural and laboratory irradiation rates may differ by at least 10^6. Measurements using unseparated minerals within the pottery are unreliable because of the wide range of minerals that would be used.

In addition to age measurements the thermoluminescence of ceramics is ideal for distinguishing between original and reproduction items (i.e. fakes). Because of the high costs associated with older items, many ceramics are routinely tested by thermoluminescence dating. There is thus the interesting

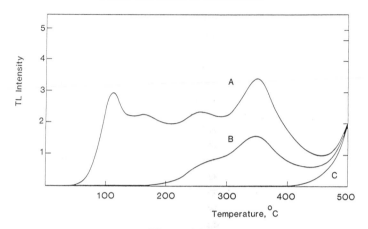

Figure 14.3.
Examples of thermoluminescence (TL) curves used in archaeological dating: curve A, after laboratory irradiation; curve B, the naturally induced signal; curve C, black-body background.

solid-state problem of how to generate fakes which pass the thermoluminescence test. At first sight this is simple and a γ-ray irradiation followed by a low-temperature anneal should leave the appropriate thermoluminescence signal. However, if the authenticity is in doubt, a comparison of the thermoluminescence from large quartz grains before and after etching off the outer few micrometres will determine the true age. The β and γ dose will be uniform throughout the grains but the outer skin will be subjected to additional short-range radiation from α-irradiation. This cannot be reproduced for individual grains, and modern forgers then resort to an alternative method.

For recent items (e.g. less than 100 years) a variation of the thermoluminescence method is to use the change in thermoluminescence sensitivity of the 110°C quartz peak to small doses. The method is frequently used but the physical model used to justify it is very contentious.

14.2.3   Geological dating

Geological applications are more difficult because except for volcanic material, aboriginal hearthstones, or minerals such as stalagmites or flint which grow from seed crystals there is no obvious method of initializing the thermoluminescence clock. In the case of wind-blown silts (Aeolian loess) it is assumed that exposure to sunlight bleaches the shallow traps and these only refill once the material is covered. Several uncertainties exist; for example the bleaching may occur in the upper atmosphere at -40°C or on the ground at ambient temperatures and the spectrum of the sunlight is different in the two cases. Both features influence the degree of bleaching and the subsequent response to ionizing radiation. *In situ* dosimetry problems are also influenced by a changing water content which acts as a radiation shield. There are further complications that the thermoluminescence may saturate over the long time scale or, as with LiF:Mg:Ti, defects may aggregate by thermal diffusion. This possibility must be considered for samples which have ages of 10^5 years or more.

Dating is similar to that for archaeological samples but for materials such as loess the level of thermoluminescence which would have survived the optical bleaching stage during deposition must also be determined. For loess-type materials, on-site measurements of the natural radiation level are required plus laboratory experiments to determine the bleaching effects and thermoluminescence sensitivity to irradiation dosage. Routes such as those shown in Figure 14.4 can be used. In Figure 14.4(a) the non-bleachable level and the linear section of the natural signal plus additive laboratory doses $(N + \beta)$ are determined. Extrapolation back to the level of the non-bleachable line (I_0) then gives the equivalent dose ED and hence an age estimate. Figure 14.4(b) shows that after the $(N + \beta)$ dose bleaching for times t_1, t_2, etc., should lead to a set of lines which all lead back to the same intercept. A third approach, (Figure 14.4(c)) shows that signal regeneration after bleaching could provide the equivalent dose value ED. In many cases the approaches agree but there are also examples where bleaching changes the sensitivity of the thermoluminescence response. Despite the potential difficulties many samples are well behaved and numerous studies of the ages of loess or river silts now exist, extending back over 200 000 years.

14.2.4 Other thermoluminescence applications

A number of more limited thermoluminescence applications are very briefly mentioned. By noting the changes in the emission spectra or number of thermoluminescence peaks it is feasible to identify the source of particular minerals since impurity variations will be related to particular sites. This approach has been attempted in order to follow the distribution of minerals along early trade routes (e.g. obsidian arrowheads, zircon, etc.). A variation on the same theme is to retrace the path of minerals along river beds or flood plains to determine the source. Potentially this is worthwhile in the case of gold-bearing quartz veins and it has recently been attempted in uranium prospecting.

In buildings in which there has been a fire the residual natural thermoluminescence of the heated concrete or brick will indicate the temperature and duration of the heating. Structural strength can then be assessed without the need to remove large sections of the building (a particularly difficult problem with pre-stressed concrete structures).

(a)

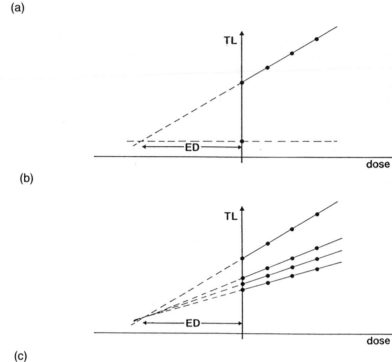

(b)

(c)

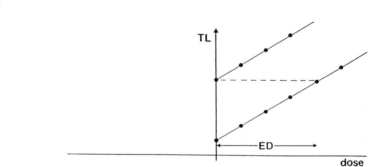

Figure 14.4.
Strategies used in sediment thermoluminescence (TL) dating to obtain the equivalent dose and the limit of optical bleaching: (a) additive dose; (b) partial bleach; (c) regeneration.

14.3   APPLICATIONS OF ION IMPLANTATION

Control of the behaviour of the surface layers of a material is a key step in a diverse range of modern technologies and ion implantation is one method for producing the control. It is convenient to consider that by "surface layer" we mean the outer few micrometres of the material; so, in order to gain a perspective of the problems, we briefly list the areas where surface properties are crucial. Foremost among these are semiconductor devices, as almost without exception the electrically interesting regions are close to the surface. Even where the devices contain many complex layers, each stage is in close proximity to the surface. By contrast, metal structures tend to be more massive in order to provide mechanical strength, but again surface properties are important as they will frequently be subject to corrosion, frictional forces between moving parts and mechanical wear. It may also be recalled that intrinsic bond strengths of metals far exceed the observed practical strengths of materials. This difference is strongly influenced by dislocations which can be generated at the surface (i.e. after machining or thermal treatment).

For insulating materials the same mechanical criteria apply but, in addition, optical properties, such as refractive index control are used in lens blooming or coatings for optical fibres. In integrated optics the methods of optical signal encoding are still under development, but all the signs are that the techniques will be controlled by surface layer properties. Surface sites are equally important in chemical engineering in work on corrosion or catalysis. Chemical features of ion implantation include changes in chemical composition or reactivity which may act as pre-cursors for enhanced lithographic etching or image recording. Further, ion implantation can enhance bonding between different layers including metal−organic, metal−insulator and metal−metal interfaces.

One of the simplest and most direct methods of controlling physical and chemical properties of surfaces is to inject impurity ions using an ion implanter (i.e. an ion accelerator). The concept is appealing because an ion accelerator produces a readily measured flux of particles and the injected ions can be placed at controlled depths beneath the surface by adjustment of their energy. Major advantages of the approach are that there are only minor constraints on the choice of the injected ion, different ions may be placed at different depths without perturbing the other impurities, only selected regions of the surface need be modified, and the processing can proceed at a selected temperature and need not be a high-temperature treatment.

For a new processing technique such as ion implantation to become accepted, it must provide properties which were not previously obtainable, improved quality control and/or lower costs. Development of semiconductor devices inevitably incorporated ion beam techniques as all the above advantages were apparent. The electrical properties require controlled additions of small (e.g. 10^{12} ions cm^{-2}) quantities of impurities in well-defined regions, and many processing steps with integrity of earlier stages. The resultant devices are small, of high intrinsic cost per surface area and have a large world market; so the cost of precision ion implanters is easily justified. Hence the use of ion implantation has been incorporated into semiconductor chip production and is now an indispensable stage in circuit manufacture.

In metallurgy there are few examples of comparable high value per unit area; so applications are commencing where we can use low-cost high-current implanters, which may lack current accuracy or isotope selectivity but are capable of implanting large surface areas (e.g. aeroplane engine components). Improvements in mechanical properties may be spectacular in a few cases, with a reduction in mechanical wear by a factor of 100 quoted as an outstanding example but, in general, the effects are more modest and friction or wear properties normally only improve by a factor of $2-10$. The economic argument for using ion-implanted components is thus less obvious and the entire manufacturing and operational costs must be appraised. Improvements of working lifetimes for inaccessible bearing components are clear examples; nevertheless if the lifetime of a cutting tool such as a thread tap is improved by a factor of 3 and the tap costs three times as much, the new technique might be ignored. However, a longer lifetime may also provide a higher tolerance for more operations and, more importantly, fewer operating breaks whilst the tap is changed by a machinist. Overall there is a significant cost advantage to be gained by the use of the ion-implanted tool.

There has been a transition from flexible research machines to dedicated industrial implanters which are fully incorporated into semiconductor production. For metallurgical uses, simple large-scale implanters are still being developed but there are now numerous commercial success stories; so we can confidently predict that implantation will become a routine feature of metallurgy during the next decade. Applications of ion implantation to insulators are less advanced but comparisons with other areas suggest that the situation will change, particularly in the development of devices for integrated optics.

Before discussing specific examples of applications, it may be convenient to summarize the advantages offered by the ion beam approach for control of the properties of surface structures. These include the following.

(1) Controllable additions of impurity ions.
(2) Ion energy control of the ion range.
(3) Multienergy implants which offer a depth tailored profile.
(4) Small lateral spread of the implanted ions.
(5) Ion ranges which are not sensitive to dislocations (by contrast with thermal diffusion).
(6) Implanted species which do not need to be in thermodynamic equilibrium.

(7)   Control of processing temperature.
(8)   Self-alignment of the implant with surface masks and/or contacts.
(9)   Implantation through existing layers or surface barrier oxide.
(10) Thermally unstable compound semiconductor faces which may be implanted via a protective barrier.
(11) Formation of surface or buried oxide isolation.
(12) Use of pulse annealing to give "solid-phase epitaxy" without perturbation of other regions of the crystal.
(13) A reduced number of wet chemical steps in complex component delineation.
(14) High vacuum cleanliness which is needed for circuit miniaturization.
(15) High vacuum compatibility of implantation and molecular beam epitaxy.
(16) Crystal amorphization (which requires relatively low ion doses for heavier implanted ions).
(17) Structures which could not be prepared by diffusion treatments.
(18) Accurate control of resistive circuit elements by both metallic and silicide layers.

The only obvious difficult in using ion implantation to form semiconductor devices is the problem of radiation damage. Present annealing techniques alleviate the problems and in some instances, e.g. in amorphization, the radiation damage may have positive features.

14.3.1   Ion implantation in silicon

A wide spread of ion doses is used to control electrical properties with injected ion doses as low as 10^{10} ions cm^{-2} (i.e. equivalent to about 10^{15} impurities cm^{-3}). Such implanted impurity doping levels are used in the conducting channel of a metal−oxide−semiconductor field effect transistor. A difficulty encountered with this low-dose example is that ion channelling may make a significant contribution. In more recent years an alternative approach has been to form an amorphized silicon layer, to implant the impurity and then to anneal the radiation damage. Amorphization of silicon with large implantation dose can in principle be achieved by self-ion (i.e. Si$^+$) bombardment but the damage path can vary from being a trail of separate point defects for light ions to gross collision cascades with heavy particles. With the lightest ions (e.g. B$^+$), it is not possible to amorphize silicon but, for heavier implants, some 12 eV per target ion needs to be delivered from the collision cascades. At room temperature with typical ion energies, amorphization requires almost 5×10^{14} P$^+$ ions cm^{-2} or about 10^{14} As$^+$ ions cm^{-2}.

The implanted impurity ions may occur in interstitial sites and some degree of thermal motion is required to cause substitution into the silicon sites. Additionally the damaged silicon lattice must be annealed or the carrier mobilities will be greatly reduced by scattering. The impurities tend to form pairs or clusters; this occurs as a strain relief mechanism and is encouraged by dislocations. To reach the necessary level of electrical activity, this aggregation is compensated for by "overdosing".

The ability of dislocations to attract impurity species has been used to getter impurities. In this case the silicon slice is initially heavily implanted on the rear face; annealing and radiation-enhanced diffusion then encourage the original defects to become trapped at dislocations at the rear of the crystal. This gettering effect is useful as it leaves a higher-purity front face which can then be implanted with a small, but controlled, number of selected impurities in a high-mobility sample.

The major problem with very large ion doses for amorphization is that the stresses which arise in real device structures from localized implantations (i.e. as opposed to academic planar demonstrations) cause dislocation motion away from the implanted region. The formation and motion are a function of dose, implant temperature and method of annealing. Recent attention has thus focused on fast annealing methods using laser, electron or ion beam pulses which can cause epitaxial regrowth in localized regions, without disturbing the remainder of the crystal. To do this, pulses as short as picoseconds have been used.

Much of the open literature that leads to the use of ion implantation in commercial methods of semiconductor device fabrication was written in the 1960s and early 1970s. However, even these older examples indicate the possibilities of the methods.

The early example of metal−oxide−semiconductor transistor (MOST) production using an ion beam demonstrates major advantages of dose and depth control of impurities coupled with excellent mask alignment and an abrupt junction region. The structure of a MOST is given in Figure 14.5. Current flow is by hole movement between source and drain when the gate is negatively biased and the current depends on the applied voltage.

The structure is built with p-type regions beneath a silicon dioxide insulator. Windows exist in the oxide to contact source and drain but, if the p-type regions were prepared by diffusion, then they must be well spaced to avoid lateral diffusion; in practice, the gate region would need to overlay both of them and this would give a capacitance layer through the SiO_2. After defining the metal layer for the source, gate and drain, the metal acts as a mask for boron implantation of a p-type layer such that there is no overlap (i.e. minimal capacitance) and an abrupt p-type edge (so that the device dimensions can be reduced).

Developments in integrated-circuit design and production have moved towards shallower devices. The value of ion implantation in the manufacture

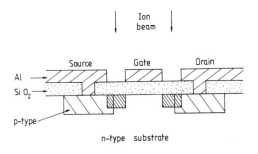

Figure 14.5.
Example of a simple MOST device formed by ion implantation through mask windows. This produces abrupt junctions and a low capacitance suitable for high-speed devices.

of shallow layers lies mainly in the ability to form a high concentration of impurities with a very abrupt termination of the dopant region. One example of a highly doped narrow-layer diode is a camel diode. It is a majority-carrier structure with a metal$-n^{++}-p^{+}-n^{-}$ configuration. The very thin p^{+} region limits the amount of band bending and thus minimizes the accumulation of minority carriers. It has a very sharp diode characteristic. A variation of the camel diode is the collector of hot electrons in the so-called monolithic hot-electron transistor.

In parallel with the trend towards shallow devices there have been improvements in high-energy ion implanters which have enabled novel device structures to be made with deeply buried layers, generally as isolation regions.

Implantation is applicable to compound semiconductors of the III$-$V, II$-$VI or IV$-$VI structures. The major disadvantage is that radiation damage products in a compound system are more varied than for an elemental semiconductor; consequently there has been an optimization of the implantation and annealing conditions. A further complexity occurs as some compound systems dissociate and may require implantation via a capping layer and control of the stabilizing vapour pressure during annealing. Numerous successful devices have been reported and there is considerable interest both for fast electronic devices and for electro-optic or microwave applications.

14.3.2 Ion implantation in insulators

This area has received least interest from the ion implantation community, despite the numerous attempts to point out the potential applications. The range of properties which are specifically associated with insulators and can be modified by implantation, include surface reflectivity, refractive index, surface acoustic wave velocity, luminescence, optical absorption, electro-optic coefficient and birefringence. Thus the implanted systems can find a role in Fourier spectrum analysers, second-harmonic phase-matching crystals, selective optical filter, solar energy conversion devices, antireflection coatings, optically flat Schmidt plates and optical waveguide circuitry.

Optical waveguide production in amorphous silica occurs as an optical beam is confined by total internal reflection in a region of high refractive index generated by densification of the glass network. Chemical changes, by N^+ implants to form an oxynitride, offer an additional index enhancement for high-dose implants in silica.

Most of the materials of specific interest to integrated optics (e.g. $LiNbO_3$, $LiTaO_3$ or quartz) must be used in the crystalline state to make electro-optic modulators or surface acoustic wave devices. Implantation destroys this crystallinity. However, in so doing, "amorphized" material is formed which is of lower density and refractive index. This is ideal for the development of optical waveguides in these materials as, if the region *adjacent* to the intended waveguide is implanted, the result is a high-index region (i.e. the crystalline waveguide) surrounded by a low-index amorphized optical barrier. Annealing of colour centres precedes crystal regrowth; so low-loss guides with strong optical confinement can be made. Although it is immediately apparent that lateral confinement of a surface guide is possible by ion implantation of the boundaries, it is slightly less obvious how to produce a buried region which optically isolates the guide from the substrate. However, reference to the mechanisms of energy transfer from the ion beam to the lattice indicates that for an energetic light ion (e.g. a megaelectron volt He^+ ion) the excitation is primarily electronic, and amorphization and hence index reduction only ensue at the end of the ion range where nuclear collision terms are important. This is ideal for an optical waveguide as the guides are of micrometre-size dimensions; so we must resort to high-energy ions to achieve the necessary penetration into the solid and the guide properties immediately ensue.

14.3.3   Implantation in metals

Surface treatments of metals are generally directed to solve the problems of wear, friction, hardness, fatigue, corrosion and oxidation. The use of ion implantation in this context is relatively recent with much of the pioneering work originating from the Harwell group of Dearnaley and Hartley. The experimental problems are very different from those of semiconductor physics and the work may be divided into that which shows direct industrial benefits and longer-term research where the implantations are used to develop new materials. In the latter context, ion beams offer a direct route to the formation of metallic alloys with accurately controlled compositions not limited by conventional solubility. Hence the properties of new materials can be explored and optimum cases for "normal" metallurgical preparation selected. Because of cost considerations, exotic alloys formed solely by implantation may be limited in use to superconductors or other high-cost technologies.

In the ion-implanted metals the surface modification gives negligible changes in physical dimensions and is therefore ideal for application to products after they have been machined to final tolerances. Implantation surface hardening does not cause distortions. The initial choice of ions has been governed by availability of implanter ion sources and prior knowledge of those ions which provide beneficial effects in conventional metal treatments; nitrogen is foremost among these. Implantation "nitriding" is marginally superior to the best chemical cases and has the advantage that it is almost universally applicable; other commonly used implants are titanium, carbon and boron nitride.

There is no general agreement about the physical process by which the implanted ions improve surface properties although there are various suggestions; nitrogen blocks dislocation motion, causes compaction of the surface, forms new metallic chemical compounds, reacts to give oxynitride compounds at the surface or initiates new metallic phases. Similarly there are conflicting data and proposed mechanisms to explain the persistence of the enhanced surface properties beyond the initial projected ion range. This is not a trivial effect as it is not uncommon to note that improved wear resistance may continue to depths of 10 or 20 times the projected range of the implanted nitrogen. Similarly, implanted traces of yttrium, to reduce oxidation, are also known to recede with the surface and to continue to inhibit corrosion.

Despite the uncertainties in the form of the physical mechanisms there is no doubt about the efficacity of the treatments. Improvements in wear lifetime obtained by nitrogen implantation are typically a factor of $2-4$ but in special cases factors of $100-1000$ are cited. For steels, demonstrations of wear reduction have frequently been made under unlubricated test conditions, and improvements recorded for a wide range of steels. Comparisons of N^+ with Ti^+ implants have given differing wear lifetime improvements in dry and lubricated tests. A factor of about 3 was achieved for both ions in lubricated wear tests for soft steels, but there was no correlation between wear and hardness measurements.

More spectacular results have been seen for ferritic and austenitic stainless steels (except those with a low chromium content) in which the wear was reduced one hundredfold by nitrogen implants. Sputtering limited the injection of N^+ at 100 keV to about 7×10^{17} ions cm^{-2} but the improved wear properties continued for about $20R_p$.

Nitrogen implantation has raised the microhardness and wear resistance of titanium alloys by a factor of 100, and a factor of 1000 has been achieved by Dearnaley in replacement hip joints. Conventional joints have a working life of under 10 years but the ion-implanted versions should survive several hundred years.

Measurements of wear rate dominated the first 10 years' work in this field for the obvious reason that there are immediate industrial applications. Associated with the reduction in wear are features of reduced friction, increases in microhardness and changes in oxidation rate. Even machine tool tips of tungsten carbide can be hardened by ion implantation, and improvements of 20% have been noted. Unexpected benefits accrue in improved machine tools; for example the surface temperature during cutting of plastics may be lower by up to 100°C. The consequence of the lower temperature is a cleaner parallel-sided hole and there is additionally a reduction in noxious vapours produced from the drilling.

Frictional forces can be reduced by ion implantation, and halving of the coefficient of friction is commonly reported. The reasons for this range from inhibition of debris breaking off the original surface to the addition of soft coating elements (e.g. lead). The role of oxide formation is important and there are studies in which metallic corrosion is inhibited, or monolayers of stable hard oxides (TiO_xN_y) provide a protective coating. In all cases, we expect to reduce the rate of oxidation by a densification of the surface layer and hence to block attack via rapid diffusion along dislocation lines. Ideally, surface dislocations can be removed by amorphization of the surface layer. Cooling rates within collision cascades exceed those used in splat-cooling methods of metallic glass production and therefore there have been numerous studies of metallic glasses.

Chemical application of metallic ion bombardment was attempted as early as 1974 where it was demonstrated that platinum implanted in monolayer concentrations could give catalytic performance equal to a pure platinum layer and we imagine the chemical possibilities of surface layers tailored by implantation will continue to develop.

14.3.4   Implantation in superconductors

Implantation has been used in alloy production of metallic superconductors and generated several new materials with critical temperatures in the range up to 23 K. Equally, for the high-temperature superconductors (i.e. $T_c >$ 77 K), implants could be used to adjust stoichiometry or, as with the formation of optical waveguides, to destroy the crystalline properties and hence to define patterns in the material. We can therefore envisage the formation of SQUID patterns by ion beams in high-T_c films in which the pattern is virgin material and the boundaries are amorphized.

14.3.5 Ion beam mixing

A feature of ion beam bombardment which is increasingly used is in the field of adhesion between two layers. The energy deposited during bombardment temporarily destroys small volumes of the lattice. Within a single crystal the majority of the ions will regrow in the original crystal structure but, if the collision cascades intersect an interface between two different layers, the cascades generate an intimate mixing of the two systems. The scale may be only tens of nanometres but the result is a greatly enhanced interface bond. The process probably includes simple kinetic disorder from nuclear collisions as well as, for insulators, electronic processes of relaxation and mixing. The interface mixing may be further aided by radiation-enhanced diffusion or the existence of thermodynamic driving forces from heats of formation of compounds within the cascade volume. Interest in ion beam mixing has rapidly escalated as nearly all metal−insulator−organic junctions show enhanced bonding. We imagine that future industrial applications will include low-cost layer deposition with ion beam mixing to give the requisite interface bond strengths.

14.4 FUTURE SOLID-STATE LASERS

The first visible laser used ruby (chromium-doped Al_2O_3) as the active element. This was a fruitful application of defect science in which higher-energy states with a wide spectrum of excitation could be pumped with a flash tube and electrons could be concentrated in a long-lived state in the decay route. This led to population inversion so that, in a high-gain crystal within a reflective cavity, there was stimulated emission and lasing. Despite the excitement which was generated by this laser, it has the inherent weakness that it is effectively a three-level system (Figure 14.6); so the ground state of the optical transition cannot be kept empty. Progress to the four-level lasers overcomes this problem; neodymium-based lasers are effective for this reason. Solid-state laser technology developed with a wide range of host materials which were doped with impurities such as neodymium or chromium. Control of the point defects lead to a lower threshold power, a higher gain and a more uniform response. This was necessary as, in the ruby example, attempts to increase the dopant level caused preferential chromium impurity pairing to reduce the lattice strain and the actual lasing frequently takes place in preferred filamentary paths rather than uniformly over the laser rod. Such difficulties result in noisy spikes of laser power and poor mode control. Although there is a good selection of lasing transitions available, most early lasers operated at a single wavelength or required a complete change of cavity mirrors.

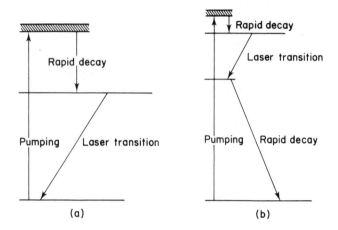

Figure 14.6.
Examples of energy level schemes for (a) three-level and (b) four-level lasers.

Solid-state lasers have been complemented by gas lasers and lasers can now be purchased which operate at wavelengths from the ultraviolet to the infrared. Many of these lasers require, and deliver, kilowatts of power, are physically large and, inevitably, are expensive. For many purposes, small lasers with wavelength tunability at low cost and modest-power (milliwatts) levels would be preferable. Semiconductor junction lasers partly fulfil this need as they operate efficiently at specific wavelengths determined by their chemical composition (and hence energy band gap) and are economically incorporated into optical fibre communication systems and video disc players. By tailoring the composition and physical dimensions of heterojunction semiconductor lasers the "quantum well" geometries which lase at energies above the band gap can even be produced, but composition control on an atomic scale during growth is required. Techniques such as molecular beam epitaxy achieve this and visible (red) lasers with semiconductor junctions of the III–V elements can be generated. However, even for the laser, the operating wavelength is pre-selected and invariant beyond the initial mode envelope.

Thus the stage has been reached at which, by combination of many lasers or laser–fluorescent dye combinations, lasers can be produced in most parts of the spectrum, but this is not the same as continuous tunability. Two

Table 14.1.
Examples of vibronic lasers

Matrix	Pump	Operation	Tuning range (nm)	Operating temperature (K)	Maximum quoted power
YL_iF_4:Ce^+	Excimer	Pulsed	309–325		
Al_2O_3:Ti^+	Argon	Continuous wave	600–1100		
Alexandrite	Mercury arc	Continuous wave	701–826	Room temperature	50 W
Emerald	Krypton	Continuous wave	729–809		320 mW
$GdScGe(SiO_4)_3$:Cr^+	Argon or krypton	Continuous wave	742–842	77	250 mW
MgF_2:V^{2+}	Mercury arc	Continuous wave	1121	77	
MgO:Ni^{2+}	Mercury arc	Continuous wave	1314	77	
MgF_2:Co^{2+}	Yttrium aluminium garnet	Continuous wave	1500–2300	77	4.3 W
MgF_2:Ni^{2+}	Mercury arc	Continuous wave	1623	77	1 W

Table 14.2.
Examples of colour centre lasers

Type	Matrix	Pump	Emission (nm)	Operating temperature (K)	Power
F_A	$KCl:Li^+$	Argon or krypton	2300−3120	77	1−10 kW
F_A	$RbCl:Li^+$	Argon or krypton	2500−3640	77	100 mW
F_B	$KCl:Na^+$	Argon or krypton	2200−2740	77	100 mW
F_2^+	$KF:Pd^+$	$LiF:F_2^+$	1240−1400	77	5 W
F_2^+	$NaF:OH^-$	Krypton	1100−1380	77	100 mW
F^+	CaO	Argon	357−420	77	15 mW

approaches are being used which are based on imperfections. The alternatives are dopants with impurity ions or colour centres. In both cases the defects luminesce over a wide spectral region so that by inclusion in a tunable Fabry−Pérot cavity the gain can be optimized at a single wavelength within this envelope of vibronic states. Work with doped tunable lasers has centred on the 3d paramagnetic ions cobalt, nickel and titanium or the rare earths samarium and cerium. Table 14.1 summarizes some of these major characteristics. It can be seen that many have a weakness in that they only operate at low temperature and/or require strong pump sources. One of the more promising crystals is $Al_2O_3:Ti^+$ as this is tunable over a very broad range with high efficiency.

Colour centre lasers based on defects of F_A, F_B or F_2^+-type centres in alkali halides or F^+ centres in oxides have produced tunable lasers covering the 0.8−3.65 μm wavelength range with high outputs and Table 14.2 cites a few examples. Again the weakness is operation at 77 K plus major pump sources. At the present stage of development many of the crystals have a further drawback in that the effects of excitation encourage charge transfer and defect migration so that the laser performance degrades. Doping with suitable cations (e.g. Tl^+) has been tried with relative success.

Several approaches might be considered to minimize the difficulties. Firstly, the total pump power could be reduced if the laser medium were strongly confined in the form of a waveguide. This has been demonstrated for optical fibres containing neodymium in silica and fibres of crystalline laser systems have recently been produced. Rather than draw an optical fibre, it is also possible to provide optical confinement by the techniques used for integrated optics.

The second possibility which then emerges if the pump power requirement is reduced is (Figure 14.7) to use a standard high-power semiconductor junction laser as an infrared source, to frequency double this in a non-linear crystal and then to pump a tunable crystal. The efficiency of each stage can be 10−50%; so the net result is a tunable miniature milliwatt laser operating on a few watts of electrical power.

Defect properties are multifaceted and range from beneficial features which offer luminescence and lasing transitions to problems of loss from absorption or scattering. In high-power operation, whether within a waveguide or a bulk crystal, the laser power can induce photorefractive effects. If the objective is second-harmonic generation, this photorefractivity must be suppressed but, as is discussed next, it may in turn be utilized in other contexts.

IR laser → frequency doubler → tunable laser

Figure 14.7.
A scheme for conversion of light from a diode laser to form a tunable visible laser: IR, infrared.

14.5 APPLICATIONS OF THE PHOTOREFRACTIVE EFFECT

As discussed in section 4.15, the photorefractive effect involves a reversible light-induced change in refractive index that can be relatively stable in the dark. The mechanism of the stabilization and the sensitivity of the laser light are clearly related to defect states as they are sample dependent. Attempts to offer defect models to explain the phenomena are being made so that the effect can be optimized or inhibited. Even at this initial stage of understanding, a number of applications for optical technology are being developed such as holographic storage of information, coherent optical amplification, optical data processing, real-time holography, interferometry and phase-conjugated devices. Some of these applications are briefly discussed.

14.5.1 Optical memories—holographic storage

Modern technology is demanding data storage devices with increasingly higher capacity and speed performances. Optical systems not requiring mechanical moving parts may offer very-fast-access capabilities by comparison with classical magnetic storage devices such as tapes or discs. The upper limit for the storage capacity of an optical memory is set by diffraction effects and is about $1/\lambda^3 = 10^{12}$ bits cm^{-3}.

There are two main technical approaches to the optical storage: the localized or bit-by-bit method and the holographic method. In localized memories, each bit is separately stored at a small location of the memory surface. Because of the small dimensions of the bit, this technique demands very strict tolerances for focusing and alignment. In the holographic memory the information is stored in the form of a hologram which is the Fourier transform of the bit pattern. The information being spatially spread over the entire hologram. Therefore, at variance with localized memories, dust and scratches do not result in the loss of information bits so that holographic memories are less sensitive to macroscopic imperfections.

One interesting feature of the holographic method is the possibility of volume storage, therefore allowing for very high packing capacities. This can be accomplished by varying the angle of incidence of the reference beam, each individual hologram being associated with a well-defined angle of incidence. Selective reconstruction of the superimposed holograms can be achieved because efficient diffraction takes place only when the hologram is addressed at the Bragg angle.

In any storage technology, three basic operations are being dealt with: writing, read-out and erasure of the information. For holographic memories based on the photorefractive effect the typical set-up for writing is shown in Figure 14.8. The beam produced by the laser is divided in two beams by a beam splitter B. One beam illuminates the object and is then focused on the photorefractive material, e.g. $LiNbO_3$. The other beam is projected onto the $LiNbO_3$ through the mirror M_2. The interference of the two beams on the $LiNbO_3$ induces a charge or electric field pattern which, via the electro-optic coupling, causes a corresponding refractive index pattern which acts as a storage element.

For read-out, the hologram is then illuminated by the reference beam, or another suitable probe laser. The diffracted wave yields a reconstructed object, whose image may be brought into a detector through an appropriate lens.

Erasure of the optical information can also be performed through high-power illumination of the hologram with the same set-up as in Figure 14.8.

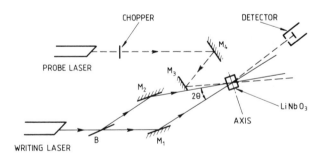

Figure 14.8.
Typical set-up for writing of a holographic grating in a photorefractive material. The beam used to probe the grating is also shown.

This erasure process is a major drawback for the holographic memory, since it limits the number of readings that can be performed on a given hologram. This limitation is particularly severe for volume holograms.

The illumination-induced degradation of the holographic memory in LiNbO$_3$ can be avoided by "fixing". This technique is based on a transformation of the electronic charge distribution into an ionic charge pattern which is not sensitive to light. This can be achieved by heating to 120°C, where ionic conductivity is appreciable. Fixing turns a reversible memory into a permanent memory. We can only speculate on the details of the defect interactions of this stage.

Main performance parameters for an LiNbO$_3$ holographic memory are summarized in Table 14.3.

14.5.2 Coherent optical amplification

The crossing of two laser beams inside a photorefractive crystal induces a refractive index pattern. During the writing of this pattern and depending on experimental conditions and the operative carrier transport mechanisms in the material, a coupling between the two beams may occur. Then, energy can be transferred from the more intense to the weaker beam. In this way, optical signals and images can be coherently amplified. Amplifications by a factor of more than 10 in intensity have been demonstrated.

Table 14.3.
Parameters for LiNbO$_3$ holograms

	Reversible	Fixed
Resolution	$>10^4$ lines mm^{-1}	Lower
Efficiency		
Single hologram	95%	95%
Multiple hologram	0.01–1%	Up to 1%
Minimum writing energy	0.1–1 mJ cm^{-2}	2–13 J cm^{-2}
Number of read-outs	10^5–10^7	10^{10}
Erasure energy	1–100 J cm^{-2}	
Number of superimposed holograms	100–500	>500
Bit capacity	10^6 – 5×10^6 bits mm^{-2}	$>5 \times 10^6$ bits mm^{-2}
Storage time at room temperature	15 min–300 days	2–3 years

14.5.3   Phase-conjugated mirrors

Non-linear materials can be used in wave-mixing experiments. In degenerate four-wave mixing (Figure 14.9), two pump beams of the same frequency ω impinge on the material in opposite directions. When another beam of the same frequency (signal wave) is sent to the crystal, the third-order optical susceptibility induces a polarization $P^3(\omega)$, causing a fourth beam, which is phase conjugated to the signal wave. In other words, this beam has an opposite K vector and a complex conjugated amplitude. This phase-conjugated beam propagates exactly as the signal beam does but reversed in direction.

The non-linear material behaves as a peculiar mirror, which does not follow the usual (snell) reflection laws. Moreover, the reflector coefficient can be higher than unity, so that a mirror with optical gain capabilities can be built. These conjugated mirrors have very promising applications for the restoration of phase-distorted images (adaptive optics). Examples include light propagation through a distorting medium, such as the atmosphere or an optical fibre, and photolithography.

14.6   DEFECT CHEMISTRY AND NUCLEAR FUELS

The majority of modern thermal reactors use uranium dioxide as the fuel; while the mixed oxide $UO_2 - PuO_2$ is employed in breeder or fast reactors. The defect chemistry of the compounds has a major influence on the chemical, physical and mechanical behaviour of the fuels, and as such is a good illustration of the way in which a knowledge of defect structure relates directly to technological problems.

UO_2 has the fluorite structure and consequently, in common with structural materials such as CaF_2, anion Frenkel disorder would be expected to

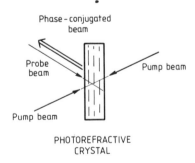

Figure 14.9.
Beam geometry in a four-wave mixing experiment. As a consequence of beam interaction, the phase-conjugated wave of the probe beam is generated.

predominate. However, the anion Frenkel formation energies are high (about 4 eV) and, since the material is non-stoichiometric, showing both oxygen excess (UO_{2+x}) and oxygen deficient (UO_{2-x}) compositions, we would expect the defects induced by deviations from stoichiometry to dominate except at the highest temperatures. As disorder is predominantly on the oxygen sublattice, oxidation of the crystal leads to creation of oxygen interstitials and introduction of oxygen vacancies. Indeed the redox reactions can be represented by

$$2U_U + \tfrac{1}{2}O_2 \rightarrow 2U_U' + O_{int}''$$

for oxidation and

$$2U_U \rightarrow 2U_U' + V_O''$$

for reduction; vacancy and interstitial concentrations are linked in all composition regions by the Frenkel disorder reaction

$$2O_O \overset{K_F}{\rightleftharpoons} V_O^{\cdot\cdot} + O_{int}''$$

One of the most important practical consequences of these equilibra on the oxygen sublattice is their effect on the minority cation defects as transport of the latter controls the rate of creep, which is of vital importance in influencing the mechanical behaviour of the fuel at high temperatures. The concentration of cation defects are controlled by the Schottky disorder reaction

$$U_U + 2O_O \overset{K_S}{\rightleftharpoons} V_U'''' + 2V_O^{\cdot\cdot} + \text{``}UO_2\text{''}$$

(where by UO_2 we indicate uranium dioxide on the surface of the crystal). The corresponding mass action equation is given by

$$c_{U_{Vac}} c_{O_{Vac}}{}^2 = K_S \qquad (14.1)$$

where $c_{U_{Vac}}$ and $c_{O_{Vac}}$ are the concentration of uranium and oxygen vacancies, respectively. Thus, in UO_{2+x} where the concentration of oxygen interstitials is enhanced, and consequently by the operation of the Frenkel disorder reaction that of the oxygen vacancies is suppressed, the cation vacancy concentration increases. Indeed since, in UO_{2+x}, $c_{O_{int}} \approx x$ (where $c_{O_{int}}$ is the concentration of oxygen interstitials) and since the Frenkel disorder reaction gives rise to the corresponding mass action equation

$$c_{O_{Vac}} c_{O_{int}} = K_F \qquad (14.2)$$

we can write, for UO_{2+x},

$$c_{U_{Vac}} \approx \frac{K_S x^2}{K_F{}^2} \qquad (14.3)$$

A similar analysis shows that in UO_{2-x} the cation vacancy concentration decreases given by

$$c_{U_{Vac}} \approx \frac{K_S}{x^2} \qquad (14.4)$$

Using similar arguments, it can be shown that the Arrhenius enthalpy h_m for diffusion decreases from

$$h_m = h_S \qquad (14.5)$$

In UO_{2-x} where h_S is the Schottky enthalpy to

$$h_m = h_S - 2h_F$$

In UO_{2+x} where h_F is the Frenkel enthalpy. These predictions are verified experimentally.

The effect of these changes is to change drastically (i.e. by several orders of magnitude) the rate of cation diffusion on passing from the substoichiometric (anion-deficient) to the hyperstoichiometric (anion-excess) regions of the phase diagrams. The consequences of these changes for the mechanical properties of the fuel are equally dramatic. The rate of plastic flow increases markedly as does that of other properties (e.g. the release of fission gas) which depend on cation diffusion rates.

The behaviour of nuclear fuels is therefore critically dependent on the oxygen-to-metal ratio, and indeed the stoichiometry of the fuel is a factor of central importance in practical fuel situations. This is directly due to the fundamental defect chemistry of the material.

14.7 SENSORS

Sensing is a broad topic which covers the detection of gases, humidity, radiation and pressure. A wide variety of materials is used. Here we concentrate on one particular application, i.e. oxygen sensing, which has industrial uses such as the control of the air—fuel mixture in "lean-burn" internal combustion engines.

The simplest types of oxygen sensor use a Nernst cell with a zirconia electrolyte, i.e. a standard oxygen atmosphere is separated by the electrolyte from the unknown atmosphere, with the electromotive force being proportional to the logarithm of the ratio of the pressures. The role of the electrolyte is to conduct oxygen ions between the two halves of the cell.

The second type of sensor which uses TiO_2 is of more interest in the present context as the operation is directly dependent on the nature of the defect structure of the material. TiO_2 is a non-stoichiometric oxide, which loses oxygen according to the reaction

$$O_O \xrightleftharpoons{K_R} V_O^{\cdot\cdot} + 2M' + \tfrac{1}{2}O_2 \qquad (14.6)$$

The reduced cations M' correspond to electrons in the conduction band and are responsible for the electrical conductivity of the material. Indeed, we can write

$$\sigma \propto c_{M'}$$

where $c_{M'}$ is the concentration of the reduced cations. However, $c_{M'} = 2c_{O_{vac}}$; hence the mass action equation governing reduction can be written

$$\frac{c_{M'}{}^3 p_{O_2}{}^{1/2}}{2} = K_R \qquad (14.7)$$

giving

$$\sigma \propto p_{O_2}{}^{-1/6}$$

The electrical conductivity is therefore a probe for the oxygen pressure.

In practice the $p_{O_2}{}^{-1/6}$ behaviour is only observed at high p_{O_2} where there are low deviations from stoichiometry. At lower p_{O_2}, different exponents from $\tfrac{1}{4}$ are observed. However, the same principle holds, i.e. the oxygen pressure is "sensed" by measuring the conductivity. Devices working on exactly this principle are currently used in the motor industry and are indeed a vital component of fuel-efficient lean-burn engines.

14.8 SUMMARY

These briefly sketched examples have the common feature that they involve materials whose property of interest is undoubtedly controlled by lattice imperfections. Although attempts to understand the defect interactions has aided the reproducibility or performance of the devices discussed, in many cases a definitive statement of the structure and role of the defects cannot be offered. Even for semiconductors, we do not understand all the effects caused by defects and have only identified a small selection of those present. The positive and hopeful feature of these examples is that, despite the lack of detailed knowledge, we can usefully develop operating systems. In-depth understanding of defects does lead to improved systems and therefore anyone working in our field is assured of an employable and useful future.

Index